T0212686

Lecture Notes in Artificial Intelligence   8791

Subseries of Lecture Notes in Computer Science

LNAI Series Editors

Randy Goebel
  University of Alberta, Edmonton, Canada
Yuzuru Tanaka
  Hokkaido University, Sapporo, Japan
Wolfgang Wahlster
  DFKI and Saarland University, Saarbrücken, Germany

LNAI Founding Series Editor

Joerg Siekmann
  DFKI and Saarland University, Saarbrücken, Germany

More information about this series at http://www.springer.com/series/1244

Laurent Besacier · Adrian-Horia Dediu
Carlos Martín-Vide (Eds.)

Statistical Language and Speech Processing

Second International Conference, SLSP 2014
Grenoble, France, October 14–16, 2014
Proceedings

 Springer

Editors

Laurent Besacier
University Joseph Fourier
Grenoble
France

Adrian-Horia Dediu
Carlos Martín-Vide
Rovira i Virgili University
Tarragona
Spain

ISSN 0302-9743
ISBN 978-3-319-11396-8
DOI 10.1007/978-3-319-11397-5

ISSN 1611-3349 (electronic)
ISBN 978-3-319-11397-5 (eBook)

Library of Congress Control Number: 2014947801

LNCS Sublibrary: SL7 – Artificial Intelligence

Springer Cham Heidelberg New York Dordrecht London

Printed on acid-free paper

Springer is part of Springer Science+Business Media (www.springer.com)

Preface

This volume contains the papers presented at the Second International Conference on Statistical Language and Speech Processing (SLSP 2014), held in Grenoble, France during October 14–16, 2014.

SLSP 2014 is the second event in a series to host and promote research on the wide spectrum of statistical methods that are currently in use in computational language or speech processing; it aims to attract contributions from both fields. The conference encourages discussion on the employment of statistical methods (including machine learning) within language and speech processing. The scope of the SLSP series is rather broad, and includes the following areas: phonology, phonetics, prosody, morphology; syntax, semantics; discourse, dialog, pragmatics; statistical models for natural language processing; supervised, unsupervised and semi-supervised machine learning methods applied to natural language, including speech; statistical methods, including biologically-inspired methods; similarity; alignment; language resources; part-of-speech tagging; parsing; semantic role labeling; natural language generation; anaphora and coreference resolution; speech recognition; speaker identification/verification; speech transcription; speech synthesis; machine translation; translation technology; text summarization; information retrieval; text categorization; information extraction; term extraction; spelling correction; text and web mining; opinion mining and sentiment analysis; spoken dialog systems; author identification, plagiarism and spam filtering.

SLSP 2014 received 53 submissions, which were reviewed by the Program Committee members, some of whom consulted with external referees as well. Among the submissions, 32 received three reviews, and 21 two reviews. After a thorough and lively discussion, the committee decided to accept 18 papers (which represents an acceptance rate of 33.96 %). The program also includes three invited talks.

Part of the success in the management of such a number of submissions is due to the excellent facilities provided by the EasyChair conference management system. We would like to thank all invited speakers and authors for their contributions, the Program Committee and the reviewers for their cooperation, and Springer for its very professional publishing work.

July 2014

Laurent Besacier
Adrian-Horia Dediu
Carlos Martín-Vide

Organization

SLSP 2014 was organized by the Study Group for Machine Translation and Automated Processing of Languages and Speech (GETALP) from the Laboratoire d'Informatique de Grenoble (LIG), in France, and the Research Group on Mathematical Linguistics (GRLMC) from Rovira i Virgili University, Tarragona, Spain.

Program Committee

Sophia Ananiadou	University of Manchester, UK
Srinivas Bangalore	AT&T Labs-Research, USA
Patrick Blackburn	University of Roskilde, Denmark
Hervé Bourlard	IDIAP Research Institute, Switzerland
Bill Byrne	University of Cambridge, UK
Nick Campbell	Trinity College Dublin, Ireland
David Chiang	University of Southern California at Marina del Rey, USA
Kenneth W. Church	Thomas J. Watson Research Center, USA
Walter Daelemans	University of Antwerpen, Belgium
Thierry Dutoit	University of Mons, Belgium
Alexander Gelbukh	National Polytechnic Institute, Mexico
James Glass	Massachusetts Institute of Technology at Cambridge, USA
Ralph Grishman	New York University, USA
Sanda Harabagiu	University of Texas at Dallas, USA
Xiaodong He	Microsoft Research, USA
Hynek Hermansky	Johns Hopkins University at Baltimore, USA
Hitoshi Isahara	Toyohashi University of Technology, Japan
Lori Lamel	CNRS-LIMSI, France
Gary Geunbae Lee	Pohang University of Science and Technology, South Korea
Haizhou Li	Institute for Infocomm Research, Singapore
Daniel Marcu	SDL Research, USA
Carlos Martín-Vide (Chair)	Rovira i Virgili University, Spain
Manuel Montes-y-Gómez	National Institute of Astrophysics, Optics and Electronics, Mexico
Satoshi Nakamura	Nara Institute of Science and Technology, Japan
Shrikanth S. Narayanan	University of Southern California at Los Angeles, USA
Vincent Ng	University of Texas at Dallas, USA
Joakim Nivre	Uppsala University, Sweden
Elmar Nöth	University of Erlangen-Nüremberg, Germany

Maurizio Omologo	Bruno Kessler Foundation, Italy
Mari Ostendorf	University of Washington at Seattle, USA
Barbara H. Partee	University of Massachusetts at Amherst, USA
Gerald Penn	University of Toronto, Canada
Massimo Poesio	University of Essex, UK
James Pustejovsky	Brandeis University at Waltham, USA
Gaël Richard	TELECOM ParisTech, France
German Rigau	University of the Basque Country, Spain
Paolo Rosso	Technical University of Valencia, Spain
Yoshinori Sagisaka	Waseda University, Tokyo, Japan
Björn W. Schuller	Imperial College London, UK
Satoshi Sekine	New York University, USA
Richard Sproat	Google, USA
Mark Steedman	University of Edinburgh, UK
Jian Su	Institute for Infocomm Research, Singapore
Marc Swerts	Tilburg University, The Netherlands
Jun'ichi Tsujii	Microsoft Research Asia, China
Gertjan van Noord	University of Groningen, The Netherlands
Renata Vieira	Pontifical Catholic University of Rio Grande do Sul, Brazil
Dekai Wu	Hong Kong University of Science and Technology, Hong Kong
Feiyu Xu	German Research Center for Artificial Intelligence, Germany
Roman Yangarber	University of Helsinki, Finland
Geoffrey Zweig	Microsoft Research, USA

External Reviewers

Banerjee, Somnath
Collovini, Sandra
De Pauw, Guy
Franco Salvador, Marc

Leitzke Granada, Roger
Lopes, Lucelene
Mohtarami, Mitra
Tanigaki, Koichi

Organizing Committee

Laurent Besacier (Co-chair), Grenoble, France
Adrian-Horia Dediu, Tarragona, Spain
Benjamin Lecouteux, Grenoble, France
Carlos Martín-Vide (Co-chair), Tarragona, Spain
Florentina-Lilica Voicu, Tarragona, Spain

Contents

Machine Learning Methods

Text Extraction and Categorization

Mining Text

Invited Talks

Syntax and Data-to-Text Generation

Claire Gardent[✉]

CNRS/LORIA, Nancy, France
claire.gardent@loria.fr

Abstract. With the development of the web of data, recent statistical, data-to-text generation approaches have focused on mapping data (e.g., database records or knowledge-base (KB) triples) to natural language. In contrast to previous grammar-based approaches, this more recent work systematically eschews syntax and learns a direct mapping between meaning representations and natural language. By contrast, I argue that an explicit model of syntax can help support NLG in several ways. Based on case studies drawn from KB-to-text generation, I show that syntax can be used to support supervised training with little training data; to ensure domain portability; and to improve statistical hypertagging.

Keywords: Computational grammars · Natural language generation · Statistical natural language processing · Hybrid symbolic/statistical approaches

1 Introduction

Given some non-linguistic input, the task of data-to-text generation consists in producing a text verbalising that input. Data-to-text generation has been used, e.g., to summarise medical data [31], to generate weather reports from numerical data [32] and to automatically produce personalised letters [11].

Earlier statistical work on data-to-text generation has mainly focused on inducing large probabilistic grammars from treebanks and on using these grammars to generate from meaning representations derived from those same treebanks. Thus, [9] induces a Probabilistic Lexical Functional Grammar (LFG) from the PTB and uses it to generate from f(unctional)-structures automatically derived from that treebank. Reference [3] uses a large scale Tree Adjoining Grammar (TAG, [34]) and a tree model trained on the derivation trees of 1 million words of the Wall Street Journal to map dependency trees to sentences. And [38] induces a probabilistic Combinatory Categorial Grammar (CCG, [33]) from the CCGBank [21] which is then used to generate from hybrid logic dependency semantics [2].

With the development of the web of data however, interest has recently shifted to data-to-text generators which can generate from less linguistic, more data oriented, meaning representations. While logical formulae and dependency

© Springer International Publishing Switzerland 2014
L. Besacier et al. (Eds.): SLSP 2014, LNAI 8791, pp. 3–20, 2014.
DOI: 10.1007/978-3-319-11397-5_1

trees may provide generic meaning representations for natural language, they typically fail to support a straightforward mapping between data and natural language (NL) expressions. This is because both the signature of the meaning representation language and the alignment between meaning and basic grammar units are specified independently of the application data. Typically, predicate names are simply lemmas (each word will be represented by a meaning representation including its lemma as a predicate symbol) and the alignment between meaning and string is determined by syntax. When generating from e.g., knowledge or database data, these assumptions generally fail to hold. That is, lemmas must be disambiguated and mapped to application-specific predicate symbols while the alignment between meaning representation sub-units and NL expressions is often at odd with grammar syntax.

To address these issues, recent statistical, data-to-text approaches have therefore focused on mapping e.g., database records or knowledge-base (KB) triples to natural language. In particular, data-to-text generators [1,10,23,24,39] were trained and developed on datasets from various domains including the air travel domain [13], weather forecasts [5,26] and sportscasting [10]. In contrast to the previous, grammar-based approaches, this more recent work systematically eschews syntax. Instead, the dominant approach consists in learning a direct mapping between meaning representations and natural language.

In this paper, we take a middleroad between these two approaches. We focus on generating from "real" data i.e., knowledge base data, but we argue that an explicit model of syntax is valuable in several ways. More specifically, we argue that syntax:

- *can help compensate for the lack of large quantities of training data*. Using an international benchmark consisting of only 207 training instances, we show that inducing a linguistically principled, non probabilistic grammar from this data, allows for the development of a data-to-text generator which shows good coverage while preserving output quality. When compared with the other two participating systems, the approach performs comparably with a rule-based, manually developed system and markedly outperforms an existing statistical generator.
- *can help ensure genericity*. Focusing on the task of verbalising user queries on knowledge bases, we show that a small hand-crafted grammar, combined with an automatically constructed lexicon, permits verbalising queries independent of which domain the queried KB bears on.
- *can help improve the performance of a statistical, hypertagging module* designed to reduce the initial search space of the generator. In particular, we show that the high level linguistic abstractions captured by the grammar permits developing a hypertagging module which improves the generator speed, supports sentence segmentation and preserves output quality.

The paper is structured as follows. In Sect. 2, we start by introducing the grammar framework which we use to support data-to-text generation namely, Feature-Based Lexicalised Tree Adjoining Grammar (FB-LTAG). We then explain the generation algorithm which permits generating sentences given some input data and an FB-LTAG. Sections 3, 4 and 5 illustrate how an explicit model of syntax can help improve generation. Section 6 concludes with pointers for further research.

2 Feature-Based Lexicalised Tree Adjoining Grammar

In this section, we start by defining the grammar formalism (Sect. 2.1) and the lexicon (Sect. 2.2) we use to mediate between data and natural language. We then describe the generation algorithm which exploits these lexicon and grammar to map data into text (Sect. 2.3).

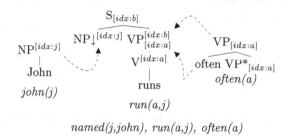

named(j,john), run(a,j), often(a)

Fig. 1. Derivation and Semantics for "John often runs"

2.1 Grammar

Following [17], we use a Feature-Based Lexicalised Tree Adjoining Grammar (FB-LTAG) augmented with a unification based semantics for generation. For a precise definition of FB-LTAG, we refer the reader to [36]. In essence, an FB-LTAG is a set of trees whose nodes are decorated with feature structures and which can be combined using either substitution or adjunction. Substitution of tree γ_1 at node n of the derived tree γ_2 rewrites n in γ_2 with γ_1. n must be a substitution node (marked with a downarrow). Adjunction of the tree β at node n of the derived tree γ_2 inserts β into γ_2 at n (n is spliced to "make room" for β). The adjoined tree must be an auxiliary tree that is a tree with a foot node (marked with a star) and such that the category of the foot and of the root node is the same.

In an FB-LTAG with unification semantics, each tree is furthermore associated with a semantics and shared variables between syntax and semantics ensure the correct mapping between syntactic and semantic arguments. As trees are combined, the semantics of the resulting derived tree is the union of their semantics modulo unification.

The semantic representation language used to represent meaning in the grammar is a flat semantics language [6, 12] which consists of a set of literals and can be used e.g., to specify first order logic formulae or RDF triples. For a precise definition of the syntax and semantics of that language, see [18].

Figure 1 shows an example toy FB-LTAG with unification semantics. The dotted arrows indicate possible tree combinations (substitution for *John*, adjunction for *often*). Thus given the grammar and the derivation shown, the semantics of *John often runs* is as shown namely, *named(j john), run(a,j), often(a)*.

2.2 Lexicon

Semantics: RUN
Tree: nx0V
Syntax: Canonical
Anchor: *runs*

Semantics: SLEEP
Tree: nx0V
Syntax: Canonical
Anchor: *sleep*

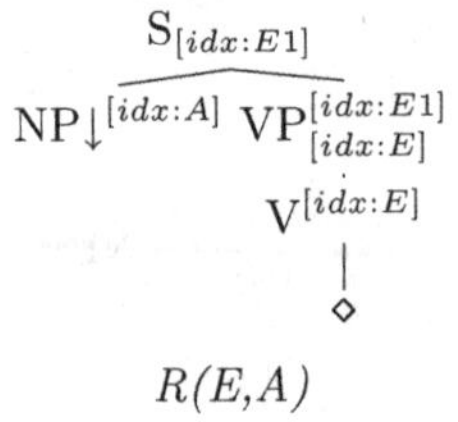

$$S_{[idx:E1]}$$

$$NP\downarrow^{[idx:A]} \quad VP^{[idx:E1]}_{[idx:E]}$$

$$V^{[idx:E]}$$

$$\diamond$$

$$R(E,A)$$

Fig. 2. FB-LTAG tree schema nx0V and two Lexical Entries associated with that tree schema.

The *Lexicon* permits abstracting over lexical and semantic information in an FB-LTAG tree and relating a single tree schema to several lexical items. For instance, the lexical entries shown on the left of Fig. 2 relates the predicate symbols RUN and SLEEP to the TAG tree nx0V shown on the right. During generation, these relation predicate symbols will be used to instantiate the predicate variable R in the semantic schema $R(E, A)$ associated with that tree; and the anchor values (*runs/sleeps*) to anchor this tree i.e., to label the terminal node marked with the anchor sign ($\diamond$). That is, the $\diamond$ node will be labelled with the terminal *runs/sleeps*.

2.3 Surface Realisation

For surface realisation, we use the chart-based algorithm described in [19]. This algorithm proceeds in four main steps as follows.

- Lexical Selection. Retrieves from the grammar all grammar units whose semantics subsumes the input semantics. For instance, given the semantics *named(j john), run(a,j), often(a)*, lexical selection will return the three trees shown in Fig. 1.
- Tree Combination. Substitution and adjunction are applied on the set of selected trees and on the resulting derived trees until no further combination is possible. For instance, the three trees selected in the previous lexical selection step will be combined to yield a complete phrase structure tree.

- Sentence Extraction. All syntactically complete trees which are rooted in S and associated with exactly the input semantics are retrieved. Their yields provide the set of generated (lemmatised) sentences e.g., *John run often* in our running example.
- Morphological Realisation. Lexical lookup and unification of the features associated with lemmas in the generated lemmatised sentences yield the final set of output sentences e.g., *John runs often*.

3 Grammar as a Means to Compensate for the Lack of Training Data

The KBGen task[1] was introduced as a new shared task at Generation Challenges 2013 [4] to evaluate and compare systems that generate text from knowledge base data. Figure 3 shows an example input and output.

```
:TRIPLES (
  (|Release-Of-Calcium646| |object| |Particle-In-Motion64582|)
  (|Release-Of-Calcium646| |base| |Endoplasmic-Reticulum64603|)
  (|Gated-Channel64605| |has-function||Release-Of-Calcium646|)
  (|Release-Of-Calcium646| |agent| |Gated-Channel64605|))
:INSTANCE-TYPES
  (|Particle-In-Motion64582| |instance-of| |Particle-In-Motion|)
(|Endoplasmic-Reticulum64603| |instance-of| |Endoplasmic-Reticulum|)
  (|Gated-Channel64605| |instance-of| |Gated-Channel|)
  |Release-Of-Calcium646| |instance-of| |Release-Of-Calcium|))
:ROOT-TYPES (
  (|Release-Of-Calcium646| |instance-of| |Event|)
  (|Particle-In-Motion64582| |instance-of| |Entity|)
  (|Endoplasmic-Reticulum64603| |instance-of| |Entity|)
  (|Gated-Channel64605| |instance-of| |Entity|)))
```

The function of a gated channel is to release particles from the endoplasmic reticulum

Fig. 3. Example KBGEN Input and Reference Sentence

One characteristic of the KBGen shared task is that the size (207 input/output pairs) of the training data is relatively small which makes it difficult to learn efficient statistical approaches. In what follows, we show that, by inducing a linguistically principled grammar from the training data, we can develop a generator which performs well on the test data. While fully automatic, our approach produces results which are comparable to those obtained by a hand written, rule based system; and which markedly outperform a data-driven, generate-and-rank approach based on an automatically induced probabilistic grammar.

Grammar induction generally relies on large syntactically annotated corpora (treebank) and results in large grammars whose combinatorics are constrained by the probability estimates derived from the treebank. In contrast, we define a grammar induction algorithm which yields compact, linguistically principled, FB-LTAG grammars. The induction process is informed by the two main principles underlying the linguistic design of an FB-LTAG namely, the extended domain of locality and the semantic principle. The *extended domain of locality*

---

[1] http://www.kbgen.org

Algorithm 1. Grammar Induction Algorithm

Require: An input semantics (set of triples) ϕ with variables V_ϕ, reference sentence S and parse tree τ_S.

1: **Variable/String Alignment** Align each variable in V_ϕ with one or more tokens in S
2: **Variable Projection** Use the Variable/String alignment and a set of hand-written rules to project the input semantics variables onto non terminal nodes in the parse tree τ_S of the reference sentence.
3: **Extracting Trees** Extract NP and relational (prepositions, verbs, conjunctions, etc.) trees from τ_S using the variables labelling the nodes of the parse tree. NP trees are NP subtrees whose root node are labelled with an input variable ($v \in V_\phi$). Relational trees are subtrees containing all and only input variables that are related to each other by relations in ϕ (cf. [20] for a more precise definition).
4: **Associating Trees with Semantics.** Each subtree is assigned a set of input triples based on the input variables it is labeled with. An NP tree labeled with input variable v is associated with all input literal whose first argument is v. Each relational tree is associated with all literals whose argument variables are variables labelling this tree.
5: **Generalising from Trees to Tree Schemas** Isomorphic trees which differ only in their semantics and lexical content are converted to a single tree schema and several lexical entries capturing the multiple possible instantiations of that tree schema.
6: **Generalising from Bigger to Smaller Trees** Large Verb trees are used to derive smaller more general trees e.g., by deriving an intransitive verb tree from a transitive one; or by splitting a tree containing a PP into one tree without that PP and a PP tree.
7: **return** An FB-LTAG with unification semantics and a lexicon mapping semantic triples to FB-LTAG trees

---

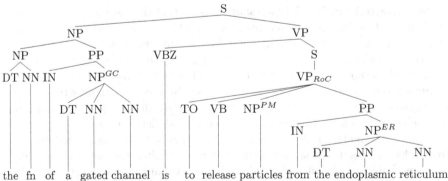

instance-of(GC,Gated-Channel),instance-of(RoC,Release-of-Calcium)
instance-of(PM,Particle-In-Motion),instance-of(ER,Endoplasmic-Reticulum)

Fig. 4. Parse tree with projected semantic variables. Input variables are first aligned with word forms (Step 1) and then projected onto parse tree nodes (Step 2).

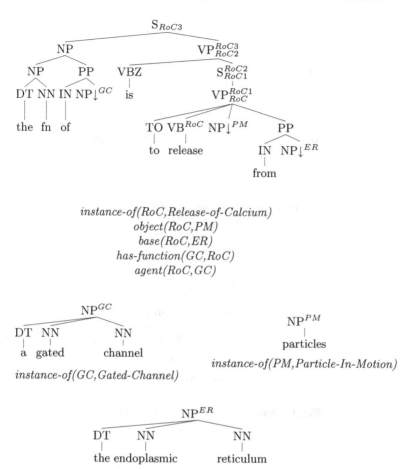

Fig. 5. Extracted Grammar for "*The function of a gated channel is to release particles from the endoplasmic reticulum*". Variable names have been abbreviated and the KBGen tuple notation converted to terms so as to fit the input format expected by our surface realiser.

principle requires that elementary TAG trees group together in a single structure a syntactic functor and its arguments while the *semantic principle* requires that each elementary tree captures a single semantic unit. Together these two principles ensure that TAG elementary trees capture basic semantic units and their dependencies.

Figure 1 gives a high level description of our grammar algorithm (see [20] for a more detailed description). In brief, the algorithm takes as input a set of (ϕ, S) pairs provided by the KBGen challenge where ϕ is a set of triples and S is a sentence verbalising ϕ. First, input variables in ϕ are aligned with word forms in S. Variables are then projected on non terminal nodes of S parse tree and

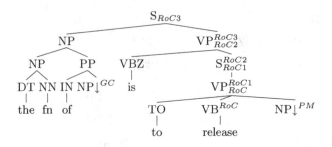

S RoC3
NP
VP RoC3 RoC2
NP
PP
VBZ
S RoC2 RoC1
DT NN IN NP↓GC
is
VP RoC1 RoC
the fn of
TO
VB RoC
NP↓PM
to
release
instance-of(RoC,Release-of-Calcium)
object(RoC,PM)
has-function(GC,RoC)
agent(RoC,GC)
VP RoC
VP*,RoC
PP
IN NP↓ER
from
base(RoC,ER)

Fig. 6. Deriving smaller from larger trees

System	All	Covered	Coverage	# Trees
IMS	0.12	0.12	100%	
UDEL	0.32	0.32	100%	
AutExp	0.29	0.29	100%	477

Fig. 7. BLEU scores and grammar size (number of elementary TAG trees)

	Fluency		Grammaticality		Meaning Similarity	
System	Mean	Homogeneous Subsets	Mean	Homogeneous Subsets	Mean	Homogeneous Subsets
UDEL	4.36	A	4.48	A	3.69	A
AutExp	3.45	B	3.55	B	3.65	A
IMS	1.91	C	2.05	C	1.31	B

Fig. 8. Human evaluation results on a scale of 0 to 5. Homogeneous subsets are determined using Tukey's Post Hoc Test with $p < 0.05$

used to constrain both tree extraction and the association of syntactic trees with semantics. Steps 5 and 6 of the algorithm generalise the extracted grammar by abstracting away from specific lexical and predicative information (Step 5) and by deriving smaller trees from extracted ones (Step 6).

Figure 4 shows a parse tree after variable projection and Fig. 5 shows the grammar produced by Step 4. Generalisation is illustrated in Fig. 6.

As illustrated by Figs. 5 and 6, the grammars extracted by our grammar induction algorithm are FB-LTAGs which conform with the semantic and the extended domain locality principle. In [20], we show that inducing such grammars help compensate the small size of the training data. Figures 7 and 8 show the results obtained on the KBGen data and compare them with those of the other two participating systems namely, a symbolic system based on hand written rules (UDEL) and a statistical system (IMS) based on a probabilistic grammar extracted from the KBGen training data. As the results show, our approach provides an interesting middle way between the two approaches. On the one hand, it produces sentences that are comparable in quality with those generated by the symbolic UDEL system but it produces them in a fully automatic manner thereby eschewing the need for highly skilled and time consuming manual labour. On the other hand, it generates sentences of much higher quality than those output by the statistical system. In sum, by extracting a linguistically principled grammar, we achieved good quality output while eschewing the need for manual grammar writing.

4   Grammar as a Means to Increase Domain Independence

We now turn to a second data-driven NLG application namely, the task of verbalising knowledge-base queries. Interfaces to knowledge bases which make use of Natural Language Generation have been shown to successfully assist the user by allowing her to formulate a query while knowing neither the formal query language nor the content of the KB being queried. In these interfaces, the user never sees the formal query. Instead, at each step in the query process, the generator verbalises all extensions of the current query which are consistent with this query and with the knowledge base. The user then chooses from among the set of generated NL queries, the query she intends.

One issue with such NLG based interfaces is that they should be domain independent. They should be usable for various KBs and various domains. Moreover, they should allow for an incremental processing of the user query. That is, they should support the user in incrementally refining her queries in a way that is consistent with the KB content.

While most previous work on generating from knowledge bases has assumed a restricted syntax and used either templates or a procedural framework (Definite Clause Grammars) to model the interaction between syntax, NL expressions and semantics, we developed an approach which uses a small hand-written FB-LTAG with unification semantics to support query verbalisation [30]. In essence, this approach consists in implementing a small hand-written FB-LTAG with unification semantics; in automatically constructing a lexicon which maps concept and relation names to trees in the FB-LTAG; and in adapting an Earley style parsing algorithm to support the incremental revision of a user queries.

In [30], we showed that this approach has several advantages.

First, because it does not rely on the existence of a training corpus, it is generic i.e., it results in a KB query system which can be used with different KBs on different domains. The key to developing a generic approach lies in combining a generic grammar which captures syntactic variations (canonical clauses, relative clauses, ellipses etc) with an automatically extracted lexicon which captures the lexicalisation of concepts and relations. While the grammar is hand written, the lexicon is automatically extracted from each ontology using the methodology described in [35]. When tested on a corpus of 200 ontologies, this approach was shown to be able to provide appropriate verbalisation templates for about 85 % of the relation identifiers present in these ontologies. 12 000 relation identifiers were extracted from the 200 ontologies and 13 syntactic templates were found to be sufficient to verbalise these relation identifiers (see [35] for more details on this evaluation).

Thus, in general, the extracted lexicons permit covering about 85 % of the ontological data. We further evaluated the coverage of our approach by running the generator on 40 queries generated from five distinct ontologies. The domains observed are cinema, wines, human abilities, disabilities, and assistive devices, e-commerce on the Web, and a fishery database for observations about an aquatic resource. The extracted lexicons contained 453 lexical entries in average and the coverage (proportion of formal queries for which the generator produced a NL query) was 87 %. Fuller coverage could be obtained by manually adding lexical entries, or by developing new ways of inducing lexical entries from ontologies (c.f. e.g. [37]).

A second advantage of the grammar based approach is that because it uses a well-defined grammar framework rather than e.g., templates or a procedural framework, it allows for the use of existing generation techniques and algorithms. In particular, we showed that the Earley style generation algorithm proposed in [8] could straightforwardly be adapted to support the incremental generation of user queries.

Finally, using a fully blown approach to syntax rather than templates or programs allows for more syntactic variability and better control on syntactic and lexical interactions (e.g., by using features to ensure number agreement or control the use of elliptical constructions). In comparison, template based approach often generate one clause per relation[2]. Thus for instance, the template-based Quelo system [16] will generate (1a) while our grammar based approach supports the generation of arguably more fluent sentences such as (1b).

(1)   a. I am looking for a car. Its make should be a Land Rover. The body style of the car should be an off-road car. The exterior color of the car should be beige.

---

[2] This is modulo aggregation of relations. Thus two subject sharing relations may be realised in the same clause.

b. I am looking for car whose make is a Land Rover, whose body style is an off-road car and whose exterior color is beige.

A human-based experiment indicate that the queries generated by our grammar based approach are perceived as more fluent than those produced by the Quelo template based approach (1.97 points[3] in average for the grammar based approach against 0.72 for the template based approach).

5   Grammar as a Means to Improve Statistical Disambiguisation

The generation algorithm described in Sect. 2.3 (i) explores the whole search space and (ii) is limited to generating a single sentence at a time. Generation from flat semantics is NP-complete however [7, 22] and existing algorithms for realization from a flat input semantics all have runtimes which, in the worst case, are exponential in the length of the input. Moreover, user queries can typically require the generation of several sentences. For instance, the query in (2) is better verbalised as (2a) than as (2b).

(2)   a. CarMake(x) isMakeOf(x y) CrashCar(y) DemonstrationCar(y) hasCar-Body(y z) OffRoad(z) soldBy(z w) CarDealer(w)

   b. *I am looking for a car make which should be the make of a crash car, a demonstration car. The body style of the crash car should be an off road and it should be sold by a car dealer.*

   c. *I am looking for a car make which should be the make of a crash car which is a demonstration car and whose body style should be an off road and should be sold by a car dealer.*

In [29], we present a hypertagger which (i) restricts the search space output by the lexical selection step and (ii) segments the input into sentence size chunks. That is, our hypertagger not only restricts the initial search space thereby reducing timeouts and generation time, but it also permits a joint modelling of surface realisation and sentence segmentation. This is possible because, in contrast to approaches such as [14, 23, 24, 27] which directly map semantics to strings, we mediate this relation using a grammar which differentiates between sentence starting (e.g., nx0VVVpnx1 in Fig. 1) and clause extending trees (e.g., relative clauses, sentence and VP coordination, PPs and elliptical clauses). Thus, a tagging sequence in effect determines sentence segmentation. For instance, given the query shown in (3a), if the hypertagger returns the sequence of tree tags shown in (3b), the output verbalisation will be (3d) because the tag Tnx0VVpnx1 indicates a sentence starting tree thereby forcing the segmentation of the input into two sentences.

---

[3] Fluency was rated on a scale from 0 to 5.

(3)   a. CarDealer(x) locatedIn(x,y) City(y) sell(x z) Car(z) runOn(z w) Diesel(w)
      b. (Trees) Tnx betanx0VPpnx1 nx ANDWHnx0VVnx1 nx nx0VVpnx1 nx
      c. (Synt.Classes) NP ParticipialOrGerund NP SubjRelAnd NP Canonical
      d. *I am looking for a car dealer located in a city and who should sell a car.*
          The car should run on a diesel.

In practice however, we do not use tree names as hypertags because of data
sparsity. The hypertagger is a model learned on a parallel corpus of NL and
formal queries. Creating such a corpus is labour intensive and we only collected
145 training instances. A first experiment which predicts tree as labels yielded
a tagging accuracy on complete inputs of 57.86 when considering the 10 best
outputs. More importantly, the model often failed to predict a sequence which
would allow for generation even when inputing to the tree combination phase
the 10 best tree sequences predicted by the hypertagger.

The tags learned by our hypertagger are therefore not tree names but more
general *syntactic classes* which capture the syntactic realisation of a semantic
token independent of its lexical class. Thus, the input query in (3a) will, in fact,
be tagged with the syntactic classes shown in (3c).

Table 1 shows the tree names and the syntactic classes associated with each
tree selected by the EQUIPPEDWITH relation while Table 2 shows example tree
names for lexical classes with distinct subcategorisation patterns. As can be
seen, while a tree name describes both the lexical and the syntactic pattern of a
lexical item (e.g., nx0VVVnx1 describes a transitive verb with canonical subject
and object NPs), syntactic classes capture syntactic generalisations which cut
across all subcategorisation patterns (e.g., the Canonical class is true of all reali-
sations with canonical subject and object NPs). Since in our grammar, each tree
is automatically associated by the grammar compilation process with its syntactic class, we first use the hypertagger to predict the syntactic class of an input
literal. We then restrict the set of trees that were lexically selected by that literal
to only those trees which have the syntactic class returned by the hypertagger.
For instance, given the literal EQUIPPEDWITH, while lexical selection will return
the set of trees shown in Table 1, if the hypertagger predicts the SubjRelAnd
class for this literal given the overall input, then the tree combination step of
the generation algorithm will only consider the tree labeled with that syntactic
class namely, the W0nx0VVVpnx1 tree.

Hypertagging is viewed as a sequence labelling task in which the sequence
of semantic input needs to be labelled with appropriate syntactic classes. The
linear order of the semantic input is deterministically given by the linearisation
process of the tree based conjunctive input (see [15] for more details).

We use a linear-chain Conditional Random Field (CRF, [25]) model to learn
the mapping between observed input features and hidden syntactic classes. This
probabilistic model defines the posterior probability of labels (syntactic classes)
$y = \{y_1, \ldots, y_n\}$ given the sequence of input literals $x = \{x_1, \ldots, x_k\}$:

$$P(y \mid x) = \frac{1}{Z(x)} \prod_{t=1}^{T} exp \sum_{k=1}^{K} \theta_k \Phi_k(y_t, y_{t-1,x_t})$$

Table 1. Verbalisations of the EQUIPPEDWITH relation captured by the lexicon and the grammar.

Example	Tree Name	Syntactic Class
NP_0 should be equipped with NP_1	nx0VVVpnx1	Canonical
It_0 should be equipped with NP_1	PRO0VVVpnx1	SubjPro
and NP_0 should be equipped with NP_1	sCONJnx0VVVpnx1	Scoord
and it_0 should be equipped with NP_1	sCONJPRO0VVVpnx1	ScoordSubjPro
NP_0 which should be equipped with NP_1	W0nx0VVVpnx1	SubjRel
NP_0 (...) and which should be equipped with NP_1	ANDWHnx0VVVpnx1	SubjRelAnd
NP_0 (...), which should be equipped with NP_1	COMMAWHnx0VVVpnx1	SubjRelComma
NP_0 equipped with NP_1	betanx0VPpnx1	ParticipialOrGerund
NP_0 (...) and equipped with NP_1	betanx0ANDVPpnx1	ParticipialOrGerundAnd
NP_0 (...), equipped with NP_1	betanx0COMMAVPpnx1	ParticipialOrGerundComma
NP_1 with which NP_0 should be equipped	W1pnx1nx0VV	PObjRel
NP_0 (equipped with X) and with NP_1	betavx0ANDVVVpnx1	SubjEllipAnd
NP_0 (equipped with X), with NP_1	betavx0COMMAVVVpnx1	SubjEllipComma

$Z(x)$ is a normalisation factor and the parameters θ_k are weights for the feature functions Φ_k.

Given a set of candidate hypertags (syntactic classes) associated with each literal, the hypertagging task consists into finding the optimal hypertag sequence y^* for a given input semantics x:

$$y^* = argmax_{y^*} P(y^* \mid x)$$

whereby the most likely hypertag sequence is computed using the Viterbi algorithm. We used the Mallet toolkit [28] for parameter learning and inference.

We train the CRF on a corpus aligning formal queries with the syntactic classes present in the FB-LTAG grammar. The corpus contains 145 training

Table 2. Example of Canonical Trees for each Subcategorisation Class

NP_0 should generate NP_1	nx0VVnx1
NP_0 should run on NP_1	nx0VVpnx1
NP_0 should be equipped with NP_1	nx0VVVpnx1
NP_0 should be the equipment of NP_1	nx0VVDNpnx1
NP_0 should have access to NP_1	nx0VVNpnx1
NP_0 should be relevant to NP_1	nx0VVApnx1
NP_0 should be an N_1 product	nx0VVDNnx1

instance with queries for 9 ontologies for different domains (car, cinema, wines, assistive devices and fishery).

All features are derived from the input semantics i.e., a sequence of inter-leaved relations and concepts. Since concepts have low lexical ambiguity (they mostly select NP trees), most of the features are associated with relations only and in the following, we write R_{i-1} to denote the relation which precedes relation R_i independently of how many concepts intervene between R_i and R_{i-1}. Features describe (i) the chaining relations between entities, (ii) the shape of relation names and correspondingly their lexicalisation properties i.e., the sequence of POS tags and indirectly the TAG tree that will be used to verbalise them, and (iii) global structural features pertaining to the overall shape of the input.

We evaluate the hypertagging module both in isolation and in interaction with the generator. The results show that hypertagging with syntactic classes[4]:

- improves hypertagging accuracy. Hypertagging with syntactic classes rather than tree names improves accuracy by up to 10.62 points for token accuracy; and by up to 20.77 points for input accuracy (token accuracy is the proportion of input literals correctly labelled while input accuracy is the proportion of correctly labelled input sequences).
- improves generation coverage by up to 17.25 points when compared with treename-based hypertagging (Generation coverage is the proportion of input for which generation yields an output).
- improves speed both with respect to both a generator using a treename-based hypertagger (-66 ms in average per input) and a symbolic generator without hypertagging. The symbolic generator repeatedly times out yielding an average generation time of 17 min on 145 inputs.
- preserves output quality. When compared with both a grammar based and a template based generation system, the output of our hybrid statistical hypertagging/grammar-based generation system is consistently perceived by human raters as clearer and more fluent. The human based evaluation involved ratings from 12 raters on a set of 30 input queries related to 9 knowledges bases. In comparison, the template based system generates one clause per

---

[4] For all results discussed, we assume a hypertagging module returning up to 20 best solutions.

Input	Flight hasCurrentDepartureDate.[Date] hasCurrentArrivalDate.[Date]
Query	hasDestination.[Airport hasFlightTo.[Airport]] hasCarrier.[Airline] hasTicket.[AirTicket hasDateOfIssue.[Date]]
Temp	I am looking for a flight. Its current departure date should be a date. The current arrival date of the flight should be a date. The destination of the flight should be an airport. The airport should have flight to an airport. The carrier of the flight should be an airline. The ticket of the flight should be an air ticket. The air ticket should have date of a date.
Hyb	I am looking for a flight whose current departure date should be a date, whose current arrival date should be a date and whose destination should be an airport. The airport should have flight to an airport. The carrier of the flight should be an airline. The ticket of the flight should be an air ticket whose date of issue should be a date.
Symb	I am looking for a flight whose current departure date should be a date and whose current arrival date should be a date and whose destination should be an airport which should have flight to an airport. Its carrier should be an airline, the ticket of the flight should be an air ticket and its date of issue should be a date.

Fig. 9. Example input and outputs. Temp is a template based system, Symb the symbolic generator described in Sect. 2.3 and Hyb the same generator augmented with the Hypertagger

relation and, on long queries, is judged unnatural (low fluency) by the raters. The symbolic generator often fails to adequately segment the input or to score the most fluent output highest. Figure 9 shows an example input and the corresponding output by each of the three systems being compared.

6   Conclusion

Syntax describes how words combine together to form complex NL expressions. Syntax is also often viewed as a scaffold for semantic construction. In other words, syntax provides both a means to abstract over lexical units and to mediate between form and meaning. While, when enough training data is available, statistical approaches can be developed which directly map meaning to form, there is, linguistically, no good reason to ignore the wealth of research that has gone into describing the syntax and the syntax/semantics interface of natural languages. In this paper, I have argues that syntax is in fact, a valuable component of natural language generation. In particular, I have shown that, by providing a higher level of abstraction, syntax permits improving a hypertagger performance; facilitates the development of a generic, domain independent, query verbaliser; and supports the induction of compact, linguistically principled grammars which are well suited for data-to-text generation.

Acknowledgments. Thanks to Laura Perez-Beltrachini and Bikash Gyawali for working with me on grammar-based generation from knowledge bases; to Eva Banik, Eric Kow and Vinay Chaudry for organising the KBGen challenge; and to Enrico Franconi for his collaboration on generating queries from Knowledge Bases.

References

1. Angeli, G., Liang, P., Klein, D.: A simple domain-independent probabilistic approach to generation. In: Proceedings of the 2010 Conference on Empirical Methods in Natural Language Processing, pp. 502–512. Association for Computational Linguistics (2010)
2. Baldridge, J., Kruijff, G.J.M.: Multi-modal combinatory categorial grammar. In: Proceedings of the Tenth Conference on European Chapter of the Association for Computational Linguistics, vol. 1, pp. 211–218. Association for Computational Linguistics (2003)
3. Bangalore, S., Rambow, O.: Using tag, a tree model, and a language model for generation. In: Proceedings of the 1st International Natural Language Generation Conference, Citeseer (2000)
4. Banik, E., Gardent, C., Kow, E., et al.: The KBGen challenge. In: Proceedings of the 14th European Workshop on Natural Language Generation (ENLG), pp. 94–97 (2013)
5. Belz, A.: Automatic generation of weather forecast texts using comprehensive probabilistic generation-space models. Nat. Lang. Eng. **14**(4), 431–455 (2008)
6. Bos, J.: Predicate logic unplugged. In: Dekker, P., Stokhof, M. (eds.) Proceedings of the 10th Amsterdam Colloquium, pp. 133–142 (1995)
7. Brew, C.: Letting the cat out of the bag: generation for shake-and-bake mt. In: Proceedings of the 14th Conference on Computational Linguistics, vol. 2, pp. 610–616. Association for Computational Linguistics (1992)
8. Gardent, C., Perez-Beltrachini, L.: Using regular tree grammar to enhance surface realisation. Nat. Lang. Eng. **17**, 185–201 (2011). (Special Issue on Finite State Methods and Models in Natural Language Processing)
9. Cahill, A., van Genabith, J.: Robust PCFG-based generation using automatically acquired LFG approximations. In: Proceedings of the 21st International Conference on Computational Linguistics and the 44th Annual Meeting of the Association for Computational Linguistics, pp. 1033–1040. Association for Computational Linguistics (2006)
10. Chen, D.L., Mooney, R.J.: Learning to sportscast: a test of grounded language acquisition. In: Proceedings of the 25th International Conference on Machine Learning, pp. 128–135. ACM (2008)
11. Coch, J.: Evaluating and comparing three text-production techniques. In: Proceedings of the 16th Conference on Computational Linguistics, vol. 1, pp. 249–254. Association for Computational Linguistics (1996)
12. Copestake, A., Lascarides, A., Flickinger, D.: An algebra for semantic construction in constraint-based grammars. In: Proceedings of the 39th Annual Meeting of the Association for Computational Linguistics, Toulouse, France (2001)
13. Dahl, D.A., Bates, M., Brown, M., Fisher, W., Hunicke-Smith, K., Pallett, D., Pao, C., Rudnicky, A., Shriberg, E.: Expanding the scope of the atis task: The atis-3 corpus. In: Proceedings of the Workshop on Human Language Technology, pp. 43–48. Association for Computational Linguistics (1994)

14. Dethlefs, N., Hastie, H., Cuayáhuitl, H., Lemon, O.: Conditional random fields for responsive surface realisation using global features. In: Proceedings of ACL, Sofia, Bulgaria (2013)
15. Dongilli, P.: Natural language rendering of a conjunctive query. KRDB Research Centre Technical Report No. KRDB08-3. Bozen, IT, Free University of Bozen-Bolzano 2, 5 (2008)
16. Franconi, E., Guagliardo, P., Trevisan, M.: An intelligent query interface based on ontology navigation. In: Proceedings of the Workshop on Visual Interfaces to the Social and Semantic Web (VISSW 2010), vol. 565. Citeseer (2010)
17. Gardent, C., Kow, E.: A symbolic approach to near-deterministic surface realisation using tree adjoining grammar. In: ACL07 (2007)
18. Gardent, C., Kallmeyer, L.: Semantic construction in FTAG. In: Proceedings of the 10th Meeting of the European Chapter of the Association for Computational Linguistics, Budapest, Hungary (2003)
19. Gardent, C., Perez-Beltrachini, L.: RTG based Surface Realisation for TAG. In: COLING'10, Beijing, China (2010)
20. Gyawali, B., Gardent, C.: Surface realisation from knowledge-base. In: ACL, Baltimore, USA June 2014
21. Hockenmaier, J.: Data and models for statistical parsing with combinatory categorial grammar. Ph.D. thesis, University of Edinburgh, College of Science and Engineering, School of Informatics (2003)
22. Koller, A., Striegnitz, K.: Generation as dependency parsing. In: Proceedings of the 40th Annual Meeting on Association for Computational Linguistics, pp. 17–24. Association for Computational Linguistics (2002)
23. Konstas, I., Lapata, M.: Concept-to-text generation via discriminative reranking. In: Proceedings of the 50th Annual Meeting of the Association for Computational Linguistics: Long Papers, vol. 1, pp. 369–378. Association for Computational Linguistics (2012)
24. Konstas, I., Lapata, M.: Unsupervised concept-to-text generation with hypergraphs. In: Proceedings of the 2012 Conference of the North American Chapter of the Association for Computational Linguistics: Human Language Technologies, pp. 752–761. Association for Computational Linguistics (2012)
25. Lafferty, J., McCallum, A., Pereira, F.C.: Conditional random fields: probabilistic models for segmenting and labeling sequence data. In: Proceedings of the Eighteenth International Conference on Machine Learning ICML '01, pp. 282–289. Morgan Kaufmann Publishers Inc., San Francisco (2001)
26. Liang, P., Jordan, M.I., Klein, D.: Learning semantic correspondences with less supervision. In: Proceedings of the Joint Conference of the 47th Annual Meeting of the ACL and the 4th International Joint Conference on Natural Language Processing of the AFNLP, vol. 1, pp. 91–99. Association for Computational Linguistics (2009)
27. Lu, W., Ng, H.T., Lee, W.S.: Natural language generation with tree conditional random fields. In: Proceedings of the 2009 Conference on Empirical Methods in Natural Language Processing, vol. 1, pp. 400–409. Association for Computational Linguistics (2009)
28. McCallum, A.K.: Mallet: A machine learning for language toolkit (2002). http://www.cs.umass.edu/~mccallum/mallet
29. Perez-Beltrachini, L., Gardent, C.: Hypertagging for query generation. May 2014 (submitted)
30. Perez-Beltrachini, L., Gardent, C., Franconi, E.: Incremental query generation. In: EACL, Gothenburg, Sweden April 2014

31. Portet, F., Reiter, E., Hunter, J., Sripada, S.: Automatic generation of textual summaries from neonatal intensive care data. In: Bellazzi, R., Abu-Hanna, A., Hunter, J. (eds.) AIME 2007. LNCS (LNAI), vol. 4594, pp. 227–236. Springer, Heidelberg (2007)
32. Reiter, E., Sripada, S., Hunter, J., Yu, J., Davy, I.: Choosing words in computer-generated weather forecasts. Artif. Intell. **167**(1), 137–169 (2005)
33. Steedman, M.: The syntactic process, vol. 35. MIT Press (2000)
34. The XTAG Research Group: A lexicalised tree adjoining grammar for english. Technical report, Institute for Research in Cognitive Science, University of Pennsylvannia (2001)
35. Trevisan, M.: A portable menuguided natural language interface to knowledge bases for querytool. Master's thesis, Free University of Bozen-Bolzano (Italy) and University of Groningen (Netherlands) (2010)
36. Vijay-Shanker, K., Joshi, A.K.: Feature structures based tree adjoining grammars. In: Proceedings of the 12th Conference on Computational linguistics, vol. 2, pp. 714–719. Association for Computational Linguistics (1988)
37. Walter, S., Unger, C., Cimiano, P.: A corpus-based approach for the induction of ontology lexica. In: Métais, E., Meziane, F., Saraee, M., Sugumaran, V., Vadera, S. (eds.) NLDB 2013. LNCS, vol. 7934, pp. 102–113. Springer, Heidelberg (2013)
38. White, M.: Efficient realization of coordinate structures in combinatory categorial grammar. Res. Lang. Comput. **4**(1), 39–75 (2006)
39. Wong, Y.W., Mooney, R.J.: Generation by inverting a semantic parser that uses statistical machine translation. In: HLT-NAACL, pp. 172–179 (2007)

Spoken Language Processing: Time to Look Outside?

Roger K. Moore[(✉)]

Speech and Hearing Research Group, University of Sheffield Regent Court,
211 Portobello, Sheffield S1 4DP, UK
r.k.moore@sheffield.ac.uk
http://www.dcs.shef.ac.uk/~roger

Abstract. Over the past thirty years, the field of spoken language processing has made impressive progress from simple laboratory demonstrations to mainstream consumer products. However, commercial applications such as *Siri* highlight the fact that there is still some way to go in creating *Autonomous Social Agents* that are truly capable of conversing effectively with their human counterparts in real-world situations. This paper suggests that it may be time for the spoken language processing community to take an interest in the potentially important developments that are occurring in related fields such as cognitive neuroscience, intelligent systems and developmental robotics. It then gives an insight into how such ideas might be integrated into a novel *Mutual Beliefs Desires Intentions Actions and Consequences* (MBDIAC) framework that places a focus on generative models of communicative behaviour which are recruited for interpreting the behaviour of others.

Keywords: Spoken language processing · Enactivism · Language grounding · Mirror neurons · Perceptual control · Cognitive architectures · Autonomous Social Agents

1 Introduction

Since the 1980s, the introduction of stochastic modelling techniques - particularly hidden Markov models (HMMs) - into the field of spoken language processing has given rise to steady year-on-year improvements in capability [1,2]. Coupled with a relentless increase in the processing power of the necessary computing infrastructure, together with the introduction of public benchmark testing, the field has developed from a specialist area of engineering research into the commercial deployment of mainstream consumer products. With the advent of smartphone applications such as Apple's Siri, Microsoft's Cortana and Google's Now, speech-based interaction with 'intelligent' devices has entered the popular imagination, and public awareness of the potential benefits of hands-free access to information is at an all-time high [3].

The gains in performance for component technologies such as automatic speech recognition and text-to-speech synthesis have accrued directly from the

© Springer International Publishing Switzerland 2014
L. Besacier et al. (Eds.): SLSP 2014, LNAI 8791, pp. 21–36, 2014.
DOI: 10.1007/978-3-319-11397-5_2

deployment of state-of-the-art machine learning techniques in which significantly large corpora of annotated speech (often thousands of hours) are used to estimate the parameters of rich context-sensitive Bayesian models. Indeed, the immense challenges posed by the need to create accurate and effective spoken language processing has meant that speech technology researchers have become acknowledged pioneers in the use of the most advanced machine learning techniques available. A recent example of this is the performance gains arising from the use of deep neural networks (DNNs) [4].

However, notwithstanding the immense progress that has been made over the past thirty or so years, it is generally acknowledged that there is still some way to go before spoken language technology systems are sufficiently reliable for the majority of envisaged applications. Whilst the performance of state-of-the-art systems is impressive, it is still well short of what is required to provide users with an effective and reliable alternative to traditional interface technologies such as keyboards and touch-sensitive screens [5]. Moreover, it is clear that spoken language capabilities of the average human speaker/listener are considerably more robust in adverse real-world situations such as noisy environments, dealing with speakers with foreign accents or conversing about entirely novel topics. This means that there is still a clear need for significant improvements in our ability to model and process speech, and hence it is necessary to ask where these gains might arise - more training data, better models, new algorithms, or from some other source [6]?

It is posited here that it is time for the spoken language processing community to look outside the relatively narrow confines of the discipline in order to understand the potentially important developments that are taking place in related areas. Fields such as cognitive neuroscience, intelligent systems and developmental robotics are progressing at an immense pace and, although some of the tools and techniques employed in spoken language processing could be of value to those fields, there is a growing understanding outside the speech area of how living systems are organised and how they interact with the world and with each other. Some of these new ideas could have a direct bearing on future spoken language systems, and could provide a launchpad for the kinds of developments that are essential if the potential of speech-based language interaction with machines is to be realised fully. This paper addresses these issues and introduces a number of key ideas from outside the field of spoken language processing which the author believes could be of some significance to future progress.

2   Looking for Inspiration Outside

It is often remarked that spoken language could be the most sophisticated behaviour of the most complex organism we know [7–9]. However, the apparent ease with which we as human beings interact using speech tends to mask the variety and richness of the mechanisms that underpin it. In fact the spoken language processing research community has become so focused on the rather obvious surface patterning - such as lexical structure (i.e. words) - that the foundational principles on which spoken *interaction* is organised has a tendency to be overlooked. In reality, long before spoken language dialogue evolved as a rich communicative behaviour, the distant ancestors of modern human beings were coordinating their activities using a variety of communicative modes and behaviours (such as the synchronisation of body postures, making explicit gestures, the laying down of markers in the environment and the use of appropriate sounds and noises). Interactivity is thus a fundamental aspect of the behaviour of living systems, and it would seem appropriate to found spoken language interaction on more primitive behaviours.

Interestingly, interactivity is not solely concerned with the behavioural relationship between one organism and another. In the general case, interactivity takes place between an organism and its physical *environment*, where that environment potentially incorporates other living systems. From an evolutionary perspective, interactive behaviour between an organism and its environment can be seen to emerge as a survival mechanism aimed at maintaining the persistence of an organism long enough for successful procreation, and these are issues that have engaged deep thinking theorists for any years. Of particular relevance here is the growth of an approach to understanding (and modelling) living systems known as *enactivism*.

2.1   Enactivism

Enactivism grew out of seminal work by Humberto Maturana and Francisco Varela [10] in which they tackled fundamental questions about the nature of living systems. In particular, they identified *autopoiesis* (a process whereby organisational structure is preserved over time) as a critical self-regulatory mechanism and *cognition* (the operation of a nervous system) as providing a more powerful and flexible autopoietic mechanism than purely chemical interactions. They defined a minimal living system such as a single cell as an *autopoietic unity* whereby the cell membrane maintains the boundary between the unity and everything else. Hence, a unity is said to be structurally *coupled* with its external environment - 1st-order coupling - and, for survival, appropriate interactive behaviours are required to take place (such as moving up a sugar gradient).

Likewise, unities may be coupled with other unities forming *symbiotic* or *metacellular* organisational structures - 2nd -order coupling - which can then be viewed organisationally as unities in their own right. The neuron is cited as a special type of cell emerging from particular symbiotic coupling, and the nervous system is thus seen as facilitating a special form of 2nd-order metacellular organisation termed a *cognitive unity*. Finally, Maturana and Varela propose that interaction between cognitive unities - 3rd-order coupling - is manifest in the form of organised *social systems* of group behaviour, and the emergence of cooperation, communication and language are posited as a consequence of 3rd-order coupling.

The enactive perspective thus establishes a simple and yet powerful framework for understanding the complexity of interaction between living systems, and it holds the promise for the investigation of computational approaches that seek to mimic these same behaviours. The emphasis on the coupling between a cognitive unity and its external environment (including other unities) is central to the approach, and this provides two clear messages - interactivity must be viewed as essentially *multimodal* in nature and that interaction is *grounded* in the context in which it takes place. Likewise, enactivism makes it clear that (spoken) language interaction is founded upon more general-purpose behaviours for continuous communicative coupling rather than simple turn-by-turn message passing [11,12].

2.2 Multimodal Interaction and Communication

In principle, the modality in which interaction between a living system and its environment (including other living systems) takes place should be irrelevant. However, in practice, the characteristics and affordances [13] of the different modes greatly influence the modalities employed. For example, it may be easier to move a heavy object by pushing it bodily rather than by blowing air at it. Similarly, it may be safer to influence the behaviour of another living system by making a loud noise from a distance rather than by approaching it and touching it physically.

Nevertheless, notwithstanding the static advantages and disadvantages of any particular mode of interaction, in a dynamic and changing world it makes sense for an organism to be able to actively distribute information across alternative modes as a function of the situational context. Hence, even a sophisticated behaviour such as language should be viewed as being essentially a multimodal activity. Given this perspective, it would be natural to assume that there exists some significant relationship between physical gestures and vocal sounds. In such a framework, the power of multimodal behaviour such as speaking and pointing would be taken for granted, and the emergence of prosody as a fundamental carrier of unimodal vocal pointing behaviour would be more obvious.

For an up-to-date review of multimodal integration in general, see [14], and for speech and gesture in particular, see [15]. The argument here is that such behaviours are not simply 'nice to have' additional features (as they tend to be treated currently), but that they represent the basic substrate on which spoken language interaction is founded. Indeed a number of authors have argued that vocal language evolved from gestural communication (freeing up the hands for tool use or grooming) [16–18]. Hence, these insights suggest that information about multimodal characteristics and affordances should be intrinsic to the computational modelling paradigms employed in spoken language systems.

2.3 Language Grounding

The notion that an organism is not only coupled with its environment, but also with other organisms in the environment, introduces another important and fundamental aspect of interactive behaviour - passive information flow versus active signalling. In the first case, almost any behaviour could have indirect consequences in the sense that the environment could be disturbed by any physical activity, and such disturbance may provide a cue to other organisms as to what has taken place. As a result, organisms could exploit the availability of such information for their own benefit; for example, a predator could track a prey by following its trail of scent. In this situation, the emergent coupled behaviour is conditioned upon passive (unintentional) information transfer between individual organisms via the environment. In this case, the information laid down in the environment has *meaning* for the receiver, but not for the sender. However, living systems may also actively manage the information flow, and this would take the form of active (intentional) signalling - signals that have meaning for the sender (and hopefully, the receiver).

Meaning and semantics have been rather latecomers to the spoken language processing party. However, from the perspective being developed here, it is clear that the significance and implications of a behaviour are fundamental to the dynamics of the coupling that takes place between one individual and another. In other words, meaning is everything! The implication of this view is that the coupling is contingent on the *communicative context* which, in general terms, consists of the characteristics of the agents involved, the physical environment in which they are placed and the temporal context in which the actions occur. In modern terminology, meaningful communication is said to be *grounded* in the real world [19], and that generating and interpreting such behaviour is only possible with reference to the *embodied* nature and *situated* context in which the interactions take place. The grounding provided by a common physical environment gives rise to the possibility of *shared* meanings and representations [20], and crucial behaviours such *joint attention* and *joint action* emerge as a direct consequence of managing the interaction [21–24].

Such a perspective has taken strong hold in the area of developmental robotics in which autonomous agents acquire communication and language skills (and in particular, meanings) not through instruction, but through interaction [25–29]. These approaches address the *symbol grounding problem* [30] by demonstrating that linguistic structure can be mapped to physical movement and sensory perception. As such, they represent the first steps towards a more general approach which hypothesises that even the most abstract linguistic expressions may be understood by the use of *metaphor* to link high-level representations to low-level perceptions and actions [31].

2.4   Mirror Neurons and Simulation

One of the drivers behind grounding language in behaviour is the discovery in the 1990s of a neural mechanism - so-called *mirror neurons* - that links action and perception [32,33]. The original experiment involved the study of neural activity in the motor cortex of a monkey grasping a small item (such as a raisin). The unexpected outcome was that neurons in the monkey's pre-frontal motor cortex fired, not only when the monkey performed the action, but also when the monkey observed a human experimenter performing the same action. As a control, it turned out that activation did not occur when the human experimenter used a tool (such as tweezers) to perform the action. The implication was that, far from being independent faculties, action and perception were somehow intimately linked.

The discovery of mirror neurons triggered an avalanche of research aimed at uncovering the implications of significant sensorimotor overlap. The basic idea was that mirror structures appeared to facilitate mental *simulations* that could be used for interpreting the actions and intentions of others [34]. Simulation not only provides a generative forward model that may be used to explain observed events, but it also facilitates the prediction of future events, the imagination of novel events and the optimal influence of future events. The mirror mechanism thus seemed to provide a basis for a number of important behaviours such as action understanding [35], imitation and learning [36], empathy and theory of mind [37] and, of most significance here, the evolution of speech and language [38–41].

Since the simulation principle suggests that generative models of spoken language production could be implicated in human speech recognition and understanding, the discovery of mirror neurons sparked a revival of interest in the *motor theory of speech perception* [42]. The jury is still out as to the precise role of the speech motor system in speech perception, but see [43–45] for examples of discussion on this topic.

The mirror neuron hypothesis has also had some impact on robotics research (see [46], for example), and the notion of mental simulation as a *forward model/predictor* mechanism has inspired new theories of language [47–49] and speech perception [50].

2.5 Perceptual Control Theory

As suggested above, the structural coupling of an agent with its environment (including other agents) could be instantiated as a one-way causal dependency. However, it is more likely that coupling would be bi-directional, and this implies the existence of a *dynamical system* with *feedback*. Feedback - in particular, *negative* feedback - provides a powerful mechanism for achieving and maintaining *stability* (static or dynamic), and feedback control systems have been posited as a fundamental property of living systems [51,52].

Founded on principles first expounded in the field of cybernetics [53], and railing against the traditional behaviourist perspective taken by mainstream psychologists, *perceptual control theory (PCT)* focuses on the consequence of a negative-feedback control architecture in which behaviour emerges, not from an external stimulus, but from an organism's internal drive to achieve desired perceptual states [54]. Unlike the traditional stimulus-response approach, PCT is able to explain how a living organism can compensate for (unpredictable) disturbances in the environment without the need to invoke complex statistical models. For example, the orientation of a foot placed on uneven ground is controlled, not by computing the consequences of an unusual joint angle, but by the need to maintain a stable body posture. Likewise, PCT suggest that the clarity of speech production is controlled, not by computing the consequences of the amount of noise in an environment, but by the need to maintain a suitable level of perceived intelligibility.

Indeed, the importance of feedback control in speech has been appreciated for some time, coupled with the realisation that living systems have to balance the effectiveness of their actions against the (physical and neural) effort that is required to perform the actions [55]. This principle has been used to good effect in a novel form of speech synthesis that regulates its pronunciation using a crude model of the listener [56].

2.6   Intentionality, Emotion and Learning

PCT provides an insight into a model of behaviour that is active/intentional rather than passive/reactive, and this connects very nicely with the observation that human beings tend to regard other human beings, animals and even inanimate objects as *intentional* agents [57]. It also links with the view of language as an intentional behaviour [58], and thus with mirror neurons as a mechanism for inferring the communicative intentions of others [59].

Intentionality already plays a major role in the field of *agent-based modelling*, in particular using the BDI *Beliefs, Desires, Intentions* paradigm [60,61]. BDI is an established methodology for modelling emergent behaviours from swarms of 'intelligent' agents, but it doesn't specify how to recognise/interpret behaviour under conditions of ambiguity or uncertainty. Nevertheless, BDI does capture some important features of behaviour, and it is useful to appreciate that beliefs equate to *priors* (which equate to memory), desires equate to goals, and intentions drive planning and action.

Viewing the behaviour of living systems as intentional with the consequences of any actions being monitored using perceptual feedback, leads to a model of behaviour that is driven by a comparison between desired and actual perceptual states (that is, by the error signal in a PCT-style feedback control process). This difference between intention and outcome can be regarded as an *appraisal* [62] of emotional valence whereby a match is regarded as positive/happy and a mismatch is regarded as negative/unhappy [63]. From this perspective, emotion can be seen as a driver of behaviour (rather than simply a consequence of behaviour) and provides the force behind adaptation and learning.

3   Bringing the Ideas Inside

The foregoing provides a wealth of insights from outside the technical field of spoken language processing that could have a direct bearing on future spoken language systems. In particular, it points to a novel computational architecture for spoken language processing in which the distinctions between traditionally independent system components become blurred. It would seem that speech recognition and understanding should be based on forward models of speech generation/production, and that those models should be the same as those used by the system to generate output itself. It turns out that dialogue management should be concerned less with turn taking and more with synchronising and coordinating its behaviours with its users.

The ideas above also suggest that a system's goals should be to satisfy users' rather than systems' needs, and this means that systems need to be able to model users and determine their needs by empathising with them. A failure to meet users' needs should lead to negative affect in the system, an internal variable which is not only used to drive the system's behaviour towards satisfying the user, but which could also be expressed visually or vocally in order to keep a user informed of the system's internal states and intentions. The previous section also points to a view of spoken language processing that is more integrated with its external environment, and to systems which are constantly adapting to compensate for the particular contextual circumstances that prevail.

3.1   Existing Approaches

A number of these ideas have already been discussed in the spoken language processing literature, and some are finding their way into practical systems. For example, the PRESENCE (*PREdictive SENsorimotor Control and Emulation*) architecture [64–66] draws together many of these principles into a unified framework in which the system has in mind the needs and intentions of its users, and a user has in mind the needs and intentions of the system. As well as the listening speech synthesiser mentioned earlier [67], PRESENCE has informed a number of developments in spoken language processing including the use of user emotion to drive dialogue [68], *AnTon* - an animatronic model of the human tongue and vocal tract [69] and the parsimonious management of interruptions in conversational systems [70, 71].

Another area of on-going work that fits well with some of the themes identified above is the powerful notion of *incremental* processing whereby recognition, dialogue management and synthesis all progress in parallel [72–75]. These ideas fit well with contemporary approaches to dialogue management using POMDPs *Partially-Observable Markov Decision Processes* [76, 77].

However, despite these important first steps, as yet there is no mathematically grounded framework that encapsulates all of the key ideas into a practical computational architecture. Of course, this is not surprising - these are complex issues that can be difficult to interpret. So, where might one start? The following is a preliminary taste of what might be required [78].

3.2   Towards a General Conceptual Framework

One of the main messages from the foregoing is that a key driver of behaviour - including speaking - for a living system seems to be *intentionality* (based on *needs*). Consider, therefore, a *world* containing just two intentional agents - *agent*1 and *agent*2. The world itself obeys the Laws of Physics, which means that the evolution of events follows a straightforward course in which actions lead to consequences (which constitute further actions) in a continuous cycle of cause and effect.

$$Actions_t \rightsquigarrow Consequences \mapsto Actions_{t+1}. \tag{1}$$

The behaviour of the world can thus be characterised as...

$$Consequences = f_{world}\left(Actions\right), \tag{2}$$

where f is some function which transforms *Actions* into *Consequences*.

The two intentional agents are each capable of (i) effecting changes in the world and (ii) inferring the causes of changes in the world.

In the first case, the intentions of an agent lead to actions which in turn lead to consequences...

$$Intentions \rightsquigarrow Actions \rightsquigarrow Consequences. \tag{3}$$

The behaviour of the agent can thus be characterised as...

$$Actions = g_{agent}\left(Intentions\right), \tag{4}$$

where g is some function that transforms *Intentions* into *Actions*.

In the second case, an agent attempts to infer the actions that gave rise to observed consequences.

$$Actions \rightsquigarrow Consequences \rightsquigarrow \widehat{Actions}. \tag{5}$$

The behaviour of the agent can thus be characterised as...

$$\widehat{Actions} = h_{agent}\left(Consequences\right), \tag{6}$$

where h is some function that transforms *Consequences* into estimated *Actions*.

This analysis becomes interesting when there is (intentional) interaction between the two agents. However, before taking that step, it is necessary to consider the interactions between the agents and the world in a little more detail.

An Agent Manipulating the World. Consider an agent attempting to manipulate the world, that is intentions are transformed into actions which are transformed into consequences. In robotics, the process of converting an intention into an appropriate action is known as *action selection*, and the relevant transformation is shown in Eq. 4. Note, however, the emphasis here is not on the *actions* that are required, but on the *consequences* of those actions.

$$Consequences = f_{world}\left(g_{agent}\left(Intentions\right)\right), \tag{7}$$

where g is a transform from intentions to actions and, as before, f is the transform from actions to consequences.

Of course, whether the intended consequences are achieved depends on the agent having the correct transforms. It is possible to discuss how f and g might be calibrated. However, there is an alternative approach that is not dependent on knowing f or g, and that is to search over possible actions to find those that create the best match between the intentions and the observed consequences.

$$\widehat{Actions} = \underset{Actions}{\arg\min}\,(Intentions - Consequences)\,, \tag{8}$$

where $Intentions - Consequences$ constitutes an error signal that reflects the agent's *appraisal* of its actions. A large value means that the actions are poor; a small value means that the actions are good. Hence, the error signal can be said to be equivalent to *emotional valence* - as discussed in Sect. 2.6. Overall, this optimisation process is a negative feedback control loop that operates to ensure that the consequences match the intentions even in the presence of unpredictable disturbances. This is exactly the type of control structure envisaged in *Perceptual Control Theory* - Sect. 2.5.

The approach works will only function if the agent can observe the consequences of its actions. However, when an agent is manipulating another agent, the consequences are likely to be changes in internal state and thus potentially unobservable. This situation is addressed below, but first it is necessary to consider an agent interpreting what's happening in the world.

An Agent Interpreting the World. The challenge facing an agent attempting to interpret what is happening in the world is to derive the actions/causes from observing their effects/consequences. If the inverse transform f^{-1} is known (from Eq. 2), then it is possible to compute the actions directly from the observed consequences...

$$Actions = f^{-1}_{world}\,(Consequences)\,. \tag{9}$$

However, in reality the inverse transform is not known. If it can be estimated $\widehat{f^{-1}}$, then it is possible to compute an estimate of the actions...

$$\widehat{Actions} = \widehat{f^{-1}_{world}}\,(Consequences)\,. \tag{10}$$

Of course the accuracy with which the causes can be estimated depends on the fidelity of the inverse transform.

An alternative approach, which aligns well with some of the ideas in the previous section, is not to use an inverse model at all, but to use a *forward model* - that is, an estimate of f ($\hat{f}$). Estimation then proceeds by searching over possible actions to find the best match between the predicted consequences ($\widehat{Consequences}$) and the observed consequences - again, a negative-feedback control loop.

$$\widehat{Actions} = \underset{Consequences}{\arg\min}\,\left(Consequences - \widehat{Consequences}\right)\,. \tag{11}$$

Of course the forward model is itself an estimate...

$$\widehat{Consequences} = \widehat{f_{world}}\,(Actions)\,, \tag{12}$$

which leads to...

$$\widehat{Actions} = \underset{Actions}{\arg\min} \left(Consequences - \widehat{f_{world}}(Actions) \right). \qquad (13)$$

As an aside, the same idea can be expressed in a Bayesian framework, but the principle is the same - interpretation is performed using search over a forward model...

$$\Pr(Actions|Consequences) = \frac{\Pr(Consequences|Actions)\Pr(Actions)}{\Pr(Consequences)}. \qquad (14)$$

Hence the estimated action is that which maximises the following...

$$\widehat{Actions} = \underset{Actions}{\arg\max} \left(\Pr(Consequences|Actions) \right), \qquad (15)$$

where $\Pr(Consequences|Actions)$ is the forward/generative model (equivalent to $\widehat{f_{world}}(Actions)$).

An Agent Communicating Its Intentions to Another Agent. Now it is possible to turn to the situation where one agent - *agent*1 - seeks to manipulate another agent - *agent*2. As mentioned above, in this case the consequences of *agent*1's actions may not be observable (because *agent*1's intention is to change the mental state of *agent*2). However, if *agent*1 can observe its own actions, then it can use a model to *emulate* the consequences of its actions. That is, *agent*1 uses an estimate of the forward transform $\widehat{h_{agent2}}$.

$$\widehat{Actions} = \underset{Actions}{\arg\min} \left(Intentions - \widehat{h_{agent2}}(Actions) \right). \qquad (16)$$

This solution is equivalent to *agent*1 actively trying out actions in order to arrive at the correct ones. However, an even better solution is for *agent*1 not to search in the real world, but to search in a *simulated* world - that is, to imagine the consequences of its actions in advance of performing the chosen ones. This is *emulation* as described in Sect. 2.4 which introduced the action of mirror neurons.

$$\widehat{Actions} = \underset{\widetilde{Actions}}{\arg\min} \left(Intentions - \widehat{h_{agent2}}(\widetilde{Actions}) \right). \qquad (17)$$

As before, a negative-feedback control loop manages the search and, interestingly, it can also be viewed as *synthesis-by-analysis*.

An Agent Interpreting the Actions of Another Agent. For an agent to interpret the actions of another agent, they are effectively inferring the intentions of that agent. In this case, *agent*2 needs to infer the intentions of *agent*1 by comparing the observed actions with the output of a forward model for *agent*1...

$$\widehat{Intentions} = \underset{Intentions}{\arg\min} \left(Actions - \widehat{g_{agent1}}(Intentions) \right). \qquad (18)$$

As before, a negative-feedback control loop manages the search to find the best match and, in this case the process can be viewed as *analysis-by-synthesis*.

3.3   Using *Self* to Model *Other*

Most of the derivation thus far is relatively straightforward in that it reflects known approaches, albeit caste in an unusual framework. This is especially obvious if one maps the arrangements into a Bayesian formulation. The overlap with some of the ideas outlined in Sect. 2 should be apparent.

However, there is one more step that serves to set this whole approach apart from more standard analyses, and that step arises from the question "where do $\widehat{g}$ and $\widehat{h}$ come from"? From the perspective of *agent*1, $\widehat{h}$ is a property of *agent*2 (and vice-versa). Likewise, From the perspective of *agent*2, $\widehat{g}$ is a property of *agent*1 (and vice-versa). The answer, drawing again on concepts emerging from Sect. 2, is that for a particular agent - *self* - the information required to model another agent - *other* - could be derived, not from modelling the behaviour of *other*, but from the capabilities of *self*! In other words, the simulation of *other* recruits information from the existing abilities of *self* - just as observed in mirror neuron behaviour (Sect. 2.4).

If spoken language is the behaviour of interest, then such an arrangement would constitute *synthesis-by-analysis-by-synthesis* for the speaker and *analysis-by-synthesis-by-analysis* for the listener.

4   Conclusion

This paper has reviewed a number of different ideas from outside the mainstream field of spoken language processing (starting from the coupling between living cells), and given an insight into how they might be integrated into a novel framework that could have some bearing on the architecture for future *intelligent interactive empathic communicative* systems [79]. The approach - which might be termed MBDIAC *Mutual Beliefs Desires Intentions Actions and Consequences* - is different from the current paradigm in that, rather than estimate model parameters off-line using vast quantities of static annotated spoken language material, it highlights an alternative developmental paradigm based on on-line interactive skill acquisition in dynamic real-world situations and environments. It also places a focus on generative models of communicative behaviour (grounded in movement and action, and generalised using metaphor) which are subsequently recruited for interpreting the communicative behaviour of others.

What is also different about the approach suggested here is that, in principle, it subsumes everything from high-level semantic and pragmatic representations down to the lowest-level sensorimotor behaviours. The approach is also neutral with respect to the sensorimotor modalities involved; hence gesture and prosody have an equal place alongside the more conventional vocal behaviours. The overall message is that it may be time to step back from worrying about the detail of contemporary spoken language systems in order to rediscover the crucial communicative context in which communicative behaviour takes place. That way we might be able to design and implement truly *Autonomous Social Agents* that are capable of conversing effectively with their human counterparts.

Finally, some readers may be tempted to think that this approach is promoting a debate between statistical and non-statistical modelling paradigms. On the contrary, the entire edifice should be capable of being caste in a probabilistic framework. The concern here is not about probability but about *priors*!

Acknowledgments. The author would like to thank colleagues in the Sheffield Speech and Hearing research group and the Bristol Robotics Laboratory for discussions relating to the content of this paper. This work was partially supported by the European Commission [grant numbers EU-FP6- 507422, EU-FP6-034434, EU-FP7-231868, FP7-ICT-2013-10-611971] and the UK Engineering and Physical Sciences Research Council [grant number EP/I013512/1].

References

1. Huang, X., Acero, A., Hon, H.-W.: Spoken Language Processing: A Guide to Theory, Algorithm, and System Development. Prentice Hall PTR, Upper Saddle River (2001)
2. Gales, M., Young, S.: The application of hidden Markov models in speech recognition. Found. Trends Sig. Process. **1**(3), 195–304 (2007)
3. Pieraccini, R.: The Voice Mach. MIT Press, Cambridge (2012)
4. Hinton, G., Deng, L., Yu, D., Dahl, G.E., Mohamed, A., Jaitly, N., Senior, A., Vanhoucke, V., Nguyen, P., Sainath, T.N., Kingsbury, B.: Deep neural networks for acoustic modeling in speech recognition: the shared views of four research groups. IEEE Sig. Process. Mag. **29**(6), 82–97 (2012)
5. Moore, R.K.: Modelling data entry rates for ASR and alternative input methods. In: INTERSPEECH 2004 ICSLP, Jeju, Korea (2004)
6. Moore, R.K.: Spoken language processing: where do we go from here? In: Trappl, R. (ed.) Your Virtual Butler. LNCS, vol. 7407, pp. 119–133. Springer, Heidelberg (2013)
7. Dawkins, R.: The Blind Watchmaker. Penguin Books, London (1991)
8. Gopnik, A., Meltzoff, A.N., Kuhl, P.K.: The Scientist in the Crib. Perennial, New York (2001)
9. Moore, R.K.: Towards a unified theory of spoken language processing. In: 4th IEEE International Conference on Cognitive Informatics, Irvine, CA (2005)
10. Maturana, H.R., Varela, F.J.: The Tree of Knowledge: The Biological Roots of Human Understanding. New Science Library/Shambhala Publications, Boston (1987)
11. Garrod, S., Pickering, M.J.: Why is conversation so easy? Trends Cogn. Sci. **8**, 8–11 (2004)
12. Fusaroli, R., Raczaszek-Leonardi, J., Tyln, K.: Dialog as interpersonal synergy. New Ideas Psychol. **32**, 147–157 (2014)
13. Gibson, J.J.: The theory of affordances. In: Shaw, R., Bransford, J. (eds.) Perceiving, Acting, and Knowing: Toward an Ecological Psychology, pp. 67–82. Lawrence Erlbaum, Hillsdale (1977)
14. Turk, M.: Multimodal interaction: a review. Pattern Recogn. Lett. **36**, 189–195 (2014)
15. Wagner, P., Malisz, Z., Kopp, S.: Gesture and speech in interaction: an overview. Speech Commun. **57**, 209–232 (2014)

16. Mithen, S.: The Prehistory of the Mind. Phoenix, London (1996)
17. MacWhinney, B.: Language evolution and human development. In: Bjorklund, D., Pellegrini, A. (eds.) Origins of the Social Mind: Evolutionary Psychology and Child Development, pp. 383–410. Guilford Press, New York (2005)
18. Tomasello, M.: Origins of Human Communication. MIT Press, Cambridge (2008)
19. Clark, H.H., Brennan, S.A.: Perspectives on socially shared cognition. In: Resnick, L.B., Levine, J.M., Teasley, S.D. (eds.) Grounding in communication, pp. 127–149. APA Books, Washington (1991)
20. Pezzulo, G.: Shared representations as coordination tools for interaction. Rev. Philos. Psychol. **2**, 303–333 (2011)
21. Tomasello, M.: The role of joint attention in early language development. Lang. Sci. **11**, 69–88 (1988)
22. Sebanz, N., Bekkering, H., Knoblich, G.: Joint action: bodies and minds moving together. Trends Cogn. Sci. **10**(2), 70–76 (2006)
23. Bekkering, H., de Bruijn, E.R.A., Cuijpers, R.H., Newman-Norlund, R., van Schie, H.T., Meulenbroek, R.: Joint action: neurocognitive mechanisms supporting human interaction. Top. Cogn. Sci. **1**, 340–352 (2009)
24. Galantucci, B., Sebanz, N.: Joint action: current perspectives. Top. Cogn. Sci. **1**, 255–259 (2009)
25. Steels, L.: Evolving grounded communication for robots. Trends Cogn. Sci. **7**(7), 308–312 (2003)
26. Roy, D., Reiter, E.: Connecting language to the world. Artif. Intell. **167**, 1–12 (2005)
27. Roy, D.: Semiotic schemas: a framework for grounding language in action and perception. Artif. Intell. **167**, 170–205 (2005)
28. Lyon, C., Nehaniv, C.L., Cangelosi, A.: Emergence of Communication and Language. Springer, London (2007)
29. Stramandinoli, F., Marocco, D., Cangelosi, A.: The grounding of higher order concepts in action and language: a cognitive robotics model. Neural Netw. **32**, 165–173 (2012)
30. Harnad, S.: The symbol grounding problem. Physica D **42**, 335–346 (1990)
31. Feldman, J.A.: From Molecules to Metaphor: A Neural Theory of Language. Bradford Books, Cambridge (2008)
32. Rizzolatti, G., Fadiga, L., Gallese, V., Fogassi, L.: Premotor cortex and the recognition of motor actions. Cogn. Brain Res. **3**, 131–141 (1996)
33. Rizzolatti, G., Craighero, L.: The mirror-neuron system. Annu. Rev. Neurosci. **27**, 169–192 (2004)
34. Wilson, M., Knoblich, G.: The case for motor involvement in perceiving conspecifics. Psychol. Bull. **131**(3), 460–473 (2005)
35. Caggiano, V., Fogassi, L., Rizzolatti, G., Casile, A., Giese, M.A., Thier, P.: Mirror neurons encode the subjective value of an observed action. Proc. Nat. Acad. Sci. **109**(29), 11848–11853 (2012)
36. Oztop, E., Kawato, M., Arbib, M.: Mirror neurons and imitation: a computationally guided review. Neural Netw. **19**, 25–271 (2006)
37. Corradini, A., Antonietti, A.: Mirror neurons and their function in cognitively understood empathy. Conscious. Cogn. **22**(3), 1152–1161 (2013)
38. Rizzolatti, G., Arbib, M.A.: Language within our grasp. Trends Neurosci. **21**(5), 188–194 (1998)
39. Studdert-Kennedy, M.: Mirror neurons, vocal imitation, and the evolution of particulate speech. In: Stamenov, M.I., Gallese, V. (eds.) Mirror Neurons and the Evolution of Brain and Language, pp. 207–227. Benjamins, Philadelphia (2002)

40. Arbib, M.A.: From monkey-like action recognition to human language: an evolutionary framework for neurolinguists. Behav. Brian Sci. **28**(2), 105–124 (2005)
41. Corballis, M.C.: Mirror neurons and the evolution of language. Brain Lang. **112**(1), 25–35 (2010)
42. Liberman, A.M., Cooper, F.S., Harris, K.S., MacNeilage, P.J.: A motor theory of speech perception. In: Symposium on Speech Communication Seminar. Royal Institute of Technology, Stockholm (1963)
43. Galantucci, B., Fowler, C.A., Turvey, M.T.: The motor theory of speech perception reviewed. Psychon. Bull. Rev. **13**(3), 361–377 (2006)
44. Lotto, A.J., Hickok, G.S., Holt, L.L.: Reflections on mirror neurons and speech perception. Trends Cogn. Sci. **13**(3), 110–114 (2009)
45. Hickok, G.: The role of mirror neurons in speech and language processing. Brain Lang.: Mirror Neurons: Prospects Probl. Neurobiol. Lang. **112**(1), 1–2 (2010)
46. Barakova, E.I., Lourens, T.: Mirror neuron framework yields representations for robot interaction. Neurocomputing **72**(4–6), 895–900 (2009)
47. Pickering, M.J., Garrod, S.: Do people use language production to make predictions during comprehension? Trends Cogn. Sci. **11**(3), 105–110 (2007)
48. Pickering, M.J., Garrod, S.: An integrated theory of language production and comprehension. Behav. Brain Sci. **36**(04), 329–347 (2013)
49. Pickering, M.J., Garrod, S.: Forward models and their implications for production, comprehension, and dialogue. Behav. Brain Sci. **36**(4), 377–392 (2013)
50. Schwartz, J.L., Basirat, A., Mnard, L., Sato, M.: The perception-for-action-control theory (PACT): a perceptuo-motor theory of speech perception. J. Neurolinguist. **25**(5), 336–354 (2012)
51. Powers, W.T.: Behavior: The Control of Perception. Hawthorne/Aldine, New York (1973)
52. Powers, W.T.: Living Control Systems III: The Fact of Control. Benchmark Publications, Escondido (2008)
53. Wiener, N.: Cybernetics or Control and Communication in the Animal and the Machine. Wiley, New York (1948)
54. Bourbon, W.T., Powers, W.T.: Models and their worlds. Int. J. Hum.-Comput. Stud. **50**, 445–461 (1999)
55. Lindblom, B.: Explaining phonetic variation: a sketch of the H&H theory. In: Hardcastle, W.J., Marchal, A. (eds.) Speech Production and Speech Modelling, pp. 403–439. Kluwer Academic Publishers, Dordrecht (1990)
56. Moore, R.K., Nicolao, M.: Reactive speech synthesis: actively managing phonetic contrast along an H&H continuum. In: 17th International Congress of Phonetics Sciences (ICPhS), Hong Kong (2011)
57. Dennett, D.: The Intentional Stance. MIT Press, Cambridge (1989)
58. Glock, H.-J.: Intentionality and language. Lang. Commun. **21**(2), 105–118 (2001)
59. Frith, C.D., Lau, H.C.: The problem of introspection. Conscious. Cogn. **15**, 761–764 (2006)
60. Rao, A., Georgoff, M.: BDI agents: from theory to practice. Australian Artificial Intelligence Institute, Melbourne (1995)
61. Wooldridge, M.: Reasoning About Ration Agents. MIT Press, Cambridge (2000)
62. Scherer, K.R., Schorr, A., Johnstone, T.: Appraisal Processes in Emotion: Theory, Methods Research. Oxford University Press, New York/Oxford (2001)
63. Marsella, S., Gratch, J., Petta, P.: Computational models of emotion. In: Scherer, K.R., Bänziger, T., Roesch, E. (eds.) A Blueprint for Affective Computing-A Sourcebook and Manual, pp. 21–46. Oxford University Press, New York (2010)

64. Moore, R.K.: Spoken language processing: piecing together the puzzle. Speech Commun. **49**(5), 418–435 (2007)
65. Moore, R.K.: PRESENCE: a human-inspired architecture for speech-based human-machine interaction. IEEE Trans. Comput. **56**(9), 1176–1188 (2007)
66. Moore, R.K.: Cognitive approaches to spoken language technology. In: Chen, F., Jokinen, K. (eds.) Speech Technology: Theory and Applications, pp. 89–103. Springer, New York (2010)
67. Nicolao, M., Latorre, J., Moore, R.K.: C2H: A computational model of H&H-based phonetic contrast in synthetic speech. In: INTERSPEECH, Portland, USA (2012)
68. Worgan, S., Moore, R.K.: Enabling reinforcement learning for open dialogue systems through speech stress detection. In: Fourth International Workshop on Human-Computer Conversation, Bellagio, Italy (2008)
69. Hofe, R., Moore, R.K.: Towards an investigation of speech energetics using AnTon: an animatronic model of a human tongue and vocal tract. Connect. Sci. **20**(4), 319–336 (2008)
70. Crook, N., Smith, C., Cavazza, M., Pulman, S., Moore, R.K., Boye, J.: Handling user interruptions in an embodied conversational agent. In: AAMAS 2010: 9th International Conference on Autonomous Agents and Multiagent Systems, Toronto (2010)
71. Crook, N.T., Field, D., Smith, C., Harding, S., Pulman, S., Cavazza, M., Charlton, D., Moore, R.K., Boye, J.: Generating context-sensitive ECA responses to user barge-in interruptions. J. Multimodal User Interfaces **6**(1–2), 13–25 (2012)
72. Allen, J.F., Ferguson, G., Stent, A.: An architecture for more realistic conversational systems. In: 6th International Conference on Intelligent User Interfaces (2001)
73. Aist, G., Allen, J., Campana, E., Galescu, L., Gallo, C.A.G., Stoness, S.C., Swift, M., Tanenhaus, M.: Software architectures for incremental understanding of human speech. In: Ninth International Conference on Spoken Language Processing: INTERSPEECH - ICSLP, Pittsburgh, PA, USA (2006)
74. Schlangen, D., Skantze, G.: A general, abstract model of incremental dialogue processing. In: 12th Conference of the European Chapter of the Association for Computational Linguistics (EACL-09), Athens, Greece (2009)
75. Hastie, H., Lemon, O., Dethlefs, N.: Incremental spoken dialogue systems: tools and data. In: Proceedings of NAACL-HLT Workshop on Future Directions and Needs in the Spoken Dialog Community, Montreal, Canada, pp. 15–16 (2012)
76. Williams, J.D., Young, S.J.: Partially observable Markov decision processes for spoken dialog systems. Comput. Speech Lang. **21**(2), 231–422 (2007)
77. Thomson, B., Young, S.J.: Bayesian update of dialogue state: a POMDP framework for spoken dialogue systems. Comput. Speech Lang. **24**(4), 562–588 (2010)
78. Moore, R.K.: Interpreting intentional behaviour. In: Mller, M., Narayanan, S.S., Schuller, B. (eds.) Dagstuhl Seminar 13451 on Computational Audio Analysis, vol. 3, Dagstuhl, Germany (2014)
79. Moore, R.K.: From talking and listening robots to intelligent communicative machines. In: Markowitz, J. (ed.) Robots That Talk and Listen. De Gruyter, Boston (in press)

Phonetics and Machine Learning: Hierarchical Modelling of Prosody in Statistical Speech Synthesis

Martti Vainio[(✉)]

Institute of Behavioural Sciences, University of Helsinki, Helsinki, Finland
martti.vainio@helsinki.fi

Abstract. Text-to-speech synthesis is a task that solves many real-world problems such as providing speaking and reading ability to people who lack those capabilities. It is thus viewed mainly as an engineering problem rather than a purely scientific one. Therefore many of the solutions in speech synthesis are purely practical. However, from the point of view of phonetics, the process of producing speech from text artificially is also a scientific one. Here I argue – using an example from speech prosody, namely speech melody – that phonetics is the key discipline in helping to solve what is arguably one of the most interesting problems in machine learning.

Keywords: Phonetics · Machine learning · Speech synthesis · Prosody

1 Introduction

Speech is a vague term that refers to a number phenomena that range from language and linguistic structure to articulatory gestures and acoustics as well as the neurophysiological features that make both its production and perception possible. Perhaps the best metaphor summarising the phenomena is the so called speech chain [21] that depicts speech as a chain of systems, events, and processes that transport ideas and intentions from the mind of a speaker to the mind of a listener via the medium of sound, or light in the case of sign language. However, what is special about speech is that it is a uniquely human behaviour emerging from human social interactions and the articulatory capabilities afforded by the human anatomy.

Speech synthesis was originally developed as a tool to study speech production; first from a mechanical point of view and later from acoustic and articulatory. The history is fairly long starting from the mechanical synthesisers of von Kempelen and Kratzenstein at the end of the 18th century [56,69,91] to modern articulatory models based on e.g., data from MRI machines [14,15,38, 72,74]. Modern synthesis systems are mostly digital, although mechanical versions still exist [33,57]. These synthesisers are mainly used in laboratories and

© Springer International Publishing Switzerland 2014
L. Besacier et al. (Eds.): SLSP 2014, LNAI 8791, pp. 37–54, 2014.
DOI: 10.1007/978-3-319-11397-5_3

were designed for research purposes; the systems available to the public are typically designed for different purposes, namely to convert textual and other data to speech. These text-to-speech synthesisers (TTS) can be roughly divided into two types: ones using prerecorded speech data that is spliced together from a various range of sizes of units [17,42] and parametric synthesisers that use statistical models to produce the signals. The various kinds of unit-selection synthesisers based on recordings are of little phonetic interest, whereas the statistical parametric synthesisers [82,83,94] provide some interest to a speech scientist.[1] For machine learning, the statistical parametric synthesis provides an ideal set of problems for testing and developing new algorithms.

The main task of a text-to-speech synthesiser is to convert a string of discrete graphical symbols (letters, numerals, punctuation marks, mathematical signs and symbols etc.) into continuous sound waveforms [24,44,73,81]. The task reflects the process of a person reading text out loud. First the text is interpreted as a linguistic entity which is then converted to a series articulatory plans resulting in gestures and, finally, an acoustic signal. Depending on the type of text, the requirements to understand its meaning varies. In the case of neutral and maximally informative text, such as e.g., found in the newspaper, the understanding can be very shallow, whereas in the case of prose with written dialogue the requirements are considerably deeper.

Typical text-to-speech systems not so much model the speech chain as emulate the process of reading aloud. That is, they have a somewhat similar relationship with human speech as airplanes have with bird flight. As such, mapping text and speech is an attractive computational task that is at the same time extremely complex and extremely easy. That is, it is fairly easy to produce intelligible speech with fairly simple models [26,29,45] or by simply cutting and splicing together prerecorded samples. The ease is, however, a mirage based on both the redundancy of the speech signal on both linguistic and phonetic levels [46,67] and the extremely good human perceptual capabilities. Two characteristics that are heavily exploited in both automatic speech recognition (ASR) and text-to-speech synthesis (TTS).

Modern machine learning techniques offer a compelling means to take the synthesis technology further without actually utilising more basic knowledge of the speech phenomena themselves. This is very much visible in the development of modern statistical synthesisers which can utilise all of the correlations and correspondences that the complexity of the speech signal and its underlying linguistic structure provide. Given enough data, these systems can in principle be trained to produce any kind of speech with high quality.

The question arises whether in the future traditional phonetic knowledge about speech will play any role in the advancement of speech synthesis. There are, in fact, systems already in existence which forgo even such basic phonetic concepts as phones (speech sounds) and phonemes [1,92]) and translate letters directly into sound without an explicit phonetic and phonological representations. The quality of such systems is poor at the moment but it is obvious that

---

[1] For a good overview of techniques used see [43].

with more data, better statistical models and learning algorithms the quality will in the foreseeable future be on par with systems that utilise more traditional representations – such as e.g., phonemes and syllables.

The important fact that the speech chain metaphor provides us is that speech is a cascade of separate processes all of which require different representations from a number of different domains. That is, the synthesis systems necessitate at least some modularity that desists monolithic modelling. It is in the internal structure of the modules and how they interact with each where the scientific knowledge is best utilised. It can be argued, that the scientifically optimal model will be the easiest and most efficient to train regardless of the algorithms used. To be phonetically relevant, the architecture of a synthesis system should resemble the outline of the speech chain. It is as if nature has already solved many of the (most difficult) optimisation problems in the field [11].

Speech synthesis has its origin in phonetics research [19] and many of the breakthrough in the science itself have come from modelling speech [51]. It is only recently in its long history that the synthesis technology has escaped phonetics laboratories into the wild; namely to laboratories dealing with electrical engineering and even more recently, to laboratories engaged in computer science and machine learning.

Phonetics is a science that takes the speech chain as a metaphor for its research programme. As such it produces knowledge that is essential for text-to-speech synthesis and speech synthesis in general. Thus, it at least potentially offers the most appropriate knowledge for the optimal conceptual structure for a synthesis system. That is, the process of turning a graphical textual representation of a linguistic sign into a time varying acoustic signal might best be done by reflecting the flow of information and actions that produce speech in humans. As such speech synthesis should be seen as a part of the greater field of artificial intelligence as its final goal is to understand and model the mechanisms that allow us to convey via speech the understanding of the world that surrounds us [12]. Phonetics deals with how speech interfaces with the different sources of information in the world and it is essential in recognising those sources. However, dealing with the sources themselves is outside the scope of phonetics. What phonetics, as an established discipline provides is the possibility to design a speech synthesis system on firm scientific grounds to reflect the theoretical knowledge of speech production. This should help the designer avoid falling into traps that are likely to lead to local minima in the optimisation process. Optimisation here refers to both the training of a separate system and to the process of designing the system itself.

Figure 1 shows the speech production side of the speech chain, where the flow of information is from the original linguistic intent via the motor system and the articulatory apparatus to the acoustic speech sound which is a product of the sound source and the vocal tract filter function. This is, of course, a simplified view, and does in no way represent the whole of speech production. Nevertheless, it is conceptually a good working model for a TTS system in terms of modularity.

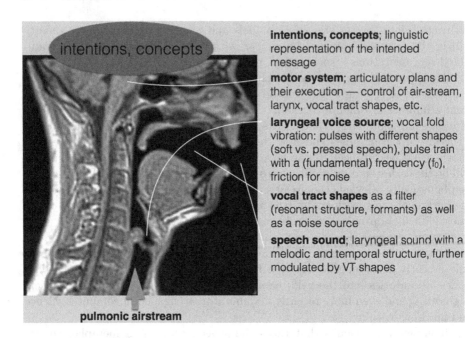

Fig. 1. The speech production chain from intention to acoustics.

The typical architecture of a TTS system does resemble the speech chain in that it is perforce divided into at least three modules: (1) the linguistic one that analyses the raw input into paragraphs, sentences, and words and turns the corresponding text into a string of phonemes or phones as opposed to letters; (2) something that could be called the phonetic module which assigns acoustic features to the linguistic input, and finally (3) a signal generation component which renders the phonetic input into sound. The phonetic components typically predict the desired prosody (phone durations and fundamental frequency contour) as well as other phonetic features. In statistical parametric synthesis this stage is the crucial one where the statistical models (usually HMMs) convert the symbolic input into continuous parameter tracks, which are then turned into an acoustic signal by a vocoder. Typically many more levels and modules are needed to produce high-quality output. A conceptually optimal synthesis system is one that avoids bottlenecks in the flow of information while at the same time reduces the dimensionality – that the complexity of speech provides – in a productive manner.

In simple terms, the synthesis system needs to model the process of speaking in conceptual terms from intentions to acoustics. It is not necessary to model the functioning of the nervous, motor, and articulatory systems per se. However, to remain scientifically interesting the modularity of the system should be such that each module could in principle be replaced with another of altogether different kind without breaking the synthesis flow in a crucial manner. For instance, changing the module mapping abstract representations of utterances to acoustics should not impact the linguistic analyses. Similarly, the linguistic analysis should not impact the acoustic modules directly.

2   Hierarchical Nature of Speech

Spoken language is organised hierarchically both structurally and phonetically. Syntactic hierarchy has been disputed lately [30], but the fact that phonology (the abstract sound structure of a language) is structured hierarchically cannot be disputed: words belong to phrases and are built up from syllables which are further divisible phonemes which stand for the actual speech sounds when the structures are realised as speech. This has many non-obvious effects on the speech signal that need to be modelled. The assumption of hierarchical structure combined with new deep learning algorithms has lead to recent breakthroughs in automatic speech recognition [23]. In synthesis the assumption has played a key role for considerably longer. The prosodic hierarchy has been central in TTS since 1970's [39, 40] and most current systems are based on some kind of a hierarchical utterance structure. Few systems go above single utterances (which typically represent sentence in written form), but some take the paragraph sized units as a basis of production [90].

Figure 2 depicts the hierarchical nature of speech as captured in a time-frequency scale-space by a continuous wavelet transform (CWT) of the signal

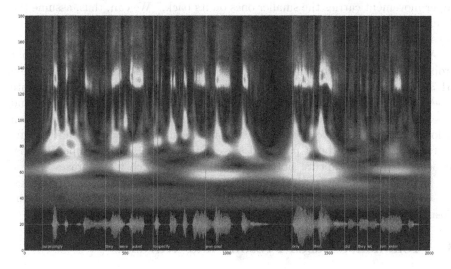

Fig. 2. A continuous wavelet transform based on the signal envelope of an English utterance showing the hierarchical structure of speech. The lower pane shows the waveform and the upper pane the CWT. In the CTW figure, the wavelet scale diminishes towards the top of the figure. The lower parts show the syllables as well as prosodic words (see text for more detail).

envelope of a typical English utterance. The upper part contains the formant structure (which is not visible due to the rectification of the signal) as well as the fundamental frequency in terms of separate pulses. Underneath the f_0 scale are the separate speech segments followed by (prominent) syllables, as well as prosodic words. The lower part including the syllables and prosodic words depicts the suprasegmental and prosodic structure which has typically not been represented in e.g., the mel-frequency cepstral coefficient (MFCC) based features in both ASR and TTS.

The utterance structure serves as a basis for modeling the prosody, e.g., speech melody, timing, and stress structure of the synthetic speech. Modeling prosody in synthesis has been based on a number of different theoretical approaches stemming from both phonological considerations as well as phonetic ones. The phonologically based ones stem from the so called Autosegmental Metrical theory [34] which is based on the three-dimensional phonology developed in [36,37] as noted in [44]. These models are sequential in nature and the hierarchical structure is only implicated in certain features of the models. The more phonetically oriented hierarchical models are based on the assumption that prosody – especially intonation – is truly hierarchical in super-positional fashion. The superpositionality was described eloquently by Bollinger [16] as follows: "The surface of the ocean responds to the forces that act upon it in movements resembling the ups and downs of the human voice. If our vision could take it all in at once, we would discern several types of motion, involving a greater and greater expanse of sea and volume of water: ripples, waves, swells, and tides. It would be more accurate to say ripples on waves on swells on tides, because each larger movement carries the smaller ones on its back." We can, thus, assume the different local prosodic phenomena have different sources – either linguistic or physiological, or both.

Actual models capturing the superpositional nature of intonation were first proposed in [58] by Öhman, whose model was further developed by Fujisaki [31,32] as a so called command-response model which assumes two separate types of articulatory commands – accents associated with stressed syllables and phrases. The accent commands produce faster changes which are superposed on a slowly varying phrase contours. Several superpositional models with a varying degree of levels have been proposed since Fujisaki [5,9,47,48]. Superpositinal models attempt to capture both the chunking of speech into phrases as well the highlighting of words within an utterance. Typically smaller scale changes, caused by e.g., the modulation of the airflow (and consequently the f_0) by the closing of the vocal tract during certain consonants, are not modelled.

3   Prominence Based Prosody and Wavelets

Prominence at the level of word is a functional phonological phenomenon that signals syntagmatic relations of units within an utterance by highlighting some parts of the speech signal while attenuating others. Thus, for instance, some of syllables within a word stand out as stressed [25]. At the level of words prominence relations can signal how important the speaker considers each word in relation to others in the same utterance. These often information based relations range from simple phrasal structures (e.g., prime minister, yellow car) to relating utterances to each other in discourse as in the case of contrastive focus (e.g., "Where did you leave your car? No, we WALKED here."). Although prominence probably works in a continuous fashion, it is relatively easily categorised in e.g., four levels where the first level stands for words that are not stressed in any fashion prosodically to moderately stressed and stressed and finally words that are emphasised (as the word WALKED in the example above). These four categories are fairly easily and consistently labeled even by non-expert listeners [6, 18, 89]. In sum, prominence functions to structure utterances in a hierarchical fashion that directs the listener's attention in a way which enables the understanding of the message in an optimal manner.

As a functional – rather than a formal – phenomenon prominence lends itself optimally to statistical synthesis. That is because the actual signalling of prosody and prominence in terms of speech parameters is extremely complex and context sensitive. As a one-dimensional feature, prominence provides for a means to reduce the complexity of a synthesis system at a juncture that is relevant in terms of both representations and data scarcity. The complex feature set that is known to effect the prosody of speech can be reduced to a few categories or a single continuum from dozens of context sensitive features, such as e.g., part-of-speech and whatever can be computed from the input text. Taken this way, word prominence can be viewed as an abstract phonological function that impacts the phonetic realisation of the speech signal. It is an essential part of the utterance structure, whereas features like part-of-speech or information content (IC) are not.

Word prominence has been shown to work well in TTS for a number of languages, even for English which has been characterised as a so called intonation language, which in principle should require a more detailed modelling scheme which requires explicit knowledge about the intonational forms [7, 10, 76–78]. The perceived prominence of a given word in an utterance is a product of many separate sources of information; mostly signal based although other linguistic factors can modulate the perception [86, 89]. Typically a prominent word is accompanied with a f_0 movement, the stressed syllable is longer in duration, and its intensity is higher. However, estimating prominences automatically is not straight-forward and a multitude of algorithms have been suggested.

Statistical speech synthesis requires relatively little data as opposed to unit-selection based synthesis. However, labelling even small amounts of speech – especially by experts – is prohibitively time consuming. In order to be practicable the labelling of any feature in the synthesis training data should be preferably attainable with automatic and unsupervised means. We have recently developed methods for automatic prominence estimation based on continuous wavelet transform (CWT) which allow for fully automatic and unsupervised means to estimate word prominences as well as boundary values from a hierarchical representation of speech [75, 79, 87].

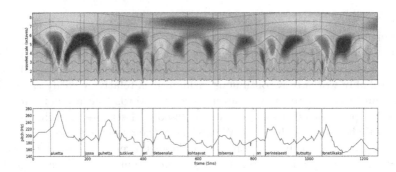

Fig. 3. CWT of the f_0 contour of a Finnish utterance. The lower pane shows the (interpolated) contour itself as well as orthographic words (word boundaries are shown as vertical lines in both panes). The upper pane shows the wavelet transform as well as eight separated scales (grey lines) ranging from segmentally influenced perturbation or microprosody (lowest scale) to utterance level phrase structure (the highest level).

Wavelets are used in many machine-learning applications to represent hierarchical properties of objects such as photographs or graphemes in both image compression and recognition. In speech research there is also a long history going back to the 1980's [4,65,66]. Summary of wavelets in speech technology can be found in [27].

The most important aspect of wavelet analysis is that it allows for time-frequency localisation for what is a priori known to exist in the speech signal; that is speech sounds, syllables, (phonological) words, phrases and so forth. This is not possible with e.g., Fourier analysis, which is painfully obvious to anyone familiar with traditional spectrograms.

Figure 3 shows a CWT of the f_0 contour of a Finnish utterance "Aluette, jossa puhetta tutkivat eri tieteenalat kohtaavat toisensa on perinteisesti kutsuttu fonetiikaksi", (The area where the sciences interested in speech meet each other has been traditionally called phonetics). The lower pane shows the (interpolated) contour itself as well as orthographic words (word boundaries are shown as vertical lines in both panes). The upper pane shows the wavelet transform as well as eight separated scales (grey lines) ranging from segmentally influenced perturbation or microprosody (lowest scale) to utterance level phrase structure (the highest level). The potentially prominent peaks in the signal are clearly visible in the scalogram.

The time-scale analysis allows for not only locating the relevant features in the signal but also estimating their relative salience, i.e., their prominence. The relative prominences of the different words are visible as positive local extrema (red in Fig. 3). There are several ways to estimate word prominences from a CWT. In [80,87] we simply used amplitude of the word prosody scale which was chosen from a discrete set of scales with ratio 2 between ascending scales as the one with the number of local maxima as close to the number of words in the corpus as possible. A more sophisticated way is presented in [88] where the lines

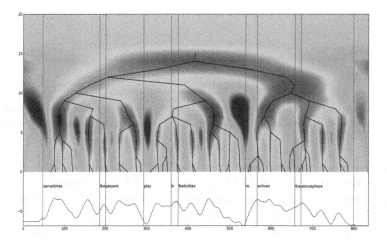

Fig. 4. CWT and LoMA based analysis of an utterance producing a hierarchical tree structure. The word prominences based on the strength of the accumulated LoMA values that reach the hierarchical level corresponding to words. See text for more detail.

of maximum amplitude (LoMA) in the wavelet image were used [35, 54, 66]. This method was shown to be on par with human estimated prominence values (on a four degree scale). However, the method still suffers from the fact that not all prominent words are identified and – more importantly – some words are estimated as prominent whereas they should be seen as non-prominent parts of either another phonological word or a phrase.

Figure 4 shows an f_0 contour of an English utterance ("Sometimes the players play in festivities to enliven the atmosphere.") analysed with CWT and LoMa. The analyses provide both an accurate measure for the locations of the prominent features in the signal as well as their magnitudes. All in all, the CWT and LoMA based analysis can be used for a fully automatic labelling of speech corpora for synthesis. The synthesis, however, cannot produce a full CWT at run time; neither does it make sense to use the full transform for training. That is, the CWT needs to be partitioned into meaningful scales for both training and producing the contours. In [75] we partitioned the CWT into five scales (two octaves apart) which roughly corresponded to the segmentally conditioned microprosody at the lowest level (smallest scale) to syllables, phonological words, phrases, and utterances (at to larges scale). In the HMM framework each level was modelled with a separate stream, which allows for the system to take properly into account the different sources responsible for the changes at all levels. At synthesis time the separate levels have to be reconstructed into a single contour. This procedure is similar to the ones used for image compression and produces a small amount of error, which is however negligible as is shown in Fig. 5. For a more technical presentation of the CWT and LoMA analyses, as well as the signal reconstruction, see [88].

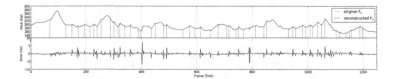

Fig. 5. An original interpolated f_0 contour and its reconstruction from five separate wavelet scales two octaves apart (upper pane) and reconstruction error (lower pane).

In earlier work, wavelets have been used in speech synthesis context mainly for parameter estimation [49,55,68] but never as a full modelling paradigm. In the HMM based synthesis framework, decomposition of f_0 to its explicit hierarchical components during acoustic modeling has been investigated in [50,93]. These approaches rely on exposing the training data to a level-dependent subset of questions for separating the layers of the prosody hierarchy. The layers can then be modelled separately as individual streams [50], or jointly with adaptive training methods [93].

In our recent work [79] we have extended the prominence based synthesis by using two-dimensional tagging of prosodic structure; in addition to the LoMA based prominence we use a boundary value of each word in order to (1) better represent the hierarchical structure of the signal, and (2) to disambiguate those prominence estimates that are estimated to be similarly prominent by the LoMA estimation alone. This brings the labelling system closer to the traditional tone-sequence models which have been widely used – with varying rates of success – in English TTS [24,41,81]. The boundary value for each word can be estimated by e.g., following the lines of minimum amplitude at word boundaries (blue areas in Fig. 4). The combination of word prominence and boundary values – together with the traditional text based utterance structure – are enough to represent the sound structure of any utterance. These utterance structures can be further modified by other functional features such as whether the utterance is a question or a statement by simply adding the feature to the top-level of the tree.

The above described scheme reduces the complexity of the symbolic representation of speech at a juncture that optimises the learning of the actual phonetic features derived from the speech events – be they parameter tracks or something else, such as e.g., articulatory gestures.

4   Implementation of Wavelet and Prominence Based Prosody in GlottHMM Synthesis

The prosody models introduced in the previous section have been implemented in the GlottHMM synthesis system, which has been developed jointly by two groups at the Aalto University (formerly Helsinki University of Technology) and University of Helsinki. It is a statistical parametric TTS system that is based on a new vocoder (signal generation component) that utilises glottal inverse filtering to estimate both the vocal tract and voice source parameters, which are modelled separately at the synthesis stage. The basic idea of glottal inverse filtering is to separate the glottal source and the vocal tract filter based on the linear speech production model [28]. This theory assumes that the production of speech can be interpreted as a linear cascade of three processes: $S(z) = G(z)V(z)L(z)$, where $S(z)$ denotes speech, and $G(z)$, $V(z)$, and $L(z)$ denote the voice source, the vocal tract filter, and the lip radiation effect, respectively. Conceptually, glottal inverse filtering corresponds to solving the glottal volume velocity $G(z)$ according to $G(z) = S(z)1/V(z)1/L(z)$. That is, the model separates the glottal excitation formed by the vocal folds vibrating in the pulmonic airflow from the acoustic filter characteristics of the upper vocal tract as it moves to form different speech sounds. In the GlottHMM system, an automatic glottal inverse filtering method, Iterative Adaptive Inverse Filtering (IAIF) [2,3] is used as a computational tool to implement glottal inverse filtering. The current TTS system utilises deep neural networks (DNN) for modelling the pulse shape [60,61]. The structure of the current version of GlottHMM system is presented in Fig. 6. The use of DNNs for modelling the glottal flow shapes has produced very promising results and we are currently looking into replacing the other components of the system, namely the decision trees and linguistic analysis components with suitable deep learning algorithms.

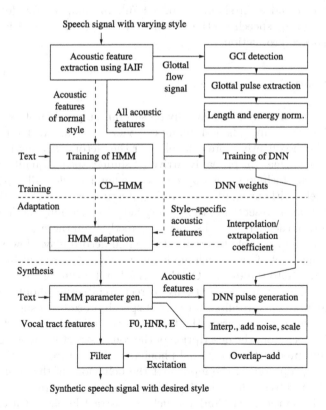

Fig. 6. The architecture of the GlottHMM synthesis system.

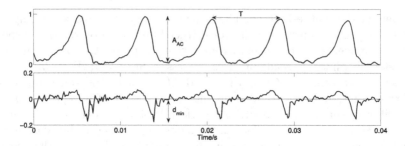

Fig. 7. A representative sample of a glottal flow signal and its derivative from the current study. The sample depicts a five glottal pulse sequence from a male speaker producing a vowel [ɑ].

Figure 7 (from [89]) show a glottal flow signal (and its derivative) acquired by the IAIF method. The measures used for computing certain characteristics of the pulse shape are also illustrated. Grounding the system on human physiology provides it with both transparency and quantifiability and consequently better control of the dynamics of speech. The system has been used successfully in modelling the whole continuum from whispery quiet speech to noise induced Lombard speech and even shouting [62–64, 76]. In sum, it provides an easily trainable system for speech synthesis which at a representational level models the human speech production.

5   Discussion

Modelling speech prosody hierarchically on several temporal levels is practicable from both linguistic and phonetic points of view as it allows for more transparency (and consequently more control) over the factors affecting the resulting speech acoustics. One can probably arrive at similar speech quality with modern deep learning techniques as opposed to – what I would call – phonetically oriented models, without similar overall architecture or the time-scale analyses and representations presented in this paper. New deep learning methods [22] can probably capture the hierarchical structure directly from the surface signal (e.g., the f_0 contour) and be able to predict the signal accordingly. However, from a purely scientific point of view, such modelling would not be as fruitful and might not carry as well into the future. On the other hand, utilising the best machine learning algorithms with a conceptually correct architecture should lead to better and more interesting results. In terms of speech prosody, as suggested in this paper, it is advisable to separate the features that influence the abstract sound structure of speech from the description of the sound structure itself. That is, the systems should predict the functional phonological structure from the linguistic and statistical properties of the input, not the behaviour of the actual phonetic parameters themselves. The phonological – or sound structural – description of an utterance is, on the other hand, on such an abstract level, that it can be used as input to a synthesiser which can be either acoustic or articulatory. That is, modularity based on phonetic principles that reflect the true nature of speech are more general should in practice work more efficiently than ones based on e.g., practical needs or constraints.

Modelling speech synthesis after human speech production analogously to the speech chain has several advantages over developing systems based on the requirements and solutions stemming from machine learning techniques alone – or needs that come from outside the scientific enterprise. Most of the advantages are not obvious and will only be relevant in the future as the methods for representing speech outside the domain of acoustic signals advance to a level where it will be practicable to build articulatory synthesisers capable of producing speech in a human-like manner, including the reductions found in spontaneous everyday talk. There are several reasons why we would prefer articulatory synthesis over ones based on acoustic parameters and signal processing. First, as a purely scientific problem, speech synthesis should model human speech production in a fashion that sheds light to the phenomenon of speech itself. For that, the current modelling as it is implemented in TTS systems is not adequate.[2] Secondly, the current methodology does not allow true interactivity between machines and humans in terms of e.g., entrainment and structuring of interactions [13].

It may be too lofty a goal, but speech synthesis (including TTS) as argued here should be viewed as modelling an integral part of the speech chain, where the other half of the whole deals with receiving and understanding the messages. Many recent findings in cognitive sciences have found evidence for the motor theory of speech perception [8,20,52], which points to the fact that at some level both speech production and speech perception could be modelled jointly even in speech technological solutions. The inclusion of the motor system in a model entails that the system be embodied in some fashion since it is via the actual articulators that the system learns to interpret the input [70,71]. Furthermore, the eventual understanding of language will require the models to include – not only the speech production and perception apparata – but arms and hands, as well [84,85]. In conclusion, when considered from the point of view of phonetics, the machine learning involved with speech synthesis should be biologically inspired [59] at all levels of representation. Moreover, as a science, phonetics is the best equipped to guide in what is arguably one of the most universally interesting machine learning problems in existence.

Acknowledgements. The research leading to these results has received funding from the European Community's Seventh Framework Programme (FP7/2007–2013) under grant agreement n° 287678 (Simple4All) and the Academy of Finland (projects 128204, 125940, and 1265610 (the MIND programme)). I would also like to thank Antti Suni, Daniel Aalto, and Juraj Šimko for their insightful discussions regarding this manuscript. Special thanks go to Paavo Alku and Tuomo Raitio for the GlottHMM collaboration.

---

[2] There are interesting developments towards more articulatory control in HMM based TTS [53]. However, this can only be seen as compromise as the units are still defined acoustically and do not necessarily correspond with the actual underlying articulatory gestures.

References

1. (2014). http://www.simple4all.org
2. Alku, P.: Glottal wave analysis with pitch synchronous iterative adaptive inverse filtering. Speech Commun. **11**(2–3), 109–118 (1992)
3. Alku, P., Tiitinen, H., Näätänen, R.: A method for generating natural-sounding speech stimuli for cognitive brain research. Clin. Neurophysiol. **110**, 1329–1333 (1999)
4. Altosaar, T., Karjalainen, M.: Multiple-resolution analysis of speech signals. In: Proceedings of IEEE ICASSP-88, New York (1988)
5. Anumanchipalli, G.K., Oliveira, L.C., Black, A.W.: A statistical phrase/accent model for intonation modeling. In: INTERSPEECH, pp. 1813–1816 (2011)
6. Arnold, D., Wagner, P., Möbius, B.: Obtaining prominence judgments from naïve listeners-influence of rating scales, linguistic levels and normalisation. In: Proceedings of Interspeech 2012 (2012)
7. Badino, L., Clark, R.A., Wester, M.: Towards hierarchical prosodic prominence generation in TTS synthesis. In: INTERSPEECH (2012)
8. Badino, L., D'Ausilio, A., Fadiga, L., Metta, G.: Computational validation of the motor contribution to speech perception. Top. Cogn. Sci. **6**(3), 461–475 (2014)
9. Bailly, G., Holm, B.: SFC: a trainable prosodic model. Speech Commun. **46**(3), 348–364 (2005)
10. Becker, S., Schröder, M., Barry, W.J.: Rule-based prosody prediction for german text-to-speech synthesis. In: Proceedings of Speech Prosody 2006, pp. 503–506 (2006)
11. Bengio, Y.: Evolving culture vs local minima. arXiv preprint arXiv:1203.2990 (2012)
12. Bengio, Y.: Deep learning of representations: looking forward. In: Dediu, A.-H., Martín-Vide, C., Mitkov, R., Truthe, B. (eds.) SLSP 2013. LNCS, vol. 7978, pp. 1–37. Springer, Heidelberg (2013)
13. Beňuš, Š.: Conversational entrainment in the use of discourse markers. In: Bassis, S., Esposito, A., Morabito, F.C. (eds.) Recent Advances of Neural Network Models and Applications, pp. 345–352. Springer, Heidelberg (2014)
14. Birkholz, P.: Modeling consonant-vowel coarticulation for articulatory speech synthesis. PloS One **8**(4), e60603 (2013)
15. Birkholz, P., Jackel, D.: A three-dimensional model of the vocal tract for speech synthesis. In: Proceedings of the 15th International Congress of Phonetic Sciences, Barcelona, Spain, pp. 2597–2600 (2003)
16. Bolinger, D.L.: Around the edge of language: intonation. Harvard Educ. Rev. **34**(2), 282–296 (1964)
17. Campbell, W.N.: CHATR: a high-definition speech re-sequencing system. In: Proceedings of 3rd ASA/ASJ Joint Meeting, pp. 1223–1228 (1996)
18. Cole, J., Mo, Y., Hasegawa-Johnson, M.: Signal-based and expectation-based factors in the perception of prosodic prominence. Lab. Phonology **1**(2), 425–452 (2010)
19. Cooper, F.S.: Speech synthesizers. In: Proceedings of 4th International Congress of Phonetic Sciences (ICPhS'61), pp. 3–13 (1962)
20. D'Ausilio, A., Maffongelli, L., Bartoli, E., Campanella, M., Ferrari, E., Berry, J., Fadiga, L.: Listening to speech recruits specific tongue motor synergies as revealed by transcranial magnetic stimulation and tissue-doppler ultrasound imaging. Philos. Trans. R. Soc. B: Biol. Sci. **369**(1644), 20130418 (2014)

21. Denes, P.B., Pinson, E.N.: The Speech Chain, p. 121. Bell Laboratory Educational Publication, New York (1963)
22. Deng, L.: A tutorial survey of architectures, algorithms, and applications for deep learning. APSIPA Trans. Signal Inf. Process. **3**, e2 (2014)
23. Deng, L., Li, X.: Machine learning paradigms for speech recognition: an overview. IEEE Trans. Audio, Speech Lang. Process. **21**(5), 1060–1089 (2013)
24. Dutoit, T.: An Introduction to Text-to-Speech Synthesis, vol. 3. Springer, New York (1997)
25. Eriksson, A., Thunberg, G.C., Traunmüller, H.: Syllable prominence: a matter of vocal effort, phonetic distinctness and top-down processing. In: Proceedings of European Conference on Speech Communication and Technology Aalborg, vol. 1, pp. 399–402, September 2001
26. Fant, C.G.M., Martony, J., Rengman, U., Risberg, A.: OVE II synthesis strategy. In: Proceedings of the Speech Communication Seminar F, vol. 5 (1962)
27. Farouk, M.H.: Application of Wavelets in Speech Processing. Springer, New York (2014)
28. Flanagan, J.L.: Speech Analysis, Synthesis and Perception, vol. 1, 2nd edn. Springer, Heidelberg (1972)
29. Flanagan, J.L.: Note on the design of "terminal-analog" speech synthesizers. J. Acoust. Soc. Am. **29**(2), 306–310 (1957)
30. Frank, S.L., Bod, R., Christiansen, M.H.: How hierarchical is language use? Proc. R. Soc. B: Biol. Sci. **279**, 4522–4531 (2012)
31. Fujisaki, H., Hirose, K.: Analysis of voice fundamental frequency contours for declarative sentences of Japanese. J. Acoust. Soc. Jpn. (E) **5**(4), 233–241 (1984)
32. Fujisaki, H., Sudo, H.: A generative model for the prosody of connected speech in japanese. Annu. Rep. Eng. Res. Inst. **30**, 75–80 (1971)
33. Fukui, K., Ishikawa, Y., Sawa, T., Shintaku, E., Honda, M., Takanishi, A.: New anthropomorphic talking robot having a three-dimensional articulation mechanism and improved pitch range. In: 2007 IEEE International Conference on Robotics and Automation pp. 2922–2927. IEEE (2007)
34. Goldsmith, J.A.: Autosegmental and Metrical Phonology, vol. 11. Blackwell, Oxford (1990)
35. Grossman, A., Morlet, J.: Decomposition of functions into wavelets of constant shape, and related transforms. Math. Phys. Lect. Recent Results **11**, 135–165 (1985)
36. Halle, M., Vergnaud, J.R.: Three dimensional phonology. J. Linguist. Res. **1**(1), 83–105 (1980)
37. Halle, M., Vergnaud, J.R., et al.: Metrical Structures in Phonology. MIT, Cambridge (1978)
38. Hannukainen, A., Lukkari, T., Malinen, J., Palo, P.: Vowel formants from the wave equation. J. Acoust. Soc. Am. **122**(1), EL1–EL7 (2007)
39. Hertz, S.R.: From text to speech with SRS. J. Acoust. Soc. Am. **72**(4), 1155–1170 (1982)
40. Hertz, S.R., Kadin, J., Karplus, K.J.: The delta rule development system for speech synthesis from text. Proc. IEEE **73**(11), 1589–1601 (1985)
41. Hirschberg, J.: Pitch accent in context: predicting intonational prominence from text. Artif. Intell. **63**(1–2), 305–340 (1993)
42. Hunt, A.J., Black, A.W.: Unit selection in a concatenative speech synthesis system using a large speech database. In: Proceedings of the 1996 IEEE International Conference on Acoustics, Speech, and Signal Processing, ICASSP-96, vol. 1, pp. 373–376. IEEE (1996)

43. King, S.: Measuring a decade of progress in text-to-speech. Loguens 1(1) (2014)
44. Klatt, D.H.: Review of text-to-speech conversion for english. J. Acoust. Soc. Am. **82**(3), 737–793 (1987)
45. Klatt, D.: Acoustic theory of terminal analog speech synthesis. In: Proceedings of 1972 International Conference on Speech Communication Processing, Boston, MA (1972)
46. Kleijn, W.B.: Principles of speech coding. In: Benesty, J., Sondhi, M.M., Huang, Y. (eds.) Springer Handbook of Speech Processing, pp. 283–306. Springer, Heidelberg (2008)
47. Kochanski, G., Shih, C.: Stem-ml: language-independent prosody description. In: INTERSPEECH, pp. 239–242 (2000)
48. Kochanski, G., Shih, C.: Prosody modeling with soft templates. Speech Commun. **39**(3), 311–352 (2003)
49. Kruschke, H., Lenz, M.: Estimation of the parameters of the quantitative intonation model with continuous wavelet analysis. In: INTERSPEECH (2003)
50. Lei, M., Wu, Y.J., Soong, F.K., Ling, Z.H., Dai, L.R.: A hierarchical f0 modeling method for HMM-based speech synthesis. In: INTERSPEECH, pp. 2170–2173 (2010)
51. Liberman, A.M., Cooper, F.S., Shankweiler, D.P., Studdert-Kennedy, M.: Perception of the speech code. Psychol. Rev. **74**(6), 431 (1967)
52. Liberman, A.M., Mattingly, I.G.: The motor theory of speech perception revised. Cognition **21**(1), 1–36 (1985)
53. Ling, Z.H., Richmond, K., Yamagishi, J.: Articulatory control of HMM-based parametric speech synthesis using feature-space-switched multiple regression. IEEE Trans. Audio Speech Lang. Process. **21**(1), 207–219 (2013)
54. Mallat, S.: A wavelet tour of signal processing. Access Online via Elsevier (1999)
55. Mishra, T., Santen, J.V., Klabbers, E.: Decomposition of pitch curves in the general superpositional intonation model. In: Speech Prosody, Dresden, Germany (2006)
56. Moro, E.B.: A 19th-century speaking machine: the tecnefón of severino perez y vazquez. Historiographia Linguistica **34**(1), 19–36 (2007)
57. Nishikawa, K., Asama, K., Hayashi, K., Takanobu, H., Takanishi, A.: Development of a talking robot. In: Proceedings of 2000 IEEE/RSJ International Conference on Intelligent Robots and Systems 2000 (IROS 2000), vol. 3, pp. 1760–1765. IEEE (2000)
58. Öhman, S.: Word and sentence intonation: a quantitative model. Speech Transmission Laboratory, Department of Speech Communication, Royal Institute of Technology (1967)
59. Pfeifer, R., Lungarella, M., Iida, F.: Self-organization, embodiment, and biologically inspired robotics. Science **318**(5853), 1088–1093 (2007)
60. Raitio, T., Lu, H., Kane, J., Suni, A., Vainio, M., King, S., Alku, P.: Voice source modelling using deep neural networks for statistical parametric speech synthesis. In: 22nd European Signal Processing Conference (EUSIPCO), Lisbon, Portugal, September 2014 (accepted)
61. Raitio, T., Suni, A., Juvela, L., Vainio, M., Alku, P.: Deep neural network based trainable voice source model for synthesis of speech with varying vocal effort. In: Proceedings of Interspeech, Singapore, accepted: September 2014
62. Raitio, T., Suni, A., Pohjalainen, J., Airaksinen, M., Vainio, M., Alku, P.: Analysis and synthesis of shouted speech. In: Interspeech, Lyon, France, pp. 1544–1548, August 2013
63. Raitio, T., Suni, A., Vainio, M., Alku, P.: Analysis of HMM-based lombard speech synthesis. In: Interspeech, Florence, Italy, pp. 2781–2784, August 2011

64. Raitio, T., Suni, A., Vainio, M., Alku, P.: Synthesis and perception of breathy, normal, and lombard speech in the presence of noise. Comput. Speech Lang. **28**(2), 648–664 (2014)
65. Ramachandran, R., Mammone, R.: Modern Methods of Speech Processing. Springer, New York (1995)
66. Riley, M.D.: Speech Time-Frequency Representation, vol. 63. Springer, New York (1989)
67. van Rooij, J.C., Plomp, R.: The effect of linguistic entropy on speech perception in noise in young and elderly listeners. J. Acoust. Soc. Am. **90**(6), 2985–2991 (1991)
68. van Santen, J.P., Mishra, T., Klabbers, E.: Estimating phrase curves in the general superpositional intonation model. In: Fifth ISCA Workshop on Speech Synthesis (2004)
69. Schroeder, M.R.: A brief history of synthetic speech. Speech Commun. **13**(1), 231–237 (1993)
70. Simko, J., Cummins, F.: Embodied task dynamics. Psychol. Rev. **117**(4), 1229 (2010)
71. Šimko, J., O'Dell, M., Vainio, M.: Emergent consonantal quantity contrast and context-dependence of gestural phasing. J. Phonetics **44**, 130–151 (2014)
72. Sondhi, M.M., Schroeter, J.: A hybrid time-frequency domain articulatory speech synthesizer. IEEE Trans. Acoust. Speech Signal Process. **35**(7), 955–967 (1987)
73. Sproat, R.W.: Multilingual Text-to-Speech Synthesis. Kluwer Academic Publishers, Boston (1997)
74. Story, B.H.: A parametric model of the vocal tract area function for vowel and consonant simulation. J. Acoust. Soc. Am. **117**(5), 3231–3254 (2005)
75. Suni, A., Aalto, D., Raitio, T., Alku, P., Vainio, M.: Wavelets for intonation modeling in HMM speech synthesis. In: 8th ISCA Speech Synthesis Workshop (SSW8), Barcelona, Spain, pp. 285–290, August-September 2013
76. Suni, A., Raitio, T., Vainio, M., Alku, P.: The GlottHMM speech synthesis entry for Blizzard Challenge 2010. In: Blizzard Challenge 2010 Workshop, Kyoto, Japan, September 2010
77. Suni, A., Raitio, T., Vainio, M., Alku, P.: The GlottHMM entry for Blizzard Challenge 2011: utilizing source unit selection in HMM-based speech synthesis for improved excitation generation. In: Blizzard Challenge 2011 Workshop, Florence, Italy, September 2011
78. Suni, A., Raitio, T., Vainio, M., Alku, P.: The GlottHMM entry for Blizzard Challenge 2012 - hybrid approach. In: Blizzard Challenge 2012 Workshop, Portland, Oregon, September 2012
79. Suni, A., Simko, J., Aalto, D., Vainio, M.: Continuous wavelet transform in text-to-speech synthesis prosody control (in preparation)
80. Suni, A.S., Aalto, D., Raitio, T., Alku, P., Vainio, M., et al.: Wavelets for intonation modeling in HMM speech synthesis. In: Proceedings of 8th ISCA Workshop on Speech Synthesis, Barcelona, 31 August-2 September 2013
81. Taylor, P.: Text-to-Speech Synthesis. Cambridge University Press, Cambridge (2009)
82. Tokuda, K., Kobayashi, T., Imai, S.: Speech parameter generation from HMM using dynamic features. In: 1995 International Conference on Acoustics, Speech, and Signal Processing, ICASSP-95, vol. 1, pp. 660–663. IEEE (1995)
83. Tokuda, K., Yoshimura, T., Masuko, T., Kobayashi, T., Kitamura, T.: Speech parameter generation algorithms for HMM-based speech synthesis. In: Proceedings of 2001 IEEE International Conference on Acoustics, Speech, and Signal Processing, ICASSP'00, vol. 3, pp. 1315–1318. IEEE (2000)

84. Vainio, L., Tiainen, M., Tiippana, K., Vainio, M.: Shared processing of planning articulatory gestures and grasping. Exp. Brain Res. **232**(7), 2359–2368 (2014)
85. Vainio, L., Schulman, M., Tiippana, K., Vainio, M.: Effect of syllable articulation on precision and power grip performance. PloS One **8**(1), e53061 (2013)
86. Vainio, M., Järvikivi, J.: Tonal features, intensity, and word order in the perception of prominence. J. Phonetics **34**, 319–342 (2006)
87. Vainio, M., Suni, A., Aalto, D.: Continuous wavelet transform for analysis of speech prosody. In: Proceedings of TRASP 2013-Tools and Resources for the Analysis of Speech Prosody, An Interspeech 2013 Satellite Event, August 30 2013, Laboratoire Parole et Language, Aix-en-Provence, France (2013)
88. Vainio, M., Suni, A., Aalto, D.: Emphasis, word prominence, and continuous wavelet transform in the control of HMM based synthesis. In: Speech Prosody in Speech Synthesis - Modeling, Realizing, Converting Prosody for High Quality and Flexible Speech Synthesis, Prosody, Phonology and Phonetics. Springer (2015)
89. Vainio, M., Suni, A., Raitio, T., Nurminen, J., Järvikivi, J., Alku, P.: New method for delexicalization and its application to prosodic tagging for text-to-speech synthesis. In: Interspeech, Brighton, UK, pp. 1703–1706, September 2009
90. Vainio, M., Suni, A., Sirjola, P.: Developing a finnish concept-to-speech system. In: Langemets, M., Penjam, P. (eds.) Proceedings of the Second Baltic Conference on Human Language Technologies, Tallinn, pp. 201–206, 4–5 April 2005
91. von Kempelen, W., de Pázmánd, W.K., Autriche, M.: Mechanismus der menschlichen Sprache nebst der Beschreibung seiner sprechenden Maschine. bei JV Degen (1791)
92. Watts, O.S.: Unsupervised learning for text-to-speech synthesis. Ph.D. thesis (2013)
93. Zen, H., Braunschweiler, N.: Context-dependent additive log f_0 model for HMM-based speech synthesis. In: INTERSPEECH, pp. 2091–2094 (2009)
94. Zen, H., Tokuda, K., Black, A.W.: Statistical parametric speech synthesis. Speech Commun. **51**(11), 1039–1064 (2009)

Machine Translation

A Hybrid Approach to Compiling Bilingual Dictionaries of Medical Terms from Parallel Corpora

Georgios Kontonatsios[(✉)], Claudiu Mihăilă, Ioannis Korkontzelos,
Paul Thompson, and Sophia Ananiadou

The National Centre for Text Mining, The University of Manchester,
131 Princess Street, Manchester M1 7DN, UK
{kontonag,mihailac,korkonti,thompsop,ananiads}@cs.man.ac.uk

Abstract. Existing bilingual dictionaries of technical terms suffer from limited coverage and are only available for a small number of language pairs. In response to these problems, we present a method for automatically constructing and updating bilingual dictionaries of medical terms by exploiting parallel corpora. We focus on the extraction of multiword terms, which constitute a challenging problem for term alignment algorithms. We apply our method to two low resourced language pairs, namely English-Greek and English-Romanian, for which such resources did not previously exist in the medical domain. Our approach combines two term alignment models to improve the accuracy of the extracted medical term translations. Evaluation results show that the precision of our method is 86 % and 81 % for English-Greek and English-Romanian respectively, considering only the highest ranked candidate translation.

1 Introduction

Bilingual dictionaries of technical terms are important resources for many *Natural Language Processing* (NLP) tasks. In *Statistical Machine Translation* (SMT), an up-to-date, bilingual dictionary can be used to dynamically update the SMT system in order to provide more accurate translations of unseen terms [12,15], i.e., current SMT approaches fail to translate terms that do not occur in the training data. In *Cross-Language Information Retrieval*, the use of a bilingual dictionary to expand queries is reported to enhance the overall performance [2]. Bilingual dictionaries can also be particularly useful for human translators who are not familiar with the domain specific terminology [10].

However, manually creating and updating bilingual dictionaries of technical terms is a labourious process. Especially in the biomedical domain, where there is a proliferation of newly coined terms [23], keeping such resources up to date can be extremely costly. For this reason, many existing bilingual dictionaries of technical terms remain incomplete and only cover a limited number of language pairs.

© Springer International Publishing Switzerland 2014
L. Besacier et al. (Eds.): SLSP 2014, LNAI 8791, pp. 57–69, 2014.
DOI: 10.1007/978-3-319-11397-5_4

The UMLS metathesaurus[1] is a popular, multilingual terminological resource for the biomedical domain. UMLS contains terms in more than 20 languages, which are linked together using common *concept* ids. However, UMLS does not index terms in Greek and in Romanian. Hence, our goals is it expand UMLS for these two low resourced languages. In the English part of UMLS, MWTs correspond to phrases that can have various syntactic forms, e.g., noun phrases (*bipolar disorder*) or adjective phrases (*abnormally high*), and which cover a wide range of biomedical concepts, e.g., qualitative concepts (*moderate to severe*), diseases or syndromes *liver cirrhosis*, pharmacological substances (*lactose monohydrate*), etc. In contrast to previous approaches, which have been concerned exclusively with identifying translations of *noun phrases* [9,19,28], we do not restrict the term alignment problem to specific syntactic categories. Without this restriction, our method is general enough to extract candidate term translations for all types of MWTs appearing in UMLS, thus providing the potential to significantly increase coverage for other languages.

Most existing term alignment algorithms suggest N ranked candidate translations for each source term (usually in the range of [1, 20]). Translation accuracy is evaluated by determining the percentage of source terms whose top N candidates contain a correct translation. Naturally, as N increases, the system performance improves, because a greater number of candidate translations is being considered. Thus, the evaluation results only tell us, if a correct translation exists, it will be somewhere amongst the N possible translation candidates. Dictionaries created using such methods (i.e., with N candidate translations for each word) are noisy, and can only be useful for a limited number of applications, e.g., in SMT systems, which use a language model to select the most probable translation out of N possible candidates. However, [8] showed that such dictionaries decreased the translation accuracy of human translators. Furthermore, automatically compiled dictionaries cannot be used to update high-quality terminological resources before human translators have removed the noisy candidates. In this paper, we aim to improve the translation accuracy on the first ranked candidate.

Our novel method firstly analyses a parallel corpus to identify terms in the source language which have a corresponding entry in UMLS. Additionally, each source term is annotated with: (a) concept id and (b) semantic category that are derived from UMLS. As a second step, we apply a term alignment method to obtain the translation equivalence in the target language. Finally, we propagate the concept id and the semantic category from the source term to the corresponding target translation.

To obtain the translation of a source term, we investigate three term alignment methods, i.e.: (a) a phrase alignment module which is part of an SMT toolkit (SPA) (b) a supervised machine learning approach that uses character n-grams, i.e., a Random Forest (RF) classifier and (c) the intersection of the above two methods, which we call the *voting* system.

For evaluation purposes, we sampled 1000 terms with their corresponding translations, and we asked bilingual speakers of both English-Greek and

---

[1] nlm.nih.gov/research/umls

English-Romanian to judge the translations. The results that we obtained showed that the voting method achieved the best translation accuracy for the *top 1* candidate translation by a significant margin. Furthermore, the same voting method exhibited competitive performance in the translation of both frequent and rare terms, in contrast to SPA, which was shown to be largely dependent on the frequency of terms in the corpus.

2   Related Work

Early works on term alignment immediately recognised the potential benefits of automatically constructed bilingual terminological resources. At AT&T labs, [6] presented *Termight*, a term alignment tool that was shown to be useful for human translators of technical manuals, despite of its fairly low accuracy (40 % translation accuracy on the top 1 candidate for English-German). *Termight* firstly performed word alignment [7] to compile a bilingual list of single words from the parallel corpus. A simple heuristic was then used to extract MWT translations, i.e., for each source MWT, candidate translations were defined as sequences whose first and last words were aligned with any of the words contained in the source MWT.

Several approaches have explored statistical methods to align MWTs in parallel corpora. Reference [25] investigated the alignment of collocations which are frequently occurring word pairs (e.g., *powerful computer*). *Champollion*, their proposed system, was iteratively building a translation by selecting words in the target language that are highly correlated (Dice coefficient) with the input source collocation. Their method achieved competitive performance of approximately 70 % translation accuracy on an English-French parallel corpus. Other approaches have investigated the use of co-occurrence frequency [9] and mutual information [5].

While statistical approaches for term alignment are frequently reported in the literature, several other techniques have been explored, including machine learning, distributional methods and hybrid approaches. Reference [19] introduced a machine learning method, an Expectation-Maximisation algorithm, for extracting translations of noun-phrases from an English-French parallel corpus. The authors reported an accuracy of 90 %, but only when considering the 100 highest ranking correspondences. Reference [3] used a simple distributional semantics approach, namely a *Vector Space model*. They constructed boolean term vectors in a source and target language of size N, where N is the number of sentences in the corpus. Each dimension of a vector corresponded to a sentence and its value indicated whether or not the term appeared in the sentence. For ranking candidate translations they used the *Jaccard Index* between the vector of a source and target term. For evaluation, they did not compute the standard precision and recall. They rather measured the effect of using the extracted bilingual dictionary within an SMT system. The results obtained showed a small improvement of +0.30 BLEU points [22] over the baseline SMT system. Reference [29] introduced a hybrid collocation alignment system that combines statistical (log likelihood ratio) and syntactic information (part-of-speech patterns). Their system achieved a competitive performance when applied on a distant language pair, namely English-Chinese.

In contrast to previous works that applied their methods to well-resourced language pairs, e.g., English-French, few approaches have focussed on low-resourced language pairs. Reference [28] applied term alignment to extend the Slovenian WordNet. Their method relied on word alignments and on lexico-syntactic patterns. In total, they were able to extract 5,597 new multi-word terms for the Slovenian WordNet, of which 2,881 were correct.

In this work, we aim at enriching UMLS with low-resourced languages that previously were absent from the thesaurus. Overall, our system retrieved 5,926 and 5,446 multi-word terms with a translation accuracy of 86 % and 81 % on the top 1 ranked candidate for English-Greek and English-Romanian respectively.

3   Methodology

We use *EMEA*, a biomedical parallel corpus from the European Medicines Agency [27]. The corpus contains approximately 1,500 sentence-aligned documents in 22 European languages. As a first step, English MWTs are identified in the corresponding part of EMEA, using a monolingual term extraction tool. For this, we use MetaMap [1], which automatically recognises biomedical terms in an input English text. Furthermore, MetaMap assigns a UMLS concept id and a semantic category to each term. In this way, the extracted English terms are mapped to the UMLS metathesaurus. This step identified 17,907 unique English MWTs in the English-Greek part of the corpus and 16,625 MWTs in the English-Romanian part of the corpus.

For the target part of the parallel corpus (Greek/Romanian), we extract candidate translations by simply considering all contiguous sequences containing up to four words. These candidates are used in both SPA and RF. Additionally, the system computes the intersection of the alignment methods (voting system) in order to improve the quality of the extracted dictionaries.

3.1   Statistical Phrase Alignment

We adopt a standard approach used in Statistical Machine Translation to align phrases from parallel corpora [16]. The *Statistical Phrase Alignment* (SPA) method builds phrase alignments using a single-word bilingual dictionary that was previously extracted from the parallel corpus. The performance of SPA is heavily dependent on the quality of the word alignment module. Word alignments are established using GIZA++ [21], an open source implementation of the 5 IBM-models [4]. GIZA++ is trained on both translation directions and extracts two word alignment tables $L_s \to L_t$ and $L_t \to L_s$ between a source L_s and target L_t language. Then we combine the two tables using the *grow-diag-final* heuristic which starts with all alignments found in the intersection of the tables and then adds neighbouring alignments of the union set. The merged translational table yields a better recall and precision of word-to-word correspondences than the intersection or union of the two tables.

Using the word alignments that are established as described above, we extract any pair of phrases (s^n, t^m), where s^n is a source phrase containing of n words and t^m a target phrase of m words, with the condition that the words in s^n are aligned to the words in t^m and not to any other words outside t^m. Finally, the candidate phrase pairs are ranked using the lexical translation probability.

3.2 Random Forest Aligner

References [17,18] introduced an Random Forest (RF) aligner that is able to automatically learn association rules of textual units between a source and target language when trained on a corpus of positive and negative examples. The method is based on the hypothesis that terms across languages are constructed using semantically equivalent textual units. Hence, if we know the translations of the basic building blocks of a term, e.g., morphemes, prefixes, suffixes, we can predict the term in a target language. Table 1 illustrates an example of the training and prediction process of the RF aligner. In this toy example, the RF aligner is trained on two English-Greek and English-Romanian instances and learns how to translate the morphemes *cardio* and *vascular*. Once trained, the model uses the previously learned associations of textual units to extract new term translations, e.g., <*cardio-vascular, καρδι-αγγειακό, cardio-vascular*>.

Table 1. Example of training and prediction process of RF Aligner

	English	Greek	Romanian
training	cardio-*myopathy*	μυο-**καρδιο**-πάθεια	cardio-*miopatie*
	extra-**vascular**	εξω-**αγγ** ειακό	*extra*-**vascular**
prediction	cardio-vascular	**καρδι-αγγειακό**	cardio-vascular

The RF aligner is a supervised, machine learning method. It uses n-grams (size of $[2, 5]$ characters) to represent a pair of terms in a source and in a target language. The feature vectors have a fixed size of $2q$. The first q n-grams are extracted from the source term while the last q n-grams from the target term. For dimensionality reduction, we considered only the $1,000$ (500 source and 500 target) most frequent character n-grams. Given a term in a source language, the model outputs a list of the top N ranked candidate translations. For ranking candidate translations, we use the prediction confidence, i.e., *classification margin*.

To train the RF aligner, we use *BabelNet* [20]. *BabelNet* is a multilingual, multi-domain, encyclopedic dictionary containing terms in 50 languages including English, Greek and Romanian. For both language pairs (English-Greek and English-Romanian) we select $10K$ positive and $10K$ negative instances to learn positive and negative associations of character grams between the source and target language. *Pseudo-negative* instances are created by randomly matching non-translation terms.

In addition to the two alignment models, we use a voted system that considers the intersection of the two models. The combined system is expected to increase the accuracy of the automatically extracted dictionaries.

4   Experiments

In this section, we evaluate the three dictionary extraction methods, namely SPA, RF and the voted system on two biomedical parallel corpora for English-Greek and English-Romanian.

We follow a standard evaluation methodology reported in the literature. We select the top N ranked translations for each source term as given by the term alignment methods and we mark a correct answer when a true translation is found among the top N candidates. In some cases both the RF and SPA failed to propose any candidate translations for a given source term (the list of ranked candidates was empty). Based upon, we use two evaluation metrics in our experiments, namely the top-N precision and top-N recall. The top-N recall is defined as the percentage of source terms whose top N candidates contain a correct translation. The top-N precision is the percentage of source terms whose list of top N candidates: (a) is not empty and (b) contains a correct translation.

4.1   Results

For evaluation purposes, we randomly sampled 1,000 English MWTs. For each English MWT, we selected the top 20 translation candidates. We asked two people whose native languages are Greek and Romanian, respectively, and who are fluent English speakers, to manually judge the translations of the English MWTs. The curators marked only exact translations as being correct.

Figures 1 and 2 show the top 20 precision of the dictionary extraction methods. We note that the performance of SPA is consistently better than RF in both datasets. SPA achieves a precision of 93.2 % and 89 % on the top 20 candidates, while RF achieves a precision of 67.1 % and 67.7 % for English-Greek and English-Romanian, respectively. A possible explanation for the poor performance of RF is that we trained the model on an *out-of-domain* lexicon, namely *BabelNet*, due to the lack of existing bilingual biomedical dictionaries for the two language pairs. Hence, the model does not explicitly learn character n-gram mappings of biomedical terms, which leads to noisy translations when we apply the model to *in-domain* datasets.

SPA achieves robust performance when considering the top 20 candidate translations. However, the performance of SPA declines steadily as fewer translation candidates are considered (49 % for English-Greek and 48 % for English-Romanian on the top 1 candidate). Thus, although SPA is able to determine

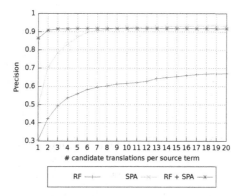

Fig. 1. Precision of top N candidates of SPA, RF and the voted method (SPA + RF) for English-Greek

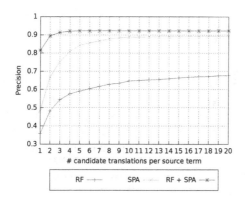

Fig. 2. Precision of top N candidates of SPA, RF and the voted method (SPA + RF) for English-Romanian

an exact translation in most cases, this could lie anywhere within the top 20 candidates.

In an attempt to improve the precision of higher ranking candidates, we considered the intersection between RF and SPA. This "hybrid" method achieved significantly better performance than either RF or SPA when considering only the highest ranked candidate (86 % and 81 % for English-Greek and English-Romanian, respectively). For the top 20 candidates, the performance of the voted system is approximately the same with SPA.

The recall which determines the coverage of the extracted dictionary is a further important feature of term alignment methods. Figures 3 and 4 illustrate the recall of SPA, RF and the voted system on an increasing number of ranked candidate translations. We observe that the recall of RF is significantly better than the recall of SPA and the voted system. For the top 20 candidate translations, RF achieves the best observed recall of approximately 67 % for both

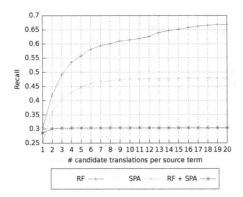

Fig. 3. Recall of top N candidates of SPA, RF and the voted method (SPA + RF) for English-Greek

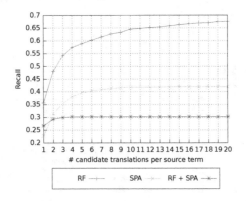

Fig. 4. Recall of top N candidates of SPA, RF and the voted method (SPA + RF) for English-Romanian

English-Greek and English-Romanian. In contrast, the voted system performs poorly, with a total coverage of 30 % on the top 20 candidates. The low recall of the voted system is caused by the significantly decreased coverage of SPA compared to RF. This impacts negatively on the voted system, since its results constitute the intersection of the two alignment methods. SPA extracted correct translations for only 48 % and 40 % of the English terms from the English-Greek and English-Romanian corpus, respectively.

4.2   Error Analysis

In this subsection, we discuss the results of an error analysis that was performed to reveal common noisy translations extracted by RF and SPA.

In the case of the RF, we identified two types of erroneous translations: (a) partial matches and (b) discontinuous translations. Partial matches refer to the cases where RF translated part of the English term but failed to retrieve an exact match. For example, the English term "urea cycle disorder" was partially translated into Romanian as "tulburărilor ciclului" (missing translation for "urea"). Additionally, in several cases, a partial translation was ranked higher than the exact match, which led to a decrease in the precision for top 1 candidate. The top ranked Greek translation for "urea cycle disorders" was " διαταραχών" (urea cycle) while the exact matched (διαταραχών του κύκλου της) was ranked fifth. In fewer cases, the translation of an English term occurred as discontinuous sequence in the target corpus. For example, the term "metabolic diseases" (*boli de metabolism*) occurred in the Romanian corpus as a discontinuous sequence with the span "boli ereditare de metabolism" (*hereditary metabolic diseases*). However, the current implementation of RF searches for candidate translations only in contiguous sequences, in order to minimise the number of classification instances. Hence, these type of translations were not captured by RF.

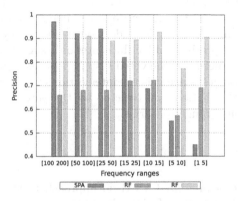

Fig. 5. Precision of top 20 candidates on different frequency ranges of SPA, RF and the voted method (SPA + RF), English-Greek dataset

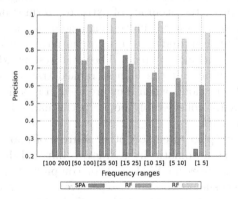

Fig. 6. Precision of top 20 candidates on different frequency ranges of SPA, RF and the voted method (SPA + RF), English-Romanian dataset

SPA is a statistical based alignment tool whose performance is largely affected by the frequency of the terms in the corpus. For high frequency terms, SPA has stronger evidence of term alignment and extracts correct translations with higher confidence. However, we predicted that the alignment quality would decrease for rarely occurring terms.

To further investigate this intuition, we evaluated the performance of SPA on terms having varying frequencies in the corpus. We segmented the 1,000 test terms for both English-Greek and English-Romanian, that were previously evaluated by the human curators, into 7 frequency ranges, from high-frequency to rare terms. Then we computed the precision for the top 20 candidates for RF, SPA and the voted system on those frequency ranges. Figures 5 and 6 show the performance of the term alignment methods. We observe that for frequent terms (i.e., those terms occurring 25 times or more in the corpus), SPA shows a robust performance for both language pairs. However, for less frequent terms, the precision of SPA steadily decreases. In contrast to SPA, RF does not exploit any corpus statistics to align terms between a source and a target language and as a result, its performance is not dependent on the frequency of occurrence of a term. Accordingly, the precision of RF fluctuates only slightly over different term frequency ranges. Furthermore, the voted system presents a stable precision over different frequency ranges, since it is the intersection of RF and SPA. Hence, we can conclude that the dictionary extracted by the voted system is robust to differences in frequency ranges of terms.

5   Conclusion

In this paper, we have presented a hybrid approach to the automatic compilation of bilingual dictionaries of biomedical terms from parallel corpora. Our novel voted system combines a supervised machine learning approach, i.e., a Random Forest (RF) aligner, with a Statistical Phrase Alignment (SPA) alignment method, to improve the accuracy of extracted term translations. We have applied our method to two low-resourced language pairs, namely English-Greek and English-Romanian, and candidate translations have been evaluated by bilingual speakers of these two language pairs. The voted system exhibits significantly better translation accuracy for the highest ranked translation candidate than either the RF or SPA methods, when they are used in isolation. In addition, the voted system achieved a considerably better performance in translating rarely occurring terms than SPA.

As future work, we plan to exploit other sources of information in order to increase the size of the automatically extracted bilingual dictionaries. Parallel corpora are useful resources for SMT and for compiling high-quality bilingual dictionaries. However, such corpora are expensive to construct because human translators need to provide the translations of the source documents. Consequently, parallel corpora are of limited size and they quickly become out-of-date (and thus are unlikely to contain neologisms). Additionally, they are not available for every domain or language pair. Due to the sparsity of parallel document collections, researchers have started to explore comparable corpora, since they are more readily available, more up-to-date and they are easier and cheaper to construct. In common with a parallel corpus, a comparable corpus is a collection of documents in a source and target language. However, in contrast to a parallel corpus, the documents in a comparable corpus are not direct translations of each other. Rather, they share common features, such as covering the same topic, domain, time period, etc. Large comparable collections can be readily constructed using freely available multilingual resources, e.g., Wikipedia [13,26,30]. This means that comparable corpora constitute a promising resource to aid in the construction and maintenance of bilingual dictionaries, especially when parallel corpora are limited or unavailable for a given language pair.

Whilst the hybrid system described in this paper cannot be applied directly to comparable corpora, since the SPA module can only process parallel data, we are planning to incorporate a context-based method into our system. This method is widely used to facilitate term alignment approaches involving comparable corpora. Context-based methods approaches (context vectors) [11,24] adapt the *distributional hypothesis* [14] to extract term translations from comparable corpora. They hypothesise that a term and its translation tend to appear in similar lexical contexts. Intuitively, the RF aligner and the context-based approach are complementary, since RF exploits the internal structure of terms, while context vectors use the surrounding lexical context. Therefore, it will be interesting to investigate how these two methods can be combined within a hybrid system.

Acknowledgements. This work was funded by the European Community's Seventh Framework Program (FP7/2007–2013) [grant number 318736 (OSSMETER)].

References

1. Aronson, A.R.: Effective mapping of biomedical text to the umls metathesaurus: the metamap program. In: Proceedings of the AMIA Symposium, p. 17. American Medical Informatics Association (2001)
2. Ballesteros, L., Croft, W.: Phrasal translation and query expansion techniques for cross-language information retrieval. In: ACM SIGIR Forum, vol. 31, pp. 84–91. ACM (1997)
3. Bouamor, D., Semmar, N., Zweigenbaum, P.: Identifying bilingual multi-word expressions for statistical machine translation. In: LREC, pp. 674–679 (2012)
4. Brown, P., Pietra, V., Pietra, S., Mercer, R.: The mathematics of statistical machine translation: parameter estimation. Comput. linguist. **19**(2), 263–311 (1993)
5. Church, K.W., Hanks, P.: Word association norms, mutual information, and lexicography. Comput. linguist. **16**(1), 22–29 (1990)
6. Dagan, I., Church, K.: Termight: identifying and translating technical terminology. In: Proceedings of the Fourth Conference on Applied Natural Language Processing, pp. 34–40. Association for Computational Linguistics (1994)

7. Dagan, I., Church, K.W., Gale, W.A.: Robust bilingual word alignment for machine aided translation. In: Proceedings of the Workshop on Very Large Corpora, pp. 1–8 (1993)

8. Delpech, E.: Evaluation of terminologies acquired from comparable corpora: an application perspective. In: Proceedings of the 18th International Nordic Conference of Computational Linguistics (NODALIDA 2011), pp. 66–73 (2011)

9. Van der Eijk, P.: Automating the acquisition of bilingual terminology. In: Proceedings of the Sixth Conference on European Chapter of the Association for Computational Linguistics, pp. 113–119. Association for Computational Linguistics (1993)

10. Fung, P., McKeown, K.: A technical word-and term-translation aid using noisy parallel corpora across language groups. Mach. Transl. 12(1), 53–87 (1997)

11. Fung, P., Yee, L.Y.: An ir approach for translating new words from nonparallel, comparable texts. In: Proceedings of the 17th International Conference on Computational linguistics, vol. 1, pp. 414–420. Association for Computational Linguistics (1998)

12. Habash, N.: Four techniques for online handling of out-of-vocabulary words in arabic-english statistical machine translation. In: Proceedings of the 46th Annual Meeting of the Association for Computational Linguistics on Human Language Technologies: Short Papers, pp. 57–60. Association for Computational Linguistics (2008)

13. Haghighi, A., Liang, P., Berg-Kirkpatrick, T., Klein, D.: Learning bilingual lexicons from monolingual corpora. In: ACL, vol. 2008, pp. 771–779 (2008)

14. Harris, Z.: Distributional structure. Word (1954)

15. Irvine, A., Callison-Burch, C.: Combining bilingual and comparable corpora for low resource machine translation. In: Proceedings of the Eighth Workshop on Statistical Machine Translation. Association for Computational Linguistics, August 2013

16. Koehn, P., Och, F.J., Marcu, D.: Statistical phrase-based translation. In: Proceedings of the 2003 Conference of the North American Chapter of the Association for Computational Linguistics on Human Language Technology, vol. 1, pp. 48–54. Association for Computational Linguistics (2003)

17. Kontonatsios, G., Korkontzelos, I., Tsujii, J., Ananiadou, S.: Using a random forest classifier to compile bilingual dictionaries of technical terms from comparable corpora. In: Proceedings of the 14th Conference of the European Chapter of the Association for Computational Linguistics: Short Papers, vol. 2, pp. 111–116. Association for Computational Linguistics, April 2014, http://www.aclweb.org/anthology/E14-4022

18. Kontonatsios, G., Korkontzelos, I., Tsujii, J., Ananiadou, S.: Using random forest to recognise translation equivalents of biomedical terms across languages. In: Proceedings of the Sixth Workshop on Building and Using Comparable Corpora, pp. 95–104. Association for Computational Linguistics, August 2013, http://www.aclweb.org/anthology/W13-2512

19. Kupiec, J.: An algorithm for finding noun phrase correspondences in bilingual corpora. In: Proceedings of the 31st Annual Meeting on Association for Computational Linguistics, pp. 17–22. Association for Computational Linguistics (1993)

20. Navigli, R., Ponzetto, S.P.: BabelNet: the automatic construction, evaluation and application of a wide-coverage multilingual semantic network. Artif. Intell. 193, 217–250 (2012)

21. Och, F.J., Ney, H.: A systematic comparison of various statistical alignment models. Comput. linguist. 29(1), 19–51 (2003)

22. Papineni, K., Roukos, S., Ward, T., Zhu, W.J.: Bleu: a method for automatic evaluation of machine translation. In: Proceedings of the 40th Annual Meeting on Association for Computational Linguistics, pp. 311–318. Association for Computational Linguistics (2002)
23. Pustejovsky, J., Castano, J., Cochran, B., Kotecki, M., Morrell, M.: Automatic extraction of acronym-meaning pairs from medline databases. Studies in health technology and informatics, pp. 371–375 (2001)
24. Rapp, R.: Automatic identification of word translations from unrelated english and german corpora. In: Proceedings of the 37th Annual Meeting of the Association for Computational Linguistics on Computational Linguistics, pp. 519–526. Association for Computational Linguistics (1999)
25. Smadja, F., McKeown, K.R., Hatzivassiloglou, V.: Translating collocations for bilingual lexicons: a statistical approach. Comput. linguist. **22**(1), 1–38 (1996)
26. Tamura, A., Watanabe, T., Sumita, E.: Bilingual lexicon extraction from comparable corpora using label propagation. In: Proceedings of the 2012 Joint Conference on Empirical Methods in Natural Language Processing and Computational Natural Language Learning, pp. 24–36. Association for Computational Linguistics (2012)
27. Tiedemann, J.: News from opus-a collection of multilingual parallel corpora with tools and interfaces. In: Recent Advances in Natural Language Processing, vol. 5, pp. 237–248 (2009)
28. Vintar, S., Fiser, D.: Harvesting multi-word expressions from parallel corpora. In: LREC (2008)
29. Wu, C.C., Chang, J.S.: Bilingual collocation extraction based on syntactic and statistical analyses. In: ROCLING (2003)
30. Yu, K., Tsujii, J.: Bilingual dictionary extraction from wikipedia. In: Proceedings of Machine Translation Summit XII, pp. 379–386 (2009)

Experiments with a PPM Compression-Based Method for English-Chinese Bilingual Sentence Alignment

Wei Liu[✉], Zhipeng Chang, and William J. Teahan

School of Computer Science, Bangor University,
Dean Street, Bangor, Gwynedd LL57 1UT, UK
{w.liu,z.chang,w.j.teahan}@bangor.ac.uk
http://www.bangor.ac.uk/cs

Abstract. Alignment of parallel corpora is a crucial step prior to training statistical language models for machine translation. This paper investigates compression-based methods for aligning sentences in an English-Chinese parallel corpus. Four metrics for matching sentences required for measuring the alignment at the sentence level are compared: the standard sentence length ratio (SLR), and three new metrics, absolute sentence length difference (SLD), compression code length ratio (CR), and absolute compression code length difference (CD). Initial experiments with CR show that using the Prediction by Partial Matching (PPM) compression scheme, a method that also performs well at many language modeling tasks, significantly outperforms the other standard compression algorithms Gzip and Bzip2. The paper then shows that for sentence alignment of a parallel corpus with ground truth judgments, the compression code length ratio using PPM always performs better than sentence length ratio and the difference measurements also work better than the ratio measurements.

Keywords: Statistical models for natural language processing · Parallel corpora · Sentence alignment · Text compression · PPM · Gzip · Bzip2

1 Background & Motivation

Accurate alignment of textual elements (e.g. paragraphs, sentences, phrase) in a parallel bilingual corpus is a crucial step for statistical machine translation. A number of different approaches have been developed over the years for aligning sentences between comparable text in a bilingual parallel corpus—for example, those based on using: sentence length; word co-occurrence; cognates; dictionaries; and parts of speech.

The assumption behind length-based approaches is that short sentences in the source language will be translated into short sentences in the target language, and the same for longer sentences, and that there is enough variation in sentence length between adjacent sentences to correct mis-alignments when they occur.

L. Besacier et al. (Eds.): SLSP 2014, LNAI 8791, pp. 70–81, 2014.
DOI: 10.1007/978-3-319-11397-5_5

Gale and Church [7] aligned sentences in English-French and English-German corpora by calculating the character length of all sentences, producing a Cartesian product of all possible alignments, then aligning the most plausible alignments iteratively until all sentences are accounted for. Their overall accuracy rate for both corpora was 96 % (97 % for English-German and 94 % for English-French). The best results were for 1:1 alignments, where one sentence in one language corresponds to one sentence in the other language. For 1:1 alignments, the error rate was only 2 %. However, there was a 10 % error rate for 2:1 alignments and 33 % error rate for 2:2 alignments. In comparison for English-Chinese corpora, Wu [19] also proposed aligning English-Chinese corpora by determining sentence length (in bytes) and also produced a high accuracy of over 95 % [19]. Length-based measurement has also had satisfactory results for evaluating the corpus extracted from China National Knowledge Infrastructure (CNKI) [6].

Brown et al. [2] calculated the length of sentences by calculating the number of words in each sentence. This generated similar results—96 to 97 %. Kay and Röscheisen [10] combined word and sentence alignment in one program. They used the dice co-efficient to calculate the probabilities of words in one language being aligned with words in the other language. Simard, Foster and Isabelle [17] pursued a cognate based approach to sentence alignment after analysing the errors produced in length-based alignment (ibid., p. 70). While they found that cognates alone cannot produce better alignments than length differences, a two-pass program, whereby strong alignments based on sentence length are made in the first pass, and cognates are used to align the more difficult sentences in the second pass, did produce better results than the simple length-based alignment. Haruno and Yamazaki [9] use both probabilistic and a bilingual dictionary to find word cognates to help align sentences. Like Kay and Röscheisen [10], this is a combined sentence and word alignment program. Haruno and Yamazaki [9] do not make use of length-based techniques because they state that these methods do not work for such structurally different languages as English and Japanese.

Papageorgiou, Cranias, and Piperidis [16] have devised a sentence alignment scheme that matches sentences on the basis of the highest matching part of speech tags, the matches restricted to content words—nouns, adjectives and verbs. With 99 % accuracy, they obtained the best results of all for sentence alignment algorithms. Melamed [13] (ibid., p. 5) however points out that *"It is difficult to compare this algorithm's performance to that of other algorithms in the literature, because results were only reported for a relatively easy bitext."*

In recent years, there have been relatively few new proposals for parallel corpora sentence alignment [21]. Existing sentence alignment algorithms are not able to link one-to-many or many-to-one mutual translations [12]. This paper will focus on adopting a novel compression-based approach as the distance measure to determine whether two sentences are aligned.

This paper is organised as follows. The next section motivates the use of compression-based methods for alignment, and describes four distance metrics for matching sentences, two based on sentence length and two based on calculating the compression code length of the sentences. The section also describes several compression algorithms used in the experiment—PPM, Gzip and Bzip2, and then describes how the compression code lengths can be calculated using a relative entropy approach and "off-the-shelf" compression software. The alignment algorithm we have used is then described next. Two experiments are then described—the first to find out which compression algorithm works best for the code length ratio metric; and the second to compare which of the four metrics perform best at aligning a corpus which was constructed with ground truth judgments concerning the alignment. Conclusions are provided in the final section.

2   Compression-Based Alignment

Our idea of using compression-based measures for alignment hinges on the premise that the compression of co-translated text (i.e. documents, paragraphs, sentences, clauses, phrases) should have similar compression code lengths [1]. This is based on the notion that the information contained in the co-translations will be similar. Since compression can be used to measure the information content, we can simply look at the ratio of the compression code lengths of the co-translated text pair to determine whether the text is aligned. That is, if you have a text string (i.e. paragraph, sentence, or phrase) in one language, and its translation in another language, then the ratio of the compression code lengths of the text string pair should be close to 1.0. Alternatively, we can use a relative entropy related measure, and use an absolute code length difference measure—in this case, a value close to 0 indicates that the text string pair are closely aligned.

Formally, given a text string of length n symbols $S^L = x_1 x_2, \ldots x_n$ in language L and a model p_L for that language, then the cross-entropy is calculated as follows:

$$H(S^L) = -\frac{1}{n} \log_2 p_L(S^L)$$

i.e. the average number of bits to encode the text string using the model.

2.1   Distance Measures

Four metrics for matching sentences required for measuring the alignment at the sentence level are compared: the standard sentence length ratio (SLR), and three new metrics, absolute sentence length difference (SLD), compression code length ratio (CR), and absolute compression code length difference (CD):

$$SLR = \max \left\{ \frac{L(S^E)}{L(S^C)}, \frac{L(S^C)}{L(S^E)} \right\} \tag{1}$$

$$SLD = \left| L(S^E) - L(S^C) \right| \tag{2}$$

$$CR = \max \left\{ \frac{H(S^E)}{H(S^C)}, \frac{H(S^C)}{H(S^E)} \right\} \tag{3}$$

$$CD = \left| H(S^E) - H(S^C) \right| \tag{4}$$

where L represents sentence length and H means code length. S^E and S^C denote English and Chinese sentences.

Sentence Length Ratio used to be used by Mújdricza-Maydt et al. [15] and has achieved good performance. The remainder of this section will describe compression schemes that we have used in the code length calculations for the experiments described below.

2.2   PPM

Prediction by Partial Matching (PPM) is an adaptive online compression scheme that predicts the next symbol or character based on a prior context with fixed length. Cleary and Witten [5] proposed PPM first using the variants of PPMA and PPMB. Then PPMC and PPMD were developed by Moffat in 1990 and Howard in 1993 [20]. The main difference between PPMA, PPMB, PPMC and PPMD is the calculation of the escape probability which is needed by the smoothing mechanism used by the algorithm for backing off to lower order models. Experiments show that PPMD in most cases performs better than PPMA, PPMB and PPMC. PPM-based methods have been widely used in natural language processing, including verification of text collections which ensures whether the collection is valid or consistent [11].

The probability p of the next symbol φ for PPMD is calculated using the following formula:

$$p(\varphi) = \frac{2c_d(\varphi) - 1}{2T_d}$$

where d denotes the current coding order, $c_d(\varphi)$ denotes the number of times that the symbol φ in the current context and T_d presents the total number of times that the current context has occurred. The calculation of the escape probability e by PPMD is as follows:

$$e = \frac{t_d}{2T_d}$$

where t_d is the total number of unique symbols that occur after the current context. When PPMD is encoding the upcoming symbol, it always starts first from the maximum order model. A maximum order of 5 is usually used in most of the experiments [18] and order 5 has also been found effective for Chinese text [20]. If the model contains the prediction for the upcoming symbol, it will be transmitted according to the order 5 distribution. If the model does not contain the prediction, the encoder will escape down to order 4. The escape process will repeat until a model is found that is able to predict the upcoming symbol, backing off if needed to a default order -1 model where symbols are equiprobable [18].

PPM code length is the size (in bytes) of the PPM-compressed output file. When using PPM as a natural language processing tool to compress text, the code length can be used to estimate the cross-entropy of the text. The cross-entropy can be calculated by the following formula:

$$H(S) = -\frac{1}{n}\log_2 p(S) = -\frac{1}{n}\sum_{i=1}^{n} -\log_2 p(x_i|x_1 \ldots x_{n-1})$$

where $H(S)$ is the average number of bits to encode the text. Order 5 PPMD models are used for both English and Chinese in this paper compressing the byte sequence of the text. i.e. for English, a single ASCII byte represents a single English character, whereas for GB-encoded Chinese text, a Chinese character is denoted by two bytes (and therefore 5 bytes will span 2.5 characters). Text compression experiments with Chinese text [20] show that compressing the byte or character sequence is noticeably better than when using the word sequence, and we also wish to avoid the problem of word segmentation for Chinese text, hence the reason for using bytes for the experiments described in this paper.

2.3 Gzip & Bzip2

Gzip (also called GNU zip) was created by the GNU project and written by Jean-Loup Gailly and Mark Adler [8]. It uses a dictionary-based Lempel-Ziv based method as opposed to the statistical context-based approach of PPM. Gzip is now a popular lossless compression utility on the Internet and Unix operating system.

Bzip2 is another lossless compression algorithm that was developed by Julian Seward [3]. It uses a block sorting compression algorithm that makes use of the Burrows Wheeler method to transform the text. Bzip2 performs better than Gzip but the speed is slower.

The reason for choosing PPM, Gzip and Bzip2 in the experiments reported below is that the three schemes represent very different compression methods—statistical (context) based, dictionary and block sorting. A primary motivation for this paper was to determine which scheme was most effective when applied to the problem of sentence alignment for parallel corpora.

2.4 Calculating Code Lengths of Gzip, Bzip2 and PPMD

We will use a relative entropy method to calculate the compression code lengths for PPM, Gzip and Bzip2. This allows us to use "off-the-shelf" software without having to re-implement the compression schemes. Since the size of the text being compressed in each sentence is relatively small, these compression schemes will not have had sufficient data to compress the text effectively since their models are uninitialised and therefore not well tuned for the languages being compressed (English and Chinese). To overcome this problem, a simple expedient is to prime the models using a large representative training sample for each language. The relative entropy technique allows us to do this in order to calculate the code length using the formula $h_t = h_{T+t} - h_T$ where h is the size of a file after it has been compressed, T represents the large training text and t is the testing text (i.e. the sentence being compressed) for which the compression code length calculation is being computed. The method simply calculates the difference in size between the compressed training text with testing text added and the compressed training text by itself.

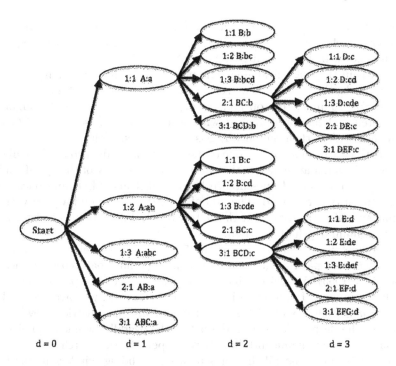

Fig. 1. 5-tree for aligning sentences.

3   Alignment Algorithm

We describe the algorithm that we used to align sentences in this section. Alignment of sentences may be one to one (1:1), one to many (e.g. 1:2, 1:3), many to one (e.g. 2:1 and 3:1) and many to many (e.g. 2:2). For efficiency reasons for our alignment algorithm, we do not consider the many to many case or the 1:n and n:1 cases where $n > 3$. In contrast, Moore [14] also proposed $n \leq 2$ because the situation of $n > 3$ is extremely rare. However, we have found that $n = 3$ did happen in English-Chinese parallel corpora, so therefore for our experiments, we use $n \in [2,3]$. Therefore, for our setting, the search for the best alignment can be considered to be a 5-tree with five branches at each node as shown in Fig. 1. The search begins at the node labelled "Start" at depth $d = 0$ in the tree where the algorithm is positioned at the beginning of each of the two list of sentences being aligned. In this example, the lists of sentences have been denoted as [$ABCDEF\ldots$] and [$abcdef\ldots$]. From the Start node there are five possible alignments to examine at depth $d = 1$—a 1:1 mapping where sentence A is aligned with sentence a, a 1:2 mapping, where sentence A is instead aligned with the first two sentences in the second list, denoted by ab, a 1:3 mapping for sentence pairs A and abc, the 2:1 mapping for the pair AB with a and the 3:1 mapping for the pair ABC and a.

For each node at depth $d = 1$, there are five child nodes at $d = 2$ that then have to be searched in turn. Note that only a subset of the set of nodes in the 5-tree are shown in Fig. 1 as it is not possible to display the full 5-tree in the diagram within the space available. The figure shows the expansion of the first two nodes at depth $d = 1$, and two selected nodes at depth $d = 2$ for illustration purposes. For example, the top node at depth $d = 3$ represents the alignment where sentence A has been aligned with sentence a, then sentences BC have been aligned with b, then sentence D has been aligned with c.

The path cost from a node to one of its child nodes is defined as a calculation result by a given distance metric that measures the quality of the specific alignment, such as ones based on sentence length (SLR and SLD) and ones based on code length (CR and CD). The aim of the search is to find a path with the minimum sum of path costs through the tree to a leaf node (which is determined by the maximum depth of the tree).

The complexity of the search for the 5-tree is 5^d. Therefore, when $d = 9$, searching the best path in the 5-tree with the minimum path cost will need to compare $5^9 = 1953125$ paths i.e. find the minimum sum of cost paths from nearly two million numbers. In our experiments described below, we have explored the case when $d \in [1, 9]$. Experiments with the four distance metrics show that in most cases, the deeper the search, the better the overall alignment quality, but this is at the cost of significantly longer time spent on the search.

In order to align the full list of sentences, a sliding window method was adopted. An alignment at a particular position is chosen using the 5-tree search which then determines the width of the window according to the alignment. The algorithm then advances to the next position after the window and so on until the entire text has been aligned.

4   Experiment 1: Comparing Different Compression Algorithms for Sentence Alignment

The purpose of the first experiment was to find out the compression scheme that is the most effective at aligning parallel corpora and also to compare the sentence length and code length metrics. For the experiment, a test corpus was needed to provide the ground truth data in order to investigate the effectiveness of the different compression algorithms. We chose 1,000 matching Chinese-English parallel sentences from the DC parallel corpus [4] at random. Table 1 shows sample calculations for the first three sentence pairs in the corpus and Fig. 2 graphs CR for all the sentences, for the three compression schemes PPMD, Gzip and Bzip2. In order to compute the code length values (as shown in bytes in Table 1), the concatenation of all of the sentences in the corpus was used as the training text T to prime the compression models, and the values were calculated by the formula for h_t (see Sect. 2.4). These values were then inserted into the formula (with $H(S) = h_t$) listed in Sect. 2.1 to calculate CR.

Note that for the second sentence (Id 0002), the h_t value for Bzip2 was 0 (the compression size in bytes of the training and testing text was exactly the same

Table 1. Comparing codelength ratio for GZIP, BZIP2 and PPMD for the test corpus.

Sent. ID	Language	GZIP (bytes)	GZIP CR	BZIP2 (bytes)	BZIP2 CR	PPMD (bytes)	PPMD CR
0001	Chinese	51	1.3784	115	3.2857	47	1.1750
	English	37		35		40	
0002	Chinese	17	2.4286	66	66.0	24	1.4118
	English	7		1		17	
0003	Chinese	85	1.6038	135	3.9706	74	1.3962
	English	53		34		53	
...	...	...	...	...	...	...	...

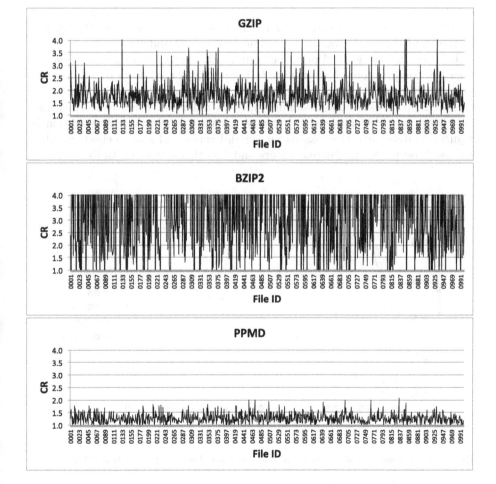

Fig. 2. Adjusted codelength ratios of the 1000 training models.

as the training text by itself). For the cases when h_t was 0, this was replaced with a value of 1 in order to avoid infinite values resulting for the CR ratio calculation. We can see from the graphs that the code length ratios of PPMD are the most stable with the largest value being 2.833. In comparison, Gzip has greater variation, with many instances when the CR value exceeds 4.0 despite the sentence pairs chosen for the corpus being accurate translations of each other. (The graphs were truncated to a maximum CR value of 4.0 in order that the three graphs could be directly compared). The widest variation clearly belongs to Bzip2, where most values are higher than 10.

Figure 3 graphs the percentages of how many sentence pairs are below a certain SLR or CR value (for PPMD, Gzip and Bzip2). From the figure, we can see that the CR values for PPMD performs better than Gzip and Bzip2 at identifying matching sentences for the lower threshold values with similar values to the SLR metric. However, the behavior for CR values calculated using Gzip and Bzip2 are noticeably different. For example, if we focus on the range between 1.0 and 1.5, there are 930 sentences out of 1,000 in this range for PPMD, but for Gzip and Bzip2, the amounts are much lower (633 and 129).

It is not clear why PPM performs significantly better at alignment than the other two compression schemes, since Gzip and Bzip2 are known to also provide good estimates of the entropy, although Gzip frequently flushes its dictionary, whereas Bzip2 uses a non-streaming approach unlike the other two algorithms and this may affect the relative entropy calculations. Further investigation is required to determine the reasons for the difference and also to check whether this result occurs for all language pairs and for other alignment tasks.

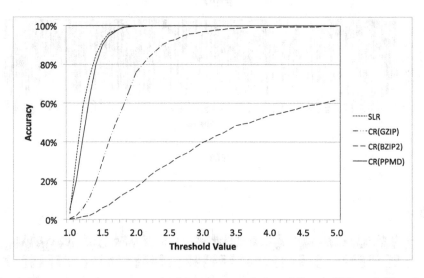

Fig. 3. Percentage of sentence pairs in text corpus below different SLR and CR values.

5 Experiment 2: Sentence Alignment for the Training Corpus

The purpose of the second experiment was to compare the four different metrics defined in Sect. 2.1. For the second experiment, another test corpus was also needed in order to verify the effectiveness of the different metrics. The test corpus includes 1000 1:1 parallel sentences, fifty 1:2 and 1:3 sentences and fifty 2:1 and 3:1 sentences placed throughout the corpus in an ad hoc manner. All the sentences were bilingual news or parallel articles downloaded from the Internet on various topics. The English part of the corpus includes 15932 words and 92508 characters, and the Chinese part has 29046 Chinese characters.

The alignment algorithm described in Sect. 3 was applied to the problem of aligning the test corpus. Table 2 compares at various search tree depths the sentence alignment accuracies that resulted using the four different metrics.

Table 2. Comparison at various search tree depths of sentence length alignment accuracies for different metrics: SLR, SLD, CR and CD.

Depth	SLR	SLD	CR	CD
1	87.5 %	90.3 %	88.4 %	**91.6 %**
2	88.4 %	**93.6 %**	90.1 %	**93.6 %**
3	88.3 %	93.2 %	91.5 %	**94.2 %**
4	87.5 %	94.3 %	92.3 %	**94.5 %**
5	86.7 %	93.8 %	92.8 %	**95.2 %**
6	87.6 %	94.1 %	93.5 %	**95.5 %**
7	88.6 %	94.4 %	93.1 %	**95.9 %**
8	88.3 %	94.7 %	93.3 %	**95.5 %**
9	88.5 %	95.2 %	93.5 %	**96.1 %**

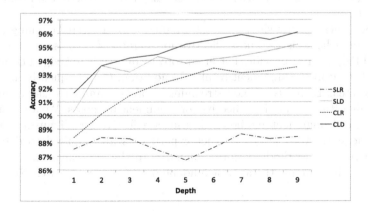

Fig. 4. Comparison at various search tree depths of sentence length alignment accuracies for the different metrics: SLR, SLD, CR and CD.

From the table, we can see that difference based metrics (SLD and CD) always performed better than their corresponding ratio based metrics (SLR and CR) and that code length metrics (CR and CD) performed better than sentence length metrics (SLR and SLD). Overall, the code length difference metric (CD) is the best performed metric in this comparison.

Figure 4 shows the performance tendencies of the four metrics where we can see significant improvements with growing depths for SLD, CR and CD. However, SLR did not show a growth trend. It is reasonable to believe that there will be more competitive results if the depth of the 5-tree is greater than 9 although this would be at a significant cost in search time. Although not optimized, the speed of code length calculation is much slower than sentence length calculation especially when depth $d \geq 6$, with it taking 7.7 s on average on a Macbook Pro laptop per 5-tree search at $d = 6$ and 69.9 s at $d = 9$. Note that there are some dips in Fig. 4, especially for depth $d = 5$ for SLR. One of the possible reasons is that sentences of the test corpus were not in a natural sequential order, and therefore the results may be affected by this.

6   Conclusion

Three new distance metrics have been introduced for matching sentences for alignment of parallel corpora. Two of the metrics are based on computing the compression code length of the sentences as this is an accurate measure of the information contained in the text. The idea is that if the sentences are aligned, then the information contained in sentences that are co-translations of each other should match. The compression-based measures will give a more accurate metric well founded in information theory than alternative metrics based on sentence length which are essentially cruder estimates of the information. Overall, the best metric for determining sentence alignment was based on absolute compression code length difference between sentence pairs. Absolute difference based metrics (including when using sentence length) were also more effective than using ratio based metrics.

Experimental results show that the Prediction by Partial Matching (PPM) compression scheme is the most effective for alignment purposes compared to Gzip and Bzip2. PPM provides better entropy estimates than Gzip of Bzip2, and this is reflected in the alignment results. In addition, Gzip frequently flushes its model, whereas Bzip2 uses a non-streaming approach, and this may contribute to these algorithms being less effective for alignment purposes. We are confident that the PPM alignment method will also be effective for alignment down to phrase and even word levels. Further experiments are required to determine how well the new methods perform compared to the approach taken by other researchers, for example the approach adopted by Gale and Church [7].

References

1. Behr, F.H., Fossum, V., Mitzenmacher, M., Xiao, D.: Estimating and comparing entropy across written natural languages using PPM compression. In: Proceedings of Data Compression Conference, p. 416 (2003)
2. Brown, P., Della Pieta, S., Della Pieta, V., Mercer, R.: The mathematics of machine translation: parameter estimation. Comput. Ling. **19**, 263–312 (1993)
3. Bzip2.: The Bzip2 Home Page (2014). http://www.bzip.org
4. Chang, Z.: A PPM-based evaluation method for Chinese-English parallel corpora in machine translation. Ph.D. thesis of Bangor University (2008)
5. Cleary, J.G., Witten, I.H.: Data compression using adaptive coding and partial string matching. IEEE Trans. Commun. **32**(4), 396–402 (1984)
6. Ding, H., Quan, L., Qi, H.: The Chinese-English bilingual sentence alignment based on length. In: International Conference on Asian Language Processing, pp. 201–204 (2011)
7. Gale, W.A., Church, K.W.: A program for aligning sentences in bilingual corpora. In: ACL'93 29th Annual Meeting, pp. 177–184 (1993)
8. Gzip.: The Gzip Home Page (2014). http://www.gzip.org
9. Haruno, M., Yamazaki, T.: High-performance bilingual text alignment using statistical and dictionary information. In: Proceedings of the 34th Annual Meeting of Association for Computational Linguistics, pp. 131–138 (1996)
10. Kay, M., Röscheisen, M.: Text-translation alignment. Comput. Ling. **19**, 121–142 (1993)
11. Khmelev, D.V., Teahan, W.J.: A repetition based measure for verification of text collections and for text categorization. In: Proceedings of the 26th Annual International ACM SIGIR Conference on Research and Development in Information Retrieval, pp. 104–110 (2003)
12. Kutuzov, A.: Improving English-Russian sentence alignment through POS tagging and Damerau-Levenshtein distance. In: Association for Computational Linguistics, pp. 63–68 (2013)
13. Melamed, I.D.: Models of translational equivalence among words. Comput. Ling. **26**(2), 221–249 (2000)
14. Moore, R.C.: Fast and accurate sentence alignment of bilingual corpora. In: Association for Machine Translation, pp. 135–144 (2002)
15. Mújdricza-Maydt, E., Körkel-Qu, H., Riezler, S., Padó, S.: High-precision sentence alignment by bootstrapping from wood standard annotations. Prague Bull. Math. Ling. **99**, 5–16 (2013)
16. Papageorgiou, H., Cranias, L., Piperidis, S.: Automatic alignment in corpora. In: Proceedings of 32nd Annual Meeting of Association of Computational Linguistic, pp. 334–336 (1994)
17. Simard, M., Foster, G.F., Isabelle, P.: Using cognates to align sentences in bilingual corpora. In: Proceedings of the Fourth International Conference on Theoretical and Methodological Issues in Machine Translation (TMI), pp. 67–81 (1992)
18. Teahan, W.J., Wen, Y., McNab, R., Witten, I.H.: A compression-based algorithm for Chinese word segmentation. Comput. Ling. **26**(3), 375–393 (2000)
19. Wu, D.: Aligning a parallel English-Chinese corpus statistically with lexical criteria. In: ACL'94 32nd Annual Meeting, pp. 80–87 (1994)
20. Wu, P.: Adaptive models of Chinese text. Ph.D. dissertation, University of Wales, Bangor (2007)
21. Yu, Q., Max, A., Yvon, F.: Revisiting sentence alignment algorithms for alignment visualization and evaluation. In: LREC Workshop, pp. 10–16 (2012)

BiMEANT: Integrating Cross-Lingual and Monolingual Semantic Frame Similarities in the MEANT Semantic MT Evaluation Metric

Chi-kiu Lo and Dekai Wu[✉]

Human Language Technology Center,
Department of Computer Science and Engineering,
Hong Kong University of Science and Technology (HKUST),
Kowloon, Hong Kong
{jackielo,dekai}@cs.ust.hk

Abstract. We present experimental results showing that integrating cross-lingual semantic frame similarity into the semantic frame based automatic MT evaluation metric MEANT improves its correlation with human judgment on evaluating translation adequacy. Recent work shows that MEANT more accurately reflects translation adequacy than other automatic MT evaluation metrics such as BLEU or TER, and that moreover, optimizing SMT systems against MEANT robustly improves translation quality across different output languages. However, in some cases the human reference translation employs different scoping strategies from the input sentence and thus standard monolingual MEANT, which only assesses translation quality via the semantic frame similarity between the reference and machine translations, fails to fairly and accurately reward the adequacy of the machine translation. To address this issue we propose a new bilingual metric, BiMEANT, that correlates with human judgment more closely than MEANT by incorporating new cross-lingual semantic frame similarity assessments into MEANT.

1 Introduction

We show that a new bilingual version of MEANT (Lo *et al.* [19]) correlates with human judgments of translation adequacy even more closely than MEANT by integrating cross-lingual semantic frame similarity assessments. We assess cross-lingual semantic frame similarity by (1) incorporating BITG constraints for word alignment within the semantic role fillers, and (2) using simple lexical translation probabilities, instead of the monolingual context vector model used in MEANT for computing the semantic role fillers similarities. We then combine this cross-lingual semantic frame similarity into the MEANT score. Our results show that integrating cross-lingual semantic frame similarity into MEANT improves its correlation with human judgment on evaluating translation adequacy.

The MEANT family of metrics (Lo and Wu [20,22]; Lo *et al.* [19]) adopt the principle that a good translation is one where a human can successfully understand the central meaning of the foreign sentence as captured by the basic

© Springer International Publishing Switzerland 2014
L. Besacier et al. (Eds.): SLSP 2014, LNAI 8791, pp. 82–93, 2014.
DOI: 10.1007/978-3-319-11397-5_6

event structure: *"who did what to whom, for whom, when, where, how and why"* (Pradhan *et al.* [31]). MEANT measures similarity between the MT output and the reference translations by comparing the similarities between the semantic frame structures of output and reference translations. Previous work indicates that the MEANT family of metrics correlates better with human adequacy judgment than commonly used MT evaluation metrics (Lo and Wu [20,22]; Lo *et al.* [19]; Lo and Wu [24]; Macháček and Bojar [25]). In addition, MEANT has been shown to be tunable—translation adequacy across different genres (ranging from formal news to informal web forum and public speech) and different languages (English and Chinese) is improved by replacing BLEU or TER with MEANT during parameter tuning (Lo *et al.* [16]; Lo and Wu [23]; Lo *et al.* [18]).

Particularly for very different languages—Chinese and English, for instance—monolingual MT evaluation strategies that compare reference and machine translations, including MEANT, often fail to properly recognize cases where alternative strategies for scoping, topicalization, and the like are employed by the input sentence and the MT output, leading to artificial differences between the reference and machine translations. As pointed out in the empirical study of Addanki *et al.* [1], this can result in drastically different semantic frame annotations. To combat this, we propose a strategy where direct bilingual comparisons of the machine translation and the original input sentence are incorporated into MEANT.

2   Related Work

2.1   MT Evaluation Metrics

A number of large scale meta-evaluations (Callison-Burch *et al.* [6]; Koehn and Monz [13]) report cases where BLEU (Papineni *et al.* [30]) strongly disagrees with human judgments of translation adequacy. Other surface-form oriented metrics such as NIST (Doddington [8]), METEOR (Banerjee and Lavie [2]), CDER (Leusch *et al.* [14]), WER (Nießen *et al.* [27]), and TER (Snover *et al.* [35]) can also suffer from similar problems because the degree of n-gram match does not accurately reflect how well the *"who did what to whom, for whom, when, where, how and why"* is preserved across translation, particularly for very different language pairs where reference translations can be extremely non-deterministic.

To address these problems of n-gram based metrics, Owczarzak *et al.* [28,29] apply LFG to extend the approach of evaluating syntactic dependency structure similarity proposed by Liu and Gildea [15]. Although they showed improved correlation with human *fluency* judgments, they did not achieve higher correlation with human *adequacy* judgments than metrics like METEOR. TINE (Rios *et al.* [32]) is a recall-oriented evaluation metric which aims to preserve the basic event structure. However, its correlation with human adequacy judgments is similar to that of BLEU and worse than that of METEOR. For a semantic MT metric to work better at the current stage of technology, we believe that it is necessary to choose (1) a suitable abstraction level for the meaning representation, and (2) the right balance of precision and recall.

Instead of prioritizing simplicity and representational transparency as Rios *et al.* [32] and Owczarzak *et al.* [28,29] do, Giménez and Màrquez [11,12] incorporate several semantic similarity features into a huge collection of n-gram and syntactic features within ULC so as to improve correlation with human adequacy judgments (Callison-Burch *et al.* [4]; Giménez and Màrquez [11]; Callison-Burch *et al.* [5]; Giménez and Màrquez [12]). However, there is no work towards tuning an SMT system using a pure form of ULC perhaps due to its expensive run time. Both SPEDE (Wang and Manning [37]), an MT evaluation metric that predicts the edit sequence needed for the MT output to match the reference via an integrated probabilistic FSM and probabilistic PDA model, and Sagan (Castillo and Estrella [7]), a semantic textual similarity metric based on a complex textual entailment pipeline, may also be susceptible to similar problems. These aggregated metrics require sophisticated feature extraction steps, contain several dozens of parameters to tune and employ expensive linguistic resources, like WordNet and paraphrase tables. Because of their expensive training, tuning and/or running times, such metrics become less useful in the MT system development cycle. We have taken the approach of keeping the representation of meaning simple and clear in MEANT, so that the resulting metric can not only be transparently understood when used in error analysis, but also be employed when scoring massive number of hypotheses for training and tuning MT systems.

2.2   The MEANT Family of Metrics

Addanki *et al.* [1] shows that for very different languages—Chinese and English, for example—there would be cases where alternative strategies for scoping or topicalization are employed by the input sentence and the MT output, leading to artificial differences between the reference and machine translations. This drives us to investigate avenues toward further improving MEANT by incorporating the cross-lingual semantic frame similarity into MEANT, so that translation output whose semantic structure is closer to the foreign input sentence than the human reference translation can be scored more fairly.

MEANT, which is a weighted f-score over the matched semantic role labels of the automatically aligned semantic frames and role fillers, has been shown to correlate with human adequacy judgments more highly than BLEU, NIST, TER, WER, CDER, and others (Lo *et al.* [19]). It is relatively easy to apply to other languages, requiring only an automatic semantic parser and a large monolingual corpus in the output language; these resources are used for identifying the semantic structures and the lexical similarity between the semantic role fillers of the reference and machine translations, respectively. MEANT is computed as follows:

1. Apply an automatic shallow semantic parser to both the reference and machine translations. (Figure 1 shows examples of automatic shallow semantic parses on both reference and machine translations.)
2. Apply maximum weighted bipartite matching to align the semantic frames between the reference and machine translations, according to the lexical similarities of the predicates.

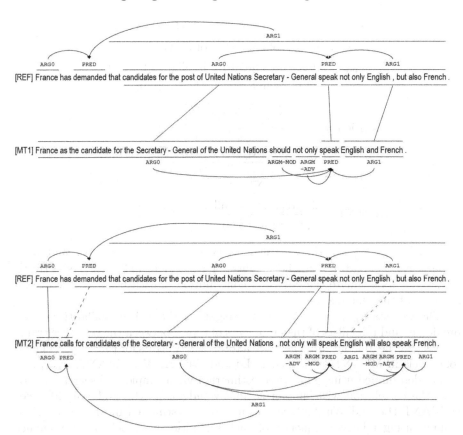

Fig. 1. Examples of monolingual semantic frame similarity captured by MEANT. MT2, the more adequate translation than MT1, is penalized by monolingual MEANT for producing translation output with more "inaccurate" semantic frames according to the reference translation. The dotted lines represent a low similarity (<0.5) semantic role alignments made by MEANT.

3. For each pair of the aligned frames, apply maximum weighted bipartite matching to align the arguments between the reference and machine translations, according to the lexical similarity of the semantic role fillers.

4. Compute the weighted f-score over the matching role labels of these aligned predicates and semantic role fillers according to the following definitions:

$$q^0_{i,j} \equiv \text{ARG j of aligned frame i in MT}$$

$$q^1_{i,j} \equiv \text{ARG j of aligned frame i in REF}$$

$$w^0_i \equiv \frac{\text{\#tokens filled in aligned frame i of MT}}{\text{total \#tokens in MT}}$$

$$w^1_i \equiv \frac{\text{\#tokens filled in aligned frame i of REF}}{\text{total \#tokens in REF}}$$

$$w_{\text{pred}} \equiv \text{weight of similarity of predicates}$$
$$w_j \equiv \text{weight of similarity of ARG j}$$
$$s_{i,\text{pred}} \equiv \text{predicate similarity in aligned frame i}$$
$$s_{i,j} \equiv \text{ARG j similarity in aligned frame i}$$

$$\text{precision} = \frac{\sum_i w_i^0 \frac{w_{\text{pred}} s_{i,\text{pred}} + \sum_j w_j s_{i,j}}{w_{\text{pred}} + \sum_j w_j |q_{i,j}^0|}}{\sum_i w_i^0}$$

$$\text{recall} = \frac{\sum_i w_i^1 \frac{w_{\text{pred}} s_{i,\text{pred}} + \sum_j w_j s_{i,j}}{w_{\text{pred}} + \sum_j w_j |q_{i,j}^1|}}{\sum_i w_i^1}$$

$$\text{MEANT} = \frac{2 \cdot \text{precision} \cdot \text{recall}}{\text{precision} \cdot \text{recall}}$$

where $q_{i,j}^0$ and $q_{i,j}^1$ are the argument of type j in frame i in MT and REF respectively. w_i^0 and w_i^1 are the weights for frame i in MT/REF respectively. These weights estimate the degree of contribution of each frame to the overall meaning of the sentence.

The weights w_{pred} and w_j are the weights of the lexical similarities of the predicates and role fillers of the arguments of type j between the reference translations and the MT output. There are a total of 12 weights for the set of semantic role labels in MEANT as defined in Lo and Wu [21]. For MEANT, w_{pred} and w_j are determined using supervised estimation via a simple grid search to optimize the correlation with human adequacy judgments (Lo and Wu [20]). For UMEANT (Lo and Wu [22]), w_{pred} and w_j are estimated in an unsupervised manner using relative frequency of each semantic role label in the reference translations. UMEANT can thus be used when human judgments on adequacy of the development set are unavailable.

$s_{i,\text{pred}}$ and $s_{i,j}$ are the lexical similarities based on a context vector model of the predicates and role fillers of the arguments of type j between the reference translations and the MT output. Lo et al. [19] and Tumuluru et al. [36] described how the lexical and phrasal similarities of the semantic role fillers are computed using geometric mean. A subsequent variant of the phrasal aggregation function that normalizes phrasal similarities according to the phrase length more accurately was proposed in Mihalcea et al. [26] and used in the work of Lo et al. [16]; Lo and Wu [23]; Lo et al. [18] and later further improved by a f-score aggregation in Lo et al. [17].

Recent studies (Lo et al. [16]; Lo and Wu [23]; Lo et al. [18]) show that tuning MT systems against MEANT produces more robustly adequate translations than the common practice of tuning against BLEU or TER across different data genres, such as formal newswire text, informal web forum text and public speech.

The promising results in evaluating and tuning with MEANT has led us to the present question: is it possible to further improve MEANT's correlation with human adequacy judgments by leveraging not only monolingual, but also cross-lingual, semantic frame similarities?

3   BiMEANT: Bilingual Semantic Frame Accuracy

Our new bilingual metric starts with monolingual MEANT's assessment of the degree of goodness of the translation, and then also integrates an assessment of roughly how well the translation captures the core semantics of the foreign input utterance. Whereas MEANT measures the lexical similarity using the monolingual context vector model and aggregates the lexical similarity into phrasal similarity using a variant of the aggregation function in Mihalcea et al. [26], for the new additional subtask of measuring semantic frame similarity cross-lingually, we propose to instead substitute simple lexical translation probabilities and a length-normalized inside probability at the root of the BITG biparse (Wu [38]; Zens and Ney [39]; Saers and Wu [34]; Adanki et al. [1]).

An example of the sorts of issues that the bilingual approach empirically helps to alleviate is shown in Fig. 1, which depicts examples of automatic shallow semantic parses on both reference and machine translations. In this case, the translation output MT2 is a more adequate translation than MT1, yet it is still too harshly penalized by monolingual MEANT for producing translation output with more "inaccurate" semantic frames as judged against the reference translation. This issue arises here because of a scoping choice in handling "not only": MT2 legitimately chooses to apply it to two separate semantic frames for the "speak" predicate, instead of the reference translation's choice of moving it inside to apply to the ARG1 of a single "speak" predicate. The dashed lines represent a low similarity (<0.5) semantic role alignments made by MEANT.

The bilingual approach, however, additionally incorporates a second way of assessing how well the semantic frames have been preserved in translation. Figure 2 shows how, by integrating a cross-lingual semantic frame similarity into MEANT, BiMEANT is able to reward the MT2 output that is closer to the semantic structure of the foreign input more fairly and accurately. To accomplish this, we compute cross-lingual semantic frame similarity in BiMEANT as follows (the differences from MEANT are underlined):

1. Apply an input language automatic shallow semantic parser to the foreign input and an output language automatic shallow semantic parser to the MT output. (Figure 2 shows examples of automatic shallow semantic parses on both foreign input and MT output. The Chinese semantic parser used in our experiments is C-ASSERT in Fung et al. [9,10].)
2. Apply the maximum weighted bipartite matching algorithm to align the semantic frames between the foreign input and MT output according to the lexical translation probabilities of the predicates.
3. For each pair of the aligned frames, apply the maximum weighted bipartite matching algorithm to align the arguments between the foreign input and MT output according to the aggregated phrasal translation probabilities of the role fillers.
4. Compute the weighted f-score over the matching role labels of these aligned predicates and role fillers according to the definitions similar to those in Sect. 2.2 except for replacing REF with IN in $q_{i,j}^1$ and w_i^1.

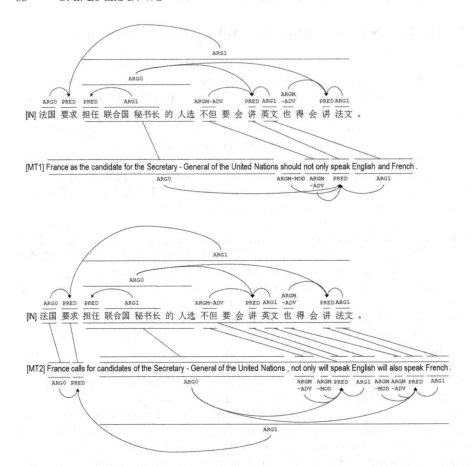

Fig. 2. Examples of bilingual semantic frame similarity captured by BiMEANT. MT2, the more adequate translation than MT1, is now fairly rewarded by the bilingual BiMEANT for producing translation with more accurate semantic frames according to the foreign input.

$$\mathbf{e}_{i,\mathrm{pred}} \equiv \text{the output side of the pred of aligned frame } i$$

$$\mathbf{f}_{i,\mathrm{pred}} \equiv \text{the input side of the pred of aligned frame } i$$

$$\mathbf{e}_{i,j} \equiv \text{the output side of the ARG } j \text{ of aligned frame } i$$

$$\mathbf{f}_{i,j} \equiv \text{the input side of the ARG } j \text{ of aligned frame } i$$

$$G \equiv \langle \{\mathrm{A}\}, \mathcal{W}^0, \mathcal{W}^1, \mathcal{R}, \mathrm{A} \rangle$$

$$\mathcal{R} \equiv \{\mathrm{A} \to [\mathrm{AA}], \mathrm{A} \to \langle \mathrm{AA} \rangle, \mathrm{A} \to e/f\}$$

$$p([\mathrm{AA}]|\mathrm{A}) = p(\langle \mathrm{AA} \rangle|\mathrm{A}) = 0.25$$

$$p(e/f|\mathrm{A}) = \frac{1}{2}\sqrt{t(e|f)\,t(f|e)}$$

$$xs_{i,\text{pred}} = \cfrac{1}{1 - \cfrac{\ln\left(P\left(A \overset{*}{\Rightarrow} e_{i,\text{pred}}/f_{i,\text{pred}}|G\right)\right)}{\max(|e_{i,\text{pred}}|,|f_{i,\text{pred}}|)}}$$

$$xs_{i,j} = \cfrac{1}{1 - \cfrac{\ln\left(P\left(A \overset{*}{\Rightarrow} e_{i,j}/f_{i,j}|G\right)\right)}{\max(|e_{i,j}|,|f_{i,j}|)}}$$

$$xp = \cfrac{\sum_i w_i^0 \frac{w_{\text{pred}} xs_{i,\text{pred}} + \sum_j w_j xs_{i,j}}{w_{\text{pred}} + \sum_j w_j |q_{i,j}^0|}}{\sum_i w_i^0}$$

$$xr = \cfrac{\sum_i w_i^1 \frac{w_{\text{pred}} xs_{i,\text{pred}} + \sum_j w_j xs_{i,j}}{w_{\text{pred}} + \sum_j w_j |q_{i,j}^1|}}{\sum_i w_i^1}$$

$$xf = \frac{2 \cdot xp \cdot xr}{xp \cdot xr}$$

$$\text{BiMEANT} = [\alpha * \text{MEANT} + (1 - \alpha) * xf]$$

where G is a bracketing ITG, whose only nonterminal is A, and where $\mathcal{R}$ is a set of transduction rules where $e \in \mathcal{W}^0 \cup \{\epsilon\}$ is an output token (or the *null* token), and $f \in \mathcal{W}^1 \cup \{\epsilon\}$ is an input token (or the *null* token). The rule probability function p is defined using fixed probabilities for the structural rules, and a translation table t trained using IBM model 1 (Brown *et al.* [3]) in both directions. A small constant (10^{-5}) is used when one of the ts is undefined. To calculate the inside probability of a pair of segments, $P\left(A \overset{*}{\Rightarrow} e/f|G\right)$, we use the algorithm described in Saers *et al.* [33]. $xs_{i,\text{pred}}$ and $xs_{i,j}$ are the length normalized BITG parsing probabilities of the predicates and role fillers of the arguments of type j between the input and the MT output. xp, xr and xf are the precision, recall and f-score of the cross-lingual semantic frame similarity computed by aggregating the BITG parsing probabilities of the predicates and role fillers in the same way as MEANT.

4   Results

Table 1 shows that BiMEANT significantly outperforms MEANT on sentence-level correlation with human adequacy judgment. This occurs despite the fact that only minimal adaptation has been done on the phrasal similarities for the cross-lingual semantic role fillers, suggesting that the performance of BiMEANT may be even better when settings are optimized.

Preliminary analysis indicates two reasons that BiMEANT improves correlation with human adequacy judgement. First, the semantic structure of the MT output often tends to be closer to that of the input sentence than that of the reference translation, due to somewhat arbitrary choices in scoping, topicalization, and similar phenomena. Secondly, the BITG constraints used in the cross-lingual assessment provide a more robust phrasal similarity aggregation function compared to the naive bag-of-words based heuristics previously employed in MEANT.

Similar results have been observed while trying to estimate word alignment probabilities where BITG constraints outperformed alignments from GIZA++ (Saers and Wu [34]).

Table 1. Sentence-level correlation with human adequacy judgement (GALE phase 2.5 evaluation data)

	Kendall
BiMEANT	**0.50**
MEANT	0.46
NIST	0.29
BLEU/METEOR/TER/PER	0.20
CDER	0.12
WER	0.10

5   Conclusion

We have presented a new bilingual automatic MT evaluation metric, BiMEANT, that correlates even more closely with human judgments of translation adequacy than standard monolingual MEANT. While previous work has established that MEANT accurately reflects translation adequacy via semantic frames and that optimizing SMT against MEANT improves translation quality, for very different languages the performance of purely monolingual metrics such as MEANT can be degraded by surface differences in choices such as scoping or topicalization, that lead to artificial differences between the reference and machine translations. The bilingual strategy employed by BiMEANT combats this by incorporating cross-lingual similarity assessments directly between the semantic frames of the input and output sentences. This is accomplished by (1) incorporating bracketing ITG constraints for aligning the lexicons in semantic role fillers, and (2) replacing the monolingual context vector model in MEANT with simple translation probabilities for computing the similarities of the semantic role fillers.

We would like to note that in this first study on a bilingual semantic frame based MT evaluation metric, we have performed minimal adaptation on the phrasal similarity assessments for the cross-lingual semantic role fillers. It is reasonable to expect that the performance of BiMEANT may further improve when the settings are optimized. The encouraging results suggest interesting potential for BiMEANT, especially across very different languages.

Acknowledgment. This material is based upon work supported in part by the Defense Advanced Research Projects Agency (DARPA) under BOLT contract nos. HR0011-12-C-0014 and HR0011-12-C-0016, and GALE contract nos. HR0011-06-C-0022 and HR0011-06-C-0023; by the European Union under the FP7 grant agreement no. 287658;

and by the Hong Kong Research Grants Council (RGC) research grants GRF620811, GRF621008, and GRF612806. Any opinions, findings and conclusions or recommendations expressed in this material are those of the authors and do not necessarily reflect the views of DARPA, the EU, or RGC. Thanks to Markus Saers, Meriem Beloucif, and Karteek Addanki for supporting work, and to Pascale Fung, Yongsheng Yang and Zhaojun Wu for sharing the maximum entropy Chinese segmenter and C-ASSERT, the Chinese semantic parser.

References

1. Addanki, K., Lo, C., Saers, M., Wu, D.: LTG vs. ITG coverage of cross-lingual verb frame alternations. In: 16th Annual Conference of the European Association for Machine Translation (EAMT-2012), Trento, Italy, May 2012
2. Banerjee, S., Lavie, A.: METEOR: an automatic metric for MT evaluation with improved correlation with human judgments. In: Workshop on Intrinsic and Extrinsic Evaluation Measures for Machine Translation and/or Summarization, Ann Arbor, Michigan, June 2005
3. Brown, P.F., Della, P., Stephen, A., Della, P., Vincent, J., Mercer, R.L.: The mathematics of machine translation: parameter estimation. Comput. Linguist. 19(2), 263–311 (1993)
4. Callison-Burch, C., Fordyce, C., Koehn, P., Monz, C., Schroeder, J.: (meta-) evaluation of machine translation. In: Second Workshop on Statistical Machine Translation (WMT-07) (2007)
5. Callison-Burch, C., Fordyce, C., Koehn, P., Monz, C., Schroeder, J.: Further meta-evaluation of machine translation. In: Third Workshop on Statistical Machine Translation (WMT-08) (2008)
6. Callison-Burch, C., Osborne, M., Koehn, P.: Re-evaluating the role of BLEU in machine translation research. In: 11th Conference of the European Chapter of the Association for Computational Linguistics (EACL-2006) (2006)
7. Castillo, J., Estrella, P.: Semantic textual similarity for MT evaluation. In: 7th Workshop on Statistical Machine Translation (WMT 2012) (2012)
8. Doddington, G.: Automatic evaluation of machine translation quality using n-gram co-occurrence statistics. In: The Second International Conference on Human Language Technology Research (HLT '02), San Diego, California (2002)
9. Fung, P., Ngai, G., Yang, Y., Chen, B.: A maximum-entropy chinese parser augmented by transformation-based learning. ACM Trans. Asian Lang. Inf. Process. (TALIP) 3(2), 159–168 (2004)
10. Fung, P., Wu, Z., Yang, Y., Wu, D.: Learning bilingual semantic frames: shallow semantic parsing vs. semantic role projection. In: The 11th International Conference on Theoretical and Methodological Issues in Machine Translation (TMI-07), Skovde, Sweden, pp. 75–84 (2007)
11. Giménez, J., Màrquez, L.: Linguistic features for automatic evaluation of heterogenous MT systems. In: Second Workshop on Statistical Machine Translation (WMT-07), Prague, Czech Republic, June 2007, pp. 256–264 (2007)
12. Giménez, J., Màrquez, L.: A smorgasbord of features for automatic MT evaluation. In: Third Workshop on Statistical Machine Translation (WMT-08), Columbus, Ohio, June 2008
13. Koehn, P., Monz, C.: Manual and automatic evaluation of machine translation between european languages. In: Workshop on Statistical Machine Translation (WMT-06) (2006)

14. Leusch, G., Ueffing, N., Ney, H.: CDer: Efficient MT evaluation using block movements. In: 11th Conference of the European Chapter of the Association for Computational Linguistics (EACL-2006) (2006)
15. Liu, D., Gildea, D.: Syntactic features for evaluation of machine translation. In: Workshop on Intrinsic and Extrinsic Evaluation Measures for Machine Translation and/or Summarization, Ann Arbor, Michigan, June 2005
16. Lo, C., Addanki, K., Saers, M., Wu, D.: Improving machine translation by training against an automatic semantic frame based evaluation metric. In: 51st Annual Meeting of the Association for Computational Linguistics (ACL 2013) (2013)
17. Lo, C., Beloucif, M., Saers, M., Wu, D.: XMEANT: better semantic MT evaluation without reference translations. In: 52nd Annual Meeting of the Association for Computational Linguistics (ACL 2014) (2014)
18. Lo, C., Beloucif, M., Wu, D.: Improving machine translation into Chinese by tuning against Chinese MEANT. In: International Workshop on Spoken Language Translation (IWSLT 2013) (2013)
19. Lo, C., Tumuluru, A.K., Wu, D.: Fully automatic semantic MT evaluation. In: 7th Workshop on Statistical Machine Translation (WMT 2012) (2012)
20. Lo, C., Wu, D.: MEANT: an inexpensive, high-accuracy, semi-automatic metric for evaluating translation utility based on semantic roles. In: 49th Annual Meeting of the Association for Computational Linguistics: Human Language Technologies (ACL HLT 2011) (2011)
21. Lo, C., Wu, D.: SMT vs. AI redux: how semantic frames evaluate MT more accurately. In: 22nd International Joint Conference on Artificial Intelligence (IJCAI-11) (2011)
22. Lo, C., Wu, D.: Unsupervised vs. supervised weight estimation for semantic MT evaluation metrics. In: Sixth Workshop on Syntax, Semantics and Structure in Statistical Translation (SSST-6) (2012)
23. Lo, C., Wu, D.: Can informal genres be better translated by tuning on automatic semantic metrics? In: 14th Machine Translation Summit (MT Summit XIV) (2013)
24. Lo, C., Wu, D.: MEANT at WMT 2013: a tunable, accurate yet inexpensive semantic frame based MT evaluation metric. In: 8th Workshop on Statistical Machine Translation (WMT 2013) (2013)
25. Macháček, M., Bojar, O.: Results of the WMT13 metrics shared task. In: 8th Workshop on Statistical Machine Translation (WMT 2013), Sofia, Bulgaria, August 2013
26. Mihalcea, R., Corley, C., Strapparava, C.: Corpus-based and knowledge-based measures of text semantic similarity. In: The 21st National Conference on Artificial Intelligence (AAAI-06), vol. 21 (2006)
27. Nießen, S., Och, F. J., Leusch, G., Ney, H.: A evaluation tool for machine translation: fast evaluation for MT research. In: The 2nd International Conference on Language Resources and Evaluation (LREC 2000) (2000)
28. Owczarzak, K., van Genabith, J., Way, A.: Dependency-based automatic evaluation for machine translation. In: Syntax and Structure in Statistical Translation (SSST) (2007)
29. Owczarzak, K., van Genabith, J., Way, A.: Evaluating machine translation with LFG dependencies. Mach. Transl. 21, 95–119 (2007)
30. Papineni, K., Roukos, S., Ward, T., Zhu, W.J.: BLEU: a method for automatic evaluation of machine translation. In: 40th Annual Meeting of the Association for Computational Linguistics (ACL-02), Philadelphia, Pennsylvania, July 2002, pp. 311–318 (2002)

31. Pradhan, S., Ward, W., Hacioglu, K., Martin, J. H., Jurafsky, D.: Shallow semantic parsing using support vector machines. In: Human Language Technology Conference of the North American Chapter of the Association for Computational Linguistics (HLT-NAACL 2004) (2004)
32. Rios, M., Aziz, W., Specia, L.: TINE: a metric to assess MT adequacy. In: 6th Workshop on Statistical Machine Translation (WMT 2011) (2011)
33. Saers, M., Nivre, J., Wu, D.: Learning stochastic bracketing inversion transduction grammars with a cubic time biparsing algorithm. In: 11th International Conference on Parsing Technologies (IWPT'09), Paris, France, October 2009, pp. 29–32 (2009)
34. Saers, M., Wu, D.: Improving phrase-based translation via word alignments from stochastic inversion transduction grammars. In: Third Workshop on Syntax and Structure in Statistical Translation (SSST-3), Boulder, Colorado, June 2009, pp. 28–36 (2009)
35. Snover, M., Dorr, B., Schwartz, R., Micciulla, L., Makhoul, J.: A study of translation edit rate with targeted human annotation. In: 7th Biennial Conference Association for Machine Translation in the Americas (AMTA 2006), Cambridge, Massachusetts, August 2006, pp. 223–231 (2006)
36. Tumuluru, A. K., Lo, C., Wu, D.: Accuracy and robustness in measuring the lexical similarity of semantic role fillers for automatic semantic MT evaluation. In: 26th Pacific Asia Conference on Language, Information, and Computation (PACLIC 26) (2012)
37. Wang, M., Manning, C.D.: SPEDE: probabilistic edit distance metrics for MT evaluation. In: 7th Workshop on Statistical Machine Translation (WMT 2012) (2012)
38. Wu, D.: Stochastic inversion transduction grammars and bilingual parsing of parallel corpora. Comput. Linguist. **23**(3), 377–403 (1997)
39. Zens, R., Ney, H.: A comparative study on reordering constraints in statistical machine translation. In: 41st Annual Meeting of the Association for Computational Linguistics (ACL-2003), Stroudsburg, Pennsylvania, pp. 144–151 (2003)

Speech and Speaker Recognition

Robust Speaker Recognition Using MAP Estimation of Additive Noise in i-vectors Space

Waad Ben Kheder[✉], Driss Matrouf, Pierre-Michel Bousquet,
Jean-François Bonastre, and Moez Ajili

LIA, University of Avignon, Avignon, France
{waad.ben-kheder,driss.matrouf,pierre-michel.bousquet,
jean-francois.bonastre,moez.ajili}@univ-avignon.fr

Abstract. In the last few years, the use of i-vectors along with a generative back-end has become the new standard in speaker recognition. An i-vector is a compact representation of a speaker utterance extracted from a low dimensional total variability subspace. Although current speaker recognition systems achieve very good results in clean training and test conditions, the performance degrades considerably in noisy environments. The compensation of the noise effect is actually a research subject of major importance. As far as we know, there was no serious attempt to treat the noise problem directly in the i-vectors space without relying on data distributions computed on a prior domain. This paper proposes a full-covariance Gaussian modeling of the clean i-vectors and noise distributions in the i-vectors space then introduces a technique to estimate a clean i-vector given the noisy version and the noise density function using MAP approach. Based on NIST data, we show that it is possible to improve up to 60 % the baseline system performances. A noise adding tool is used to help simulate a real-world noisy environment at different signal-to-noise ratio levels.

Keywords: i-vectors · MAP adaptation · Speaker recognition · Additive noise

1 Introduction

Recent work on the robustness of i-vector -based speaker recognition systems has been carried out at different levels in order to track and compensate the additive noise effect without altering the speaker-related information. After the success of VTS (Vector Taylor Series) in robust ASR applications [1], a VTS-based i-vectors extractor was proposed in [7,8] and then developed in [9] using "Unscented transforms" trying to model non-linear distortions in the mel-cepstral domain based on a non-linear noise model in order to compensate both convolutive and additive noises. This compensation scheme tackles the problem on an early stage by computing the "clean i-vector" directly by fitting the corresponding noisy GMM to a given noisy speech segment. That requires information about spectral

© Springer International Publishing Switzerland 2014
L. Besacier et al. (Eds.): SLSP 2014, LNAI 8791, pp. 97–107, 2014.
DOI: 10.1007/978-3-319-11397-5_7

data distribution to do the link between the two domains. The biggest weakness of this technique is the complexity of the estimation model and the number of imposed constraints which makes it extremely rigid and hardly extendable. The integration of many interesting techniques (like feature warping [11] for robust channel mismatch) becomes a hard task and requires to rebuild the whole model.

This motivates the development of a new kind of noise models which operates directly in the i-vectors space. We show in this paper that it's possible to reach far better results than VTS-based techniques based on an additive noise model in the i-vectors space using only noise and clean i-vectors distributions. We start by assuming that both clean i-vectors and noise can be modeled by full-covariance Gaussian distributions. Then, we present an i-vectors "cleaning" technique that uses the MAP approach to estimate a clean i-vector given a noisy i-vector version and a normal noise distribution model.

This paper is structured as follows. Section 2 describes the i-vector framework for speaker recognition. Section 3 details the proposed approach. Section 4 presents the experimental protocol, the experiments and the corresponding results.

2   The i-vectors Framework

In this section we present the i-vectors framework along with the scoring procedure that will be used further in our experiments.

2.1   The Total-Variability Subspace

In this approach, an i-vector extractor converts a sequence of acoustic vectors into a single low-dimensional vector representing the whole speech utterance. The speaker- and session-dependent super-vector s of concatenated Gaussian Mixture Model (GMM) means is assumed to obey a linear model of the form:

$$s = m + Tw \qquad (1)$$

where:

- m is the mean super-vector of the Universal Background Model (UBM)
- T is the low-rank variability matrix obtained from a large dataset by MAP estimation [6]. It represents the total variability subspace.
- w is a standard-normally distributed latent variable called "i-vector".

Extracting an i-vector from the total variability subspace is essentially a maximum a-posteriori adaptation of w in the space defined by T. The algorithms for the estimation of T and the extraction of i-vectors are described in [10].

2.2   The i-vectors Scoring System

Many dimensionality reduction techniques (such as LDA) and generative models (like PLDA, and the Two-covariance model) have been developed in order to improve the i-vectors comparison in speaker verification trials. The speaker verification score given two i-vectors w_1 and w_2 is the likelihood ratio described by:

$$score = log \frac{P(w_1, w_2 | \theta_{tar})}{P(w_1, w_2 | \theta_{non})} \tag{2}$$

where the hypothesis θ_{tar} states that inputs w_1 and w_2 are from the same speaker and the hypothesis θ_{non} states they are from different speakers.

We focus in the following on the generative model that we used in our work: the two-covariance scoring model.

The Two-Covariance Scoring Model: This model is a particular case of the Probabilistic Linear Discriminant Analysis (PLDA) described in [12]. It can be seen as a scoring method and a convolutive noise compensation technique. It consists of a simple linear-Gaussian generative model in which an i-vector w of a speaker s can be decomposed in:

$$w = y_s + \varepsilon \tag{3}$$

where the speaker model y_s is a vector of the same dimensionality as an i-vector, ε is Gaussian noise and:

$$P(y_s) = \mathcal{N}(\mu, B) \tag{4}$$

$$P(w|y_s) = \mathcal{N}(y_s, W) \tag{5}$$

$\mathcal{N}$ denotes the normal distribution, μ represents the overall mean of the training data set, B and W are the between- and within-speaker covariance matrices defined as:

$$B = \sum_{s=1}^{S} \frac{n_s}{n} (y_s - \mu)(y_s - \mu)^t \tag{6}$$

$$W = \frac{1}{n} \sum_{s=1}^{S} \sum_{i=1}^{n_s} (w_i^s - y_s)(w_i^s - y_s)^t \tag{7}$$

where n_s is the number of utterances for speaker s, n is the total number of utterances, w_i are the i-vectors of sessions of speaker s, y_s is the mean of all the i-vectors of speaker s and μ represents the overall mean of the training data set. Under assumptions (6) and (7), the score from Eq. (2) can be expressed as:

$$s = \frac{\int \mathcal{N}(w_1|y, W)\mathcal{N}(w_2|y, W)\mathcal{N}(y|\mu, B)dy}{\prod_{i=1,2} \int \mathcal{N}(w_i|y, W)\mathcal{N}(y|\mu, B)dy} \tag{8}$$

the explicit solution of (8) is given in [3].

3   MAP Estimation of Clean i-vectors

Given a noisy i-vector Y_0, the goal of this section will be to estimate the corresponding clean version $\hat{X}_0$. We will work exclusively in the i-vectors space and build a clean i-vectors estimator based solely on "i-vector space"-related data using a MAP approach.

Let's start by defining two random variables in the i-vectors space:

- X which corresponds the clean i-vectors.
- Y which corresponds the noisy i-vectors.

To model the additive noise in the i-vectors space, we define a third random variable N that links X and Y according to the following expression:

$$N = Y - X \tag{9}$$

We assume that both clean i-vectors (X) and noise data (N) can be represented by two normal distributions in the i-vectors space. We can then define the corresponding probability distribution functions $f(X)$ and $f(N)$ as:

$$f(X) = \mathcal{N}(\mu_X, \Sigma_X) \tag{10}$$

$$f(N) = \mathcal{N}(\mu_N, \Sigma_N) \tag{11}$$

where $\mathcal{N}(\mu_i, \Sigma_i)$ denotes a normal distribution with mean μ_i and full covariance matrix Σ_i.

Referring to hypothesis (9), (10) and (11) we can express $f(Y_0|X)$ for a given Y_0 as:

$$f(Y_0|X) = \frac{1}{(2\pi)^{\frac{p}{2}}|\Sigma|^{\frac{1}{2}}} \, exp\{(Y_0 - X - \mu_N)^t \Sigma_N^{-1}(Y_0 - X - \mu_N)\} \tag{12}$$

Based on the noise model (9) and the two previously defined distributions, we can estimate for a given noisy i-vector Y_0 its clean version $\hat{X}_0$ using a MAP estimator:

$$\hat{X}_0 = \underset{X}{\mathrm{argmax}}\{\ln f(X/Y_0)\} \tag{13}$$

Using the Bayesian rule, we can write $f(X/Y_0)$ as:

$$f(X/Y_0) = \frac{f(Y_0/X)f(X)}{f(Y_0)} \tag{14}$$

After combining (13) and (14):

$$\hat{X}_0 = \underset{X}{\mathrm{argmax}}\{\ln f(Y_0/X)f(X)\} \tag{15}$$

Finding $\hat{X}_0$ becomes equivalent to solving:

$$\frac{\partial}{\partial X}\{\ln f(Y_0/X) + \ln f(X)\} = 0 \tag{16}$$

By developing (16) using (10) and (12), we end up with:

$$\frac{\partial}{\partial X}\{(Y_0 - X - \mu_N)^t \Sigma_N^{-1}(Y_0 - X - \mu_N) + (X - \mu_X)^t \Sigma_X^{-1}(X - \mu_X)\} = 0 \quad (17)$$

After the derivation, we have:

$$- \Sigma_N^{-1}(Y_0 - \hat{X}_0 - \mu_N) + \Sigma_X^{-1}(\hat{X}_0 - \mu_X) = 0 \quad (18)$$

then, we find the final expression of the clean i-vector $\hat{X}_0$ given the noisy version Y_0 and both X and N distributions parameters:

$$\hat{X}_0 = (\Sigma_N^{-1} + \Sigma_X^{-1})^{-1}(\Sigma_N^{-1}(Y_0 - \mu_N) + \Sigma_X^{-1}\mu_X) \quad (19)$$

The estimation of $f(X)$ and $f(N)$ are done as so:

- $f(X)$: μ_X and Σ_X are estimated once and for all over a large set of clean i-vectors. Since this distribution is independent from the noise, there is no constraints on the number of i-vectors to be used.
- $f(N)$: In real-world conditions, the available amount of noisy data is generally limited. Possible improvements of this technique could be proposed in future publications to deal with this constraint. Based on a set of clean and noisy i-vectors pairs corresponding to the same clean utterances, the noise data set in the i-vectors space is firstly computed using $N = Y - X$. Then μ_N and Σ_N are estimated as any regular normal distribution parameters.

In i-vector -based speaker recognition systems, length-normalization was proved to improve the overall system performance [4]. In our case, it's important to mention that all used noisy and clean i-vectors in the estimation process of $\hat{X}_0$, $f(X)$ and $f(N)$ were initially length-normalized.

4   Experimental Protocol and Results

In this section, we present the configuration used in the LIA speaker recognition system along with the training and test data sets. Then, the noise adding procedure and the realized experiments are detailed.

4.1   The LIA Speaker Recognition Baseline System

Our experiments operate on 19 Mel-Frequency Cepstral Coefficients (plus energy) augmented with 19 first (Δ) and 11 second ($\Delta\Delta$) derivatives. A mean and variance normalization (MVN) technique is applied on the MFCC features estimated using the speech portion of the audio file. The low-energy frames (corresponding mainly to silence) are removed.

A gender-dependent 512 diagonal component UBM (male model) and a total variability matrix of low rank 400 are estimated using 15660 utterances corresponding to 1147 speakers (using NIST SRE 2004, 2005, 2006 and Switchboard data). The LIA_SpkDet package of the LIA_RAL/ALIZE toolkit is used for the estimation of the total variability matrix and the i-vectors extraction.

The implemented algorithms are described in [10]. Finally a two-covariance-based scoring scheme is applied.

4.2 Noise Adding

We will use two different noises in our analysis:

- A crowd-noise
- An air-cooling noise

The open-source toolkit FaNT [5] (Filtering and Noise Adding Tool) was used to add these noises at different SNR levels generating new noisy audio files.

In order to have a good estimation of the clean normal i-vectors distribution, we have selected the 6000 utterances from the training data having an SNR greater than 30 dB.

For each test condition, we used 3000 pairs of clean and noisy i-vectors to estimate the normal noise distribution model. N is firstly computed with $N = Y - X$ then $f(N)$ is estimated by computing μ_N and Σ_N.

At the end, six trial conditions will be evaluated for each noise:

- Noisy test/target data with "Crowd-noise" at SNR levels 10 db, 5 db and 0 db.
- Noisy test/target data with "Air-cooling noise" at SNR levels 10 db, 5 db and 0 db.

4.3 Test Data and Performance Evaluation

The equal-error rate (EER) over the NIST SRE 2008 test data will be used as a reference to monitor the performance improvement compared to the baseline system in noisy conditions. We will be only focused on the "short2/short3" task under the "det7" conditions [2]. In order to help visualize the improvement in the error-rate in each test configuration, the relative improvement measure (RI%) will be added.

The two studied noises have been used to create noisy versions of the test and target data over 10 db, 5 db and 0 db SNR levels.

4.4 Experiments and Results

The LIA speaker verification baseline system reaches EER=1.59 % in clean conditions. This error-rate will be the lower bound that helps evaluate the gain of the proposed technique compared to the noisy baseline performance.

In the following tables, the estimated clean i-vectors corresponding to noisy test or target i-vectors will be referred to as "I-MAP" vectors.

The system performances will be presented in two different configurations:

- Clean target i-vectors and noisy test i-vectors.
- Noisy target i-vectors and noisy test i-vectors.

First, we evaluate the baseline system performances before and after the application of our method when all noisy data (test and target noisy i-vectors) are produced by the same noise.

Clean Target i-vectors and Noisy Test i-vectors (Crowd-Noise): The Table 1 summarizes the baseline system performance while used with noisy test i-vectors (Crowd-noise) and clean target i-vectors compared to the proposed method performance:

Table 1. System performance using noisy test data (Crowd-noise)

	EER (%)		RI (%)
	Baseline system	with I-MAP test	
SNR=10 db	5.86	3.18	**45.73**
SNR=5 db	9.53	4.34	54.46
SNR=0 db	17.08	8.43	50.64

We observe more than 50 % relative improvement in average at the three SNR levels. This encourages the use of clean target models when available with noisy test data.

Clean Target i-vectors and Noisy Test i-vectors (Air-Cooling Noise): The Table 2 summarizes the baseline system performance while used with noisy test and target i-vectors (Air-cooling noise) and clean target i-vectors compared to the proposed method performance:

Table 2. System performance using noisy test data (Air-cooling noise)

	EER (%)		RI (%)
	Baseline system	with I-MAP test	
SNR=10 db	7.47	4.78	**36.01**
SNR=5 db	15.68	7.3	53.44
SNR=0 db	27.33	13.89	49.18

We observe more than 46 % relative improvement in average at the three SNR levels. The overall performance is comparable to the previous one and validates the proposed method for different noisy test conditions.

Noisy Target i-vectors and Noisy Test i-vectors (Crowd-Noise): In real speaker recognition applications, clean target data could not be available, so it's important to check the validity of the proposed method in noisy target i-vectors conditions.

The Table 3 summarizes the baseline system performance while used with noisy test and target i-vectors (Crowd-noise) compared to the proposed method performance:

Table 3. System performance using noisy test and target data (Crowd-noise)

	EER (%)		RI (%)
	Baseline system	with I-MAP target and I-MAP test	
SNR=10 db	10.72	4.34	59.51
SNR=5 db	17.79	8.15	54.19
SNR=0 db	24.77	13.44	**45.74**

We observe more than 53 % relative improvement in average at the three SNR levels. It's important to note that our method keeps its efficiency even with noisy target i-vectors.

Noisy Target i-vectors and Noisy Test i-vectors (Air-Cooling Noise): The Table 4 summarizes the baseline system performance while used with noisy test and target i-vectors (Air-cooling noise) compared to the proposed method performance:

Table 4. System performance using noisy test and target data (Air-cooling noise)

	EER (%)		RI (%)
	Baseline system	with I-MAP target and I-MAP test	
SNR=10 db	16.14	6.83	57.68
SNR=5 db	20.73	10.5	49.35
SNR=0 db	32.89	20.5	**37.67**

We observe more than 48 % relative improvement in average at the three SNR levels. The relative improvement with this noise is also comparable with the "clean target - noisy test" performance. This validates the robustness of the proposed method in different noisy target and test conditions.

It's easy to see the considerable leap between the baseline system performance and the one obtained after the MAP estimation of the clean i-vectors in all previous conditions. For each of the two noises, the average relative improvement exceeds 48 % in 10 dB and 5 dB SNR levels conditions. One of the most interesting results is the efficiency of this method even on very low SNR levels (0 dB).

Noisy data in real-world applications could be affected by different noise sources. Based on this idea, it's interesting to evaluate the performance of this technique in test conditions where more than one noise is present. To test this possibility, we mixed evenly for each SNR level the noisy i-vectors coming from both noises. This way, for every SNR level, 50 % of the noisy test i-vectors are related to the "crowd-noise" and the other 50 % is related to the "air-cooling" noise. The same mixing scheme is done on noisy target i-vectors in the "noisy test - noisy target" configuration.

The following tables summarizes the baseline system performance before and after the application of our technique.

Clean Target i-vectors and Noisy Test i-vectors (Two Noises): The Table 5 summarizes the system performance before and after the application of our technique for clean target data and mixed noisy test i-vectors (coming from two different noises):

Table 5. System performances for clean target and mixed noisy test i-vectors

	EER (%)		RI (%)
	Baseline system	with I-MAP target and I-MAP test	
SNR=10 db	7.06	3.92	**44.47**
SNR=5 db	13.24	5.92	55.28
SNR=0 db	22.55	11.86	47.40

We observe more than 50 % relative improvement in average at the three SNR levels. The overall performance is maintained compared to the first configurations when we used only one noise. These results validate the efficiency of the proposed method.

Noisy Target i-vectors and Noisy Test i-vectors (Two Noises in Both): The Table 6 summarizes the system performance before and after the application of our technique for mixed noisy target and test i-vectors (coming from two different noises):

Table 6. System performances for mixed noisy test and target i-vectors

	EER (%)		RI (%)
	Baseline system	with I-MAP target and I-MAP test	
SNR=10 db	15.49	6.16	60.23
SNR=5 db	24.15	9.79	59.46
SNR=0 db	34.16	22.78	**33.31**

Similar performance is also observed in this condition (51 % average relative improvement) showing the validity of the used method in the "noisy target - noisy test" configuration.

5   Conclusion

In this work, we introduced a new clean i-vector estimation technique referring to a noisy version based on a normal distribution model of both clean i-vectors and noise in the i-vectors space using a MAP approach. The observed improvement compared to the baseline system performance reaches 60 % in low SNR test conditions and outperforms recently developed robust speaker recognition techniques (like VTS-based i-vector extractors).

Further improvements could be achieved by extending the noise distribution model in the i-vectors space (using Gaussian mixtures instead of unimodal Gaussian distributions for example). The use of a factor analysis -based technique like PLDA could also be explored to improve the quality of the i-vectors used to build the noise distribution model.

References

1. Acero, A., Deng, L., Kristjansson, T.T., Zhang, J.: Hmm adaptation using vector taylor series for noisy speech recognition. In: INTERSPEECH, pp. 869–872 (2000)
2. The NIST year 2008 speaker recognition evaluation plan (2008). http://www.itl.nist.gov/iad/mig/tests/sre/2008/sre08_evalplan_release4.pdf. Accessed 15 May 2014
3. Brümmer, N., De Villiers, E.: The speaker partitioning problem. In: Odyssey, p. 34 (2010)
4. Garcia-Romero, D., Espy-Wilson, C.Y.: Analysis of i-vector length normalization in speaker recognition systems. In: Interspeech, pp. 249–252 (2011)
5. Hirsch, H.G.: FaNT - Filtering and Noise Adding Tool. http://dnt.kr.hsnr.de/download.html. Accessed 15 May 2014
6. Kenny, P.: Joint factor analysis of speaker and session variability: theory and algorithms. CRIM, Montreal, (Report) CRIM-06/08-13 (2005)
7. Lei, Y., Burget, L., Scheffer, N.: A noise robust i-vector extractor using vector taylor series for speaker recognition. In: 2013 IEEE International Conference on Acoustics, Speech and Signal Processing (ICASSP), pp. 6788–6791. IEEE (2013)
8. Lei, Y., McLaren, M., Ferrer, L., Scheffer, N.: Simplified vts-based i-vector extraction in noise-robust speaker recognition. Submitted to ICASSP, Florence, Italy (2014)
9. Martinez, D., Burget, L., Stafylakis, T., Lei, Y., Kenny, P., Lleida, E.: Unscented transform for ivector-based noisy speaker recognition. Submitted to ICASSP, Florence, Italy (2014)
10. Matrouf, D., Scheffer, N., Fauve, B.G., Bonastre, J.F.: A straightforward and efficient implementation of the factor analysis model for speaker verification. In: INTERSPEECH, pp. 1242–1245 (2007)

11. Pelecanos, J., Sridharan, S.: Feature warping for robust speaker verification. In: Speaker Odyssey, Crete, Greece (2001)
12. Prince, S.J., Elder, J.H.: Probabilistic linear discriminant analysis for inferences about identity. In: IEEE 11th International Conference on Computer Vision, ICCV 2007, pp. 1–8. IEEE (2007)

Structured GMM Based on Unsupervised Clustering for Recognizing Adult and Child Speech

Arseniy Gorin[1,2,3](✉) and Denis Jouvet[1,2,3]

[1] Speech Group, LORIA, Inria, 615 Rue du Jardin Botanique,
54600 Villers-lès-Nancy, France
[2] Universitè de Lorraine, LORIA, UMR 7503, 54600 Villers-lès-Nancy, France
[3] CNRS, LORIA, UMR 7503, 54600 Villers-lès-Nancy, France
{arseniy.gorin,denis.jouvet}@inria.fr

Abstract. Speaker variability is a well-known problem of state-of-the-art Automatic Speech Recognition (ASR) systems. In particular, handling children speech is challenging because of substantial differences in pronunciation of the speech units between adult and child speakers. To build accurate ASR systems for all types of speakers Hidden Markov Models with Gaussian Mixture Densities were intensively used in combination with model adaptation techniques.

This paper compares different ways to improve the recognition of children speech and describes a novel approach relying on Class-Structured Gaussian Mixture Model (GMM).

A common solution for reducing the speaker variability relies on gender and age adaptation. First, it is proposed to replace gender and age by unsupervised clustering. Speaker classes are first used for adaptation of the conventional HMM. Second, speaker classes are used for initializing structured GMM, where the components of Gaussian densities are structured with respect to the speaker classes. In a first approach mixture weights of the structured GMM are set dependent on the speaker class. In a second approach the mixture weights are replaced by explicit dependencies between Gaussian components of mixture densities (as in stranded GMMs, but here the GMMs are class-structured).

The different approaches are evaluated and compared on the TIDIGITS task. The best improvement is achieved when structured GMM is combined with feature adaptation.

Keywords: Speech recognition · Unsupervised clustering · Speaker class modeling · Stochastic trajectory modeling

1 Introduction

Hidden Markov Models with Gaussian Mixture observation densities (HMM-GMM) are successfully applied in automatic speech recognition systems, despite

© Springer International Publishing Switzerland 2014
L. Besacier et al. (Eds.): SLSP 2014, LNAI 8791, pp. 108–119, 2014.
DOI: 10.1007/978-3-319-11397-5_8

their inability to accurately model the dynamic properties of speech coming from different speakers and recording conditions. The accuracy is usually improved by applying various tuning techniques and more advanced feature processing.

Children speech is a good example of the data that is hard to recognize with conventional HMM-GMM because of the variability of the acoustic features of the same phonetic units spoken by adult and child speakers. Such variability comes from the differences in the size of the vocal tract and mispronunciation of certain phones by children [2,14]. For example, children have shorter vocal tract, than adults, which leads to higher F0 (fundamental frequency) [12].

The task becomes more complicated as the amount of available annotated children speech is not large enough for training separate models for children data. Also, frequently, the information about speaker age is available neither for test, nor for training data.

An effective strategy for handling child speech (or speaker variability in general) consists in adapting the ASR systems. These techniques either modify the acoustic features (VTLN [17], fMLLR [5]), or the model parameters (MLLR, MAP [6]) to maximize the likelihood of the adaptation data. A review paper [13] discusses various improvements and applications of VTLN-based algorithms for improving automatic recognition of children speech.

The conventional approach for handling speaker variability assumes age and gender known at least for the training data. In this case separate models are constructed for different age and gender classes by adapting the Speaker-Independent (SI) model trained on the full training dataset. In decoding the corresponding model is selected for each utterance based on knowledge of the speaker age and gender (if available), or on an automatic classification. A different approach relying on interpolation of several models was proposed in [16] and demonstrated significant improvements also on children speech data.

The main part of this work focuses on the general situation, when the dataset contains speakers of different age and gender, but the speaker age and gender are known neither for testing, nor for training. In such case unsupervised clustering is applied at the utterance level, assuming that the speaker class is not changing within the sentence [1]. Increasing the number of classes decreases the number of available training utterances associated with each class. This problem can be partially handled by soft clustering techniques, such as eigenvoice approach, where the parameters of an unknown speaker are determined as a combination of class models [10], or by explicitly enlarging the class data by allowing one utterance to belong to several classes [7,9].

Furthermore, a novel approach is proposed in this work for using speaker classes to structure an HMM-GMM. The idea is to include the speaker class information into the structure of a single HMM-GMM instead of building separate models for each class. To do this, the components of GMMs are composed from GMMs with a smaller number of components per density and trained (or adapted) on class data. Speaker class structuring leads to GMM, in which each kth component of the density (or a subset of components) is associated with a given class in contrast to conventional GMM, where the components are trained independently.

When the components are structured, the speaker class is represented as a subspace of the structured GMM (k^{th} component, or subset of components of each GMM corresponds to k^{th} speaker class). To select the corresponding subspace, additional modifications are proposed in the form of dependencies added on weights of the Gaussian components.

Class-structured GMM was first used with mixture Weights dependent on the speaker class in addition to the associated HMM state. Such a model with Speaker class-dependent Weights (SWGMM) was originally investigated in a radio broadcast transcription system [8]. In this model, the mixture weights are class-dependent and the Gaussian means and variances are class-independent, but class-structured.

Another way of using class-structured GMM is to replace state and class-dependent mixture weights by only state-dependent Mixture Transition Matrices (MTMs) of Stranded Gaussian Mixture Model (SGMM). SGMM is similar to conditional Gaussian model [15], which was recently extended, re-formulated and investigated for robust ASR [18]. In SGMM the Mixture Transition Matrix (MTM) defines the dependencies between the components of adjacent Gaussian mixture observation densities.

In [18] it was originally proposed to initialize SGMM from the conventional HMM-GMM. Instead, here, for a class-Structured SGMM (SSGMM), the SGMM is initialized from SWGMM and each GMM component (or each set of components) mainly represents a different speaker class. MTM in SSGMM is used to model the probabilities of either keeping the same component (speaker class) over time, or to dynamically switch between dominating components (classes).

The advantage of using explicit component dependencies over class-dependent mixture weights is that the weights are no more fixed at the utterance level (determined by the speaker class), but rather change depending on the observation from the previous frame. As a result, explicit trajectory modeling improves the recognition accuracy. Moreover, it does not require an additional classification step to determine the class of the utterance in decoding.

The paper is organized as follows. Section 2 describes the system and discusses the conventional adaptation-based approach. Section 3 discusses unsupervised class-based-adaptation approach for ASR (CA-GMM). Section 4 introduces class-structured GMM with Speaker class-dependent Weights (SWGMM) and describes the corresponding experiments. Section 5 recaps Stranded GMM (SGMM) framework, describes the initialization of the class-Structured SGMM (SSGMM) from SWGMM and explains the corresponding experiments. The paper ends with conclusion and future work.

2 Adaptation for Handling Age and Gender Variability

The section describes conventional approaches based on gender and age adaptation with MLLR, MAP and VTLN. Unlike the main objective of the work (use no prior information about speakers), within this section the speaker classes (adult/child and male/female) are assumed to be known for the training data.

2.1 TIDIGITS Baselines

The experiments in this paper are conducted on the TIDIGITS connected digits task [11]. The full training data set consists of 41224 digits (28329 for adult and 12895 for child speech). The test set consists of 41087 digits (28554 for adult and 12533 for child). Similarly to other work with TIDIGITS [3] the signal is down sampled to 8 kHz in order to roughly model the telephone-quality data.

The Sphinx3 toolkit [4] is used for modeling. The digits are modeled as sequences of word-dependent phones. Each phone is modeled by a 3-state HMM without skips. Each state density is modeled by 32 Gaussian components. The front-end computes 13 standard MFCC (12 cepstral + log energy) plus the first and second derivatives and a cepstral mean normalization (CMN) is applied.

Two speaker-independent (SI) models are trained from the adult subset only and from the full training set. The corresponding Word Error Rates (WER) for baseline models are shown in Table 1.

Table 1. Baseline WERs on TIDIGITS data

	Adult	Child
Training on adult data	0.64	9.92
Training on adult+child data	1.66	1.88

Training on adult data provides the best results for adult speakers, but shows a weak performance on the child subset. When child data are included in the training set, the conventional HMM-GMM improves on child, but degrades on adult subset.

2.2 Model Adaptation

Better baselines are achieved when age-gender classes are used for adapting the SI baselines with MLLR for GMM mean values followed by MAP for all model parameters.

With class-based modeling, decoding is usually done in 2 passes. In the 1st pass, for each utterance, the corresponding class is determined using a GMM classifier trained on age-gender labels of the training data. In 2nd pass the standard decoding is done with the corresponding class-based model.

In addition, the recognition hypothesis can be used for applying rapid adaptation of the features (VTLN) using only the utterance data. After such VTLN-based feature transformation a 3rd pass decoding is done.

Word Error Rates for baselines, 2-pass and 3-pass decoding of TIDIGITS data are summarized in Table 2.

Although for SI baseline using all data in training provides better results, the adaptation is more efficient when initial SI model is trained on adult data. In all cases additional VTLN pass in decoding further improves the model accuracy.

Table 2. Baseline WERs for SI and Gender-Age adapted models

	Decoding	Adaptation in decoding	WER	
			Adult	Child
Training on adult data	1 pass	–	**0.64**	**9.92**
+Gender-Age adaptation	2 pass	–	0.54	1.08
+Utterance Rapid adaptation	3 pass	VTLN	0.54	0.97
Training on adult+child data	1 pass	–	**1.66**	**1.88**
+Gender-Age adaptation	2 pass	–	1.34	1.45
+Utterance Rapid adaptation	3 pass	VTLN	1.29	1.41

3   Unsupervised Clustering for Multi-model ASR

Let us consider a set of training utterances without any knowledge about the speaker identity or class (age, gender, etc.). The objective is to automatically group the training data into classes of acoustically similar data.

A GMM-based utterance clustering algorithm is applied [9]. In this approach, a single GMM with a large number of components is first trained on the full dataset. Then, the GMM is duplicated and the mean values are perturbed. Next, the data are classified with Maximum Likelihood criterion and the GMMs are trained from the corresponding classes. The classification and training steps are repeated until convergence. This split-classification-training process is repeated until the desired number of classes is achieved. The class data are then used for adapting the SI HMM-GMM model parameters. The same classification GMMs are used in decoding to identify the class for selecting the best model for each utterance of the test set.

Although clustering of the utterances is not exactly equivalent to speaker clustering, here and later we assume that the main source of variability comes from the speaker and we will refer to the described process as speaker clustering and to the resulting classes of utterances as speaker classes.

Analyzing data clustering for mixed adult-child data. This unsupervised clustering is applied on the TIDIGITS train data. The classification GMMs consist of 256 components. The corresponding distributions of Age-Gender over these classes are summarized in Fig. 1.

The first clustering step (2 classes) mainly splits male speakers from female and child speakers. The second split (4 classes) allows to separate female speakers from child speakers. It seems impossible to distinguish boys from girls, even with more classes.

After clustering, the SI acoustic model (32 Gaussian per density) trained on full train data (adult and child) is adapted using each class data with MLLR+MAP. The bars "*CA-GMM*" in Fig. 4 illustrate WERs with the associated 95 % confidence intervals. The best result is achieved with 4 classes, for which the WER (see details in the "*4 classes CA-GMM*" row of the Table 3) is similar

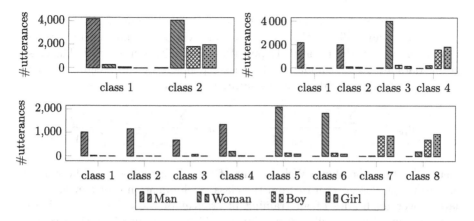

Fig. 1. Number of training utterances for each Age-Gender in the resulting 2, 4 and 8 classes

to the supervised Gender-Age adaptation of the mixed Adult-Child SI model results (see Table 1). After 4 classes, the performance degrades, because there is not enough data to adapt the class-based models.

4   Class-Structured GMM with Class-Dependent Weights

Instead of adapting all GMM parameters for each class of data, a more efficient and compact parameterization was investigated: structured GMM with Speaker class-dependent Weights (SWGMM) [8]. GMM components of this model are shared and structured with respect to speaker classes and only the mixture weights are class-dependent.

The SWGMM pdf for an HMM state j and a given speaker class c has the following form:

$$b_j^{(c)}(o_t) = \sum_{k=1}^{M} w_{jk}^{(c)} \mathcal{N}(o_t, \mu_{jk}, U_{jk}) \tag{1}$$

where M is the number of components per mixture, o_t is the observation vector at time t and $\mathcal{N}(o_t, \mu_{jk}, U_{jk})$ is the Gaussian pdf with the mean vector μ_{jk} and the covariance matrix U_{jk}.

In decoding, each utterance to be recognized is firstly automatically assigned to some class c. After that, the Viterbi decoding with the corresponding set of mixture weights is performed.

The class structuring consists in concatenating the components of GMMs of smaller dimensionality, separately trained from different classes. For example, to train a target model with mixtures of M Gaussian components from Z classes, first Z models with $L = M/Z$ components per density are trained. Then, these components are merged into a single mixture as follows:

$$\left[\mu_{j1}^{(c_1)}, \ldots, \mu_{jL}^{(c_1)}\right] \cdots \left[\mu_{j1}^{(c_Z)}, \ldots, \mu_{jL}^{(c_Z)}\right] \Rightarrow \left[\mu_{j1}, \ldots, \mu_{jL}, \ldots, \mu_{M-L+1}, \ldots, \mu_M\right]$$

For the combined (structured) model, mixture weights are also concatenated, copied and re-normalized. Finally, the means, variances and mixture weights are re-estimated in the iterative Expectation-Maximization manner. The class-specific data are used for updating the class-dependent mixture weights, whereas the whole data set is used for re-estimating the means and variances:

$$\omega_{jk}^{(c_i)} = \frac{\sum_{t=1}^{T} \gamma_{jk}^{(c_i)}(t)}{\sum_{t=1}^{T} \sum_{l=1}^{M} \gamma_{jl}^{(c_i)}(t)} \qquad \mu_{jk} = \frac{\sum_{i=1}^{Z} \sum_{t=1}^{T} \gamma_{jk}^{(c_i)}(t)o_t}{\sum_{i=1}^{Z} \sum_{t=1}^{T} \gamma_{jk}^{(c_i)}(t)} \qquad (2)$$

where $\gamma_{jk}^{(c_i)}(t)$ is the Baum-Welch count of the kth component of the state j, generating the observation o_t from the class c_i. Summation over t means summation over all frames of all training utterances of the class. The variances are re-estimated in a similar way as means. Means can also be estimated in a Bayesian way (MAP) to take into account the prior distribution.

After such re-estimation the class-dependent mixture weights are larger for the components that are associated with the corresponding classes of data (Fig. 2 shows the examples of class-dependent mixture weights of structured GMM, averaged over HMM states, for classes c_7, c_{17} and c_{27}).

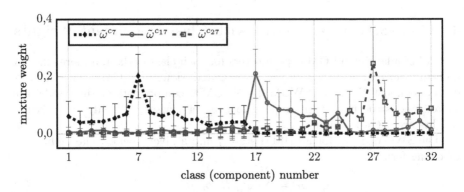

Fig. 2. Example of class-dependent mixture weights of structured GMM after joint re-estimation. Here mixture weights are averaged over HMM states with corresponding standard deviation in bars (here Z = 32, M = 32)

Experiments with class-structured SWGMM. The previous GMM-based unsupervised clustered data were used to build the proposed SWGMM. In order to build models with 32 Gaussians per density, smaller class-dependent models are combined: 2 classes modeled with 16 Gaussians per density, or 4 classes with 8 Gaussians per density, and so on up to 32 classes.

Once the SWGMM is initialized, the model is re-estimated. ML estimation (MLE) is used for mixture weights and MAP for means and variances. The corresponding results are described by the bars "*SWGMM*" in Fig. 4.

This parameterization allows to use the information from all classes for a robust estimation of the means and variances, and significantly reduces the WER with a limited number of parameters, due to the sharing of the Gaussian parameters. This model achieves the best result of 0.80 % for adult and 1.05 % for child data (see *8* and *32 classes SWGMM* rows in Table 3).

5 Class-Structured Stranded Gaussian Mixture Model

Stranded GMM was proposed [18] in the robust ASR framework. The corresponding extended training and decoding algorithms were also introduced in the original paper. This model expands the observation densities of HMM-GMM and explicitly adds dependencies between GMM components of the adjacent states.

Originally, an SGMM is initialized from an HMM-GMM. In this section after briefly recalling the conventional Stranded GMM approach, a class-Structured SGMM (SSGMM) is proposed.

5.1 Conventional Stranded GMM

The conventional SGMM consists of the state sequence $\mathcal{Q} = \{q_1, ..., q_T\}$, the observation sequence $\mathcal{O} = \{o_1, ..., o_T\}$, and the sequence of components of the observation density $\mathcal{M} = \{m_1, ..., m_T\}$, where every $m_t \in \{1, ..., M\}$ is the component of the observation density at the time t, and M denotes the number of such components in the mixture.

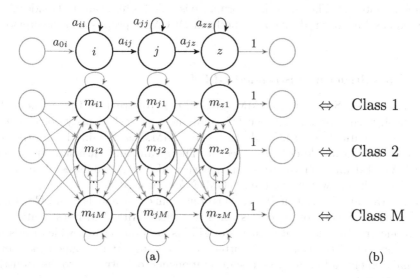

(a)   (b)

Fig. 3. (a) Stranded GMM with schematic representation of the component dependencies; (b) the idea of Structured SGMM, i.e., associating each kth component with some class of data

The difference of SGMM from HMM-GMM is that an additional dependency between the components of the mixture at the current frame m_t and at the previous frame m_{t-1} is introduced (Fig. 3-a). The joint likelihood of the observation, state and component sequences is defined by:

$$P(\mathcal{O}, \mathcal{Q}, \mathcal{M} | \lambda) = \prod_{t=1}^{T} P(o_t | m_t, q_t) P(m_t | m_{t-1}, q_t, q_{t-1}) P(q_t | q_{t-1}) \qquad (3)$$

where $P(q_t = j | q_{t-1} = i) = a_{ij}$ is the state transition probability, $P(o_t | m_t = l, q_t = j) = b_{jl}(o_t)$ is the probability of the observation o_t with respect to the single density component $m_t = l$ in the state $q_t = j$ and $P(m_t = l | m_{t-1} = k, q_t = j, q_{t-1} = i) = c_{kl}^{(ij)}$ is the mixture transition probability.

The set of component transition probabilities corresponds to the mixture transition matrices (MTMs) $C^{(ij)} = \{c_{kl}^{(ij)}\}$, where $\sum_{l=1}^{M} c_{kl}^{(ij)} = 1, \forall i, j, k$.

Experiments with conventional SGMM. In conventional SGMM, MTM rows are initialized from the mixture weights of convention HMM-GMM, and the model parameters are re-estimated with MLE. Such initialization and training processes are applied in this section. In addition, to reduce the number of parameters, only 2 MTMs are used for each state (i.e., cross-phone MTMs are shared). The WERs for SGMM are shown in the bar "*SGMM*" in Fig. 4 and in the corresponding row of Table 3.

Compared to the conventional HMM-GMM trained on all data (adult+child), SGMM improves from 1.66 % to 1.11 % on adult and from 1.88 % to 1.27 % on child speech. Both improvements are statistically significant with respect to 95 % confidence interval. The SGMM performance is even better than the Gender-Age adapted baseline, but it does not outperform SWGMM, proposed in the previous section.

5.2   Class-Structured Stranded GMM

The idea of class-Structured SGMM (SSGMM) is to structure the components of SGMM, such that initially the kth component of each density corresponds to a class of data (Fig. 3-b). To do this, the SSGMM is initialized from the re-estimated SWGMM, described in Sect. 4. The means and variances are taken from SWGMM and MTMs are defined with uniform probabilities. The class-dependent mixture weights of the SWGMM are not used.

When the initialization of SWGMM is done from class-models with 1 Gaussian per density, each component corresponds to a class. After EM re-estimation of all parameters, the diagonal elements of MTMs are dominating, which leads to the consistency of the class within utterance decoding. At the same time, non-diagonal elements allow other Gaussian components to contribute to the acoustic score computation.

The advantage of SSGMM is that it explicitly parameterizes speech trajectories and allows to automatically switch between different components (speaker classes). Therefore, the classification algorithm is no more needed in decoding.

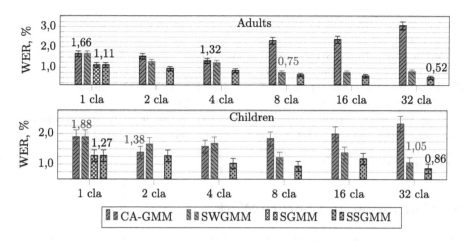

Fig. 4. WER for adult (top) and child (bottom) sets, computed with full Class-Adapted model (CA-GMM), class-structured GMM with Speaker-class dependent Weights (SWGMM), conventional Stranded GMM and class-Structured Stranded GMM built from 32 classes (SSGMM)

Table 3. Summary of the best results and the number of model parameters. Compared the baseline (SI GMM), 4 full Class-Adapted model (CA-GMM), 8 and 32 class-structured GMM with Speaker-class dependent Weights (SWGMM), conventional Stranded GMM and class-Structured Stranded GMM built from 32 classes (SSGMM) without and with additional VTLN pass in decoding)

Model	Decoding	Parameters/state	Adult	Child
SI GMM	1 pass	78*32+32=2528	**1.66**	**1.88**
4 classes CA-GMM	2 pass	4*(78*32+32)=10112	1.32	1.57
8 classes SWGMM	2 pass	78*32+8*32=2752	0.75	1.21
32 classes SWGMM	2 pass	78*32+32*32=3520	0.80	1.05
SGMM	1 pass	78*32+2*32*32=4544	1.11	1.27
SSGMM	1 pass	78*32+2*32*32=4544	**0.52**	**0.86**
SSGMM+VTLN	2 pass	78*32+2*32*32=4544	**0.52**	**0.81**

Experiments with class-Structured Stranded GMM. In the experimental study, the SSGMM is initialized from SWGMM, which was constructed using 32 classes with 1 Gaussian per class and re-estimated with ML for mixture weights and MAP for Gaussian means and variances (corresponds to the result *32 classes SWGMM* in Table 3). Two MTMs per states are defined with uniform probabilities. Then, the parameters of SSGMM are re-estimated with MLE.

The WERs for such SSGMM are described with the bars "*SSGMM*" in Fig. 4 and in the corresponding rows of Table 3. Initializing SSGMM from SWGMM with different number of classes (2, 4, 8 and 16) was always leading to accuracy improvement, compared to SGMM. Only the best result, corresponding to 32 classes, is reported.

While conventional SGMM improves from 1.66 % to 1.11 % on adult and from 1.88 % to 1,27 % on child data, compared to the SI GMM trained on full train data (adult+child), the proposed Class-Structured SGMM (*SSGMM*) further improves by achieving 0.52 % WER on adult and 0.86 % on child data.

The key improvements from all proposed techniques are summarized in Table 3. Notice, that SSGMM can be further combined with rapid feature adaptation to further slightly improve the recognition result on child data (see row *SSGMM+VTLN*).

6   Conclusion and Future Work

This paper investigated an efficient unsupervised approach for handling heterogeneous speech data without prior knowledge about speaker age and gender. Unsupervised clustering does not allow to build many speaker class models, when the amount of training data is limited. To address this problem, an efficient class-structured parameterization of GMM components has been proposed.

The structuring consists in associating subsets of Gaussian components with given speaker classes. Two models, which include this class-structured parameterization, have been investigated and lead to significant improvements of the ASR accuracy.

The first model uses Speaker class-dependent Weights (SWGMM). Unlike standard class model adaptation, the performance does not degrade, when the number of classes increases and when the number of class-associated data decreases. The class structuring approach was also applied for Stranded GMM - an explicit trajectory model with additional dependencies between the components of the observation densities. Class-Structured SGMM is initialized from SWGMM, in which Gaussian components are structured with respect to speaker classes. Mixture Transition Matrices (MTMs) were then used to replace class-dependent mixture weights and to model dependencies between components (speaker classes). SSGMM provides very promising results for both child and adult data. Moreover, it does not require classification algorithm before utterance decoding. SSGMM combined with VTLN achieves overall best performance, outperforming even the strong 3-pass MLLR+MAP age-gender adapted baseline with VTLN pass in decoding.

In the future the proposed techniques should be applied for large vocabulary speech recognition task including adult and child speakers.

References

1. Beaufays, F., Vanhoucke, V., Strope, B.: Unsupervised discovery and training of maximally dissimilar cluster models. In: Proceedings of the INTERSPEECH, Makuhari, Japan, pp. 66–69 (2010), http://www.isca-speech.org/archive/interspeech_2004/i04_0377.html

2. Benzeghiba, M., De Mori, R., Deroo, O., Dupont, S., Erbes, T., Jouvet, D., Fissore, L., Laface, P., Mertins, A., Ris, C., Tyagi, V., Wellekens, C.: Automatic speech recognition and speech variability: a review. Speech Commun. **49**(10), 763–786 (2007)
3. Burnett, D.C., Fanty, M.: Rapid unsupervised adaptation to children's speech on a connected-digit task. In: Proceedings of the ICSLP, vol. 2, pp. 1145–1148. IEEE (1996)
4. CMU: Sphinx toolkit (2014), http://cmusphinx.sourceforge.net
5. Gales, M.J.: Maximum likelihood linear transformations for HMM-based speech recognition. Comput. Speech Lang. **12**(2), 75–98 (1998)
6. Gauvain, J.L., Lee, C.H.: Maximum a posteriori estimation for multivariate Gaussian mixture observations of Markov chains. IEEE Trans. Speech Audio Process. **2**(2), 291–298 (1994)
7. Gorin, A., Jouvet, D.: Class-based speech recognition using a maximum dissimilarity criterion and a tolerance classification margin. In: 2012 IEEE Proceedings of the Spoken Language Technology Workshop (SLT), pp. 91–96. IEEE (2012)
8. Gorin, A., Jouvet, D.: Efficient constrained parametrization of GMM with class-based mixture weights for automatic speech recognition. In: Proceedings of the LTC-6th Language & Technologies Conference, pp. 550–554 (2013)
9. Jouvet, D., Gorin, A., Vinuesa, N.: Exploitation d'une marge de tolérance de classification pour améliorer l'apprentissage de modèles acoustiques de classes en reconnaissance de la parole. In: JEP-TALN-RECITAL, pp. 763–770 (2012)
10. Kuhn, R., Nguyen, P., Junqua, J.C., Goldwasser, L., Niedzielski, N., Fincke, S., Field, K., Contolini, M.: Eigenvoices for speaker adaptation. In: Proceedings of the ICSLP, vol. 98, pp. 1774–1777 (1998)
11. Leonard, R.G., Doddington, G.: Tidigits speech corpus. Texas Instruments, Inc. (1993)
12. O'Shaughnessy, D.: Acoustic analysis for automatic speech recognition. Proc. IEEE **101**(5), 1038–1053 (2013)
13. Panchapagesan, S., Alwan, A.: Frequency warping for vtln and speaker adaptation by linear transformation of standard mfcc. Computer Speech Lang. **23**(1), 42–64 (2009)
14. Stern, R.M., Morgan, N.: Hearing is believing: Biologically inspired methods for robust automatic speech recognition. IEEE Signal Process. Mag. **29**(6), 34–43 (2012), http://ieeexplore.ieee.org/xpls/abs_all.jsp?arnumber=6296528
15. Wellekens, C.J.: Explicit time correlation in hidden Markov models for speech recognition. In: Proceedings of the ICASSP, pp. 384–386 (1987)
16. Wenxuan, T., Gravier, G., Bimbot, F., Soufflet, F.: Rapid speaker adaptation by reference model interpolation. In: Proceedings of the INTERSPEECH, pp. 258–261 (2007)
17. Zhan, P., Waibel, A.: Vocal tract length normalization for large vocabulary continuous speech recognition. Technical report. DTIC Document (1997)
18. Zhao, Y., Juang, B.H.: Stranded Gaussian mixture hidden Markov models for robust speech recognition. In: Proceedings of the ICASSP, pp. 4301–4304 (2012)

Physiological and Cognitive Status Monitoring on the Base of Acoustic-Phonetic Speech Parameters

Gábor Kiss(✉) and Klára Vicsi

Department of Telecommunication and Media Informatics, Budapest University
of Technology and Economics, Budapest, Hungary
{kiss.gabor,vicsi}@tmit.bme.hu

Abstract. In this paper the development of an online monitoring system is shown in order to track physiological and cognitive condition of crew members of the Concordia Research Station in Antarctica, with specific regard to depression. Follow-up studies were carried out on recorded speech material in such a way that segmental and supra-segmental speech parameters were measured for individual researchers weakly, and the changes of these parameters were detected over time. Two kind of speech were recorded weekly by crew members in their mother tongue: a diary and a tale ("North Wind and The Sun"). An automatic language independent program was used to segment the records in phoneme level for the measurements. Such a way **Concordia Speech Databases** were constructed. Those acoustic-phonetic parameters were selected for the follow up study at Concordia, which parameters were statistically selected during a research on the base of the analysis of **Seasonal Affective Disorder Databases** gathered separately in Europe.

Keywords: Acoustic-phonetic speech analysis · Seasonal affective depression · Cognitive status monitoring · Statistical analysis · Two sample t test

1 Introduction

Physiological and cognitive condition of humans is reflected in the speech production, therefore, in the acoustic phonetic-parameters of speech as well. It is well known, that in case of vocal disorders the acoustical parameters of speech significantly differ from the normal speech. Speech production is a complicated process involving coordination of several brain areas and peripheral muscle controls. This is the reason why the acoustic-phonetic parameters of speech are sensitive to many neurological defects (cognitive dysfunctions). These defects can cause changes also in speech [1–3]. The defects generally occur in areas of phonation, articulation, prosody, and the fluency of speech. In the case of phonation defects, due to improper working of muscles, vocal cords do not work perfectly. In this case the fluctuation and scattering of pitch frequency (F0), which is described by standard deviation or by jitter (a measure of period by period variation of F0) - different from the typical one [4]. The articulation defects are reflected primarily in the pronunciation of consonants [5]. The biggest problems [5, 6] occur mainly in the pronunciation of plosives. This is again due to the improper

© Springer International Publishing Switzerland 2014
L. Besacier et al. (Eds.): SLSP 2014, LNAI 8791, pp. 120–131, 2014.
DOI: 10.1007/978-3-319-11397-5_9

working of muscles involved in speech production, in this case the active articulatory organs (lips, tongue tip, the center of the tongue, tongue base, epiglottis, and larynx). During the pronunciation of plosives the position of articulation organs changes rapidly. In some cases, patients are not able to perform these rapid changes [7]. The Voice Onset Time (VOT) defined as the time interval between the release of the plosive and the beginning of the vocal fold vibration associated with the subsequent vowel, is another frequently used acoustical parameter to determine speech production problems [8]. For example it was proved that average VOT was shortened in case of hypoxia [9] during a 48-h exposure, in a 4300 m simulated altitude.

Resonance occurs in the larynx, oral and nasal cavities. The frequencies at which these resonances occur are called formants. Formant values and corresponding spaces in the cavity are affected by the position of speech organs. Formant frequencies can be estimated by frequency analyses of speech. A disadvantage of the formant frequencies is their strong dependence on age and sex. Speaker prosody is partly based on phonation, due to the fact that the three basic acoustic features are pitch frequency (F0), speech intensity and speech rate. In addition to these basic parameters the prosodic features also include rhythm, intonation, accent, timing [10, 11], etc. These parameters are particularly sensitive to changes in mood states and emotions [12, 13]. These three areas of the continuous speech production: the phonation, the articulation, and the prosody are closely related to each other. The relationships between depression and different acoustic phonetic parameters were examined in the last years [14–17]. Data indicates that more depressed patients take longer to express themselves. They don't vocalize more; rather they speak with greater hesitation, producing more cumulative and variable pauses. Consequently, voice acoustic measures such as the percentage of pause time, vocalization or pause ratio, and speaking rates reflect depression severity. Moreover pitch variability about F0 and first and second formants correlate significantly with overall depression severity.

We are participating in an international ESA project, AO-11-Concordia, titled: Psychological Status Monitoring by Computerized Analysis of Language phenomena (COALA). The research is planned to take for two major periods: the year of 2013 and 2014. The duty of our laboratory is phonetic data collection and processing. The aim is the examination of the sensitivity of acoustic-phonetic parameters of speech regarding hypoxia and Seasonal Affective Disorder (SAD), furthermore the development of a system and a metric that alerts crew members at early stage of cognitive dysfunction (Automatic detection). For this research purposes we have developed two types of databases.

For the determination of those acoustic-phonetic parameters which are sensitive to the SAD symptoms, the Seasonal Affective Disorder Database was developed and used. The database development is described in Sect. 2, the methods and the obtained results are described in the Sect. 3.

For the physiological and cognitive status monitoring of the crew members in Concordia research station, the Concordia Speech Database has been developed. The database development is described in Sect. 2 and methods and the obtained results are described in the Sect. 4.

2 Databases

Recorded Material for the Concordia Speech Database. Speech data was collected from all crew members using their mother tongue, weekly during their stay at the Concordia Station. Baseline data was collected in normal circumstances once in Europe before arrival to Antarctica. This way the inflections due to hypoxia and occasional occurrence of SAD symptoms could be monitored. This collection of speech samples is called the Concordia Speech Database. The Concordia Speech Database consists of two parts: the first is spontaneous speech obtained from the diaries used for content analysis too. In the second part, the participants had to read a text, a standard phonetically balanced short folk tale (about 6 sentences all together), "The North Wind and the Sun". This story is from the booklet "The Principles of the International Phonetic Association" (International Phonetic Association, 1999). In this booklet phonetic transcriptions of the story are given in 50-odd languages, in order to illustrate the use of the International Phonetic Alphabet (IPA). The tale is frequently used in the phoniatric practice for all European languages, and commonly used for acoustic analysis in the framework of international, multi-lingual clinical studies.

In the Concordia Station the crew members had either French, Italian or Greek mother tongue (both seasons the doctor's mother tongue was Greek, but they were speaking English during the whole project and on the recordings as well). In the experiment, two seasons was planned: 2013 and 2014 (Tables 1 and 2). The recordings were recorded at a sampling rate of 44 100 Hz, using 16-bits.

Table 1. The Concordia speech database (2013)

Language	Number of persons	Diary length	Tale length	Total length
French	8	17 h	3 h 40 min	20 h 40 min
Italian	6	7 h 20 min	2 h 2 min	9 h 22 min
English	1	2 h 50 min	331 min	3 h 21 min

Table 2. The Concordia speech database (2014 still in progress)

Language	Number of persons	Diary length	Tale length	Total length
French	5	2 h	39 min	2 h 39 min
Italian	6	5 h	1 h 18 min	6 h 18 min
English	1	1 h 30 min	23 min	1 h 53 min

Recorded Material for the Seasonal Affective Disorder Database. Parallel with the data collection in the Concordia Station another data collection was also performed on the seasonal affective disorder patients in normal atmospheric conditions, practically in the doctor's consulting room. This, so called Seasonal Affective Disorder Database is necessary for the development of a good metric to detect SAD. Certainly, speech recording is necessary in symptomatic as well as in symptom free period of the same patient, to detect the differences between the examined parameters. For the Seasonal Affective Disorder Database we are collecting records from Hungarian patients (Table 3). (Italian SAD collection is also going on, but until now we have only small data, thus here we speak only about the Hungarian SAD databases.) We gather data from people who are suffering from depression. The Seasonal Affective Disorder database consists of two parts: the first is spontaneous speech obtained from the discussion between the patient and the doctor, the second one is the reading of the same text as before in the Concordia Speech Database, "The North Wind and the Sun". The recordings were recorded on 44 100 Hz and 16 bits per sample.

Table 3. The seasonal affective disorder database

Language	Persons	Diary length	Tale length	Total length
Hungarian	45	3 h 45 min	41 min	4 h 26 min

To measure depression we used Beck Depression Inventory (BDI), this is to specify the severity of depression in the range 0 to 63 [18]. The distribution of the BDI indices of patients for Hungarian patients is shown in Fig. 1. The distribution of the Ages of the patients for Hungarian patients is shown in Fig. 2.

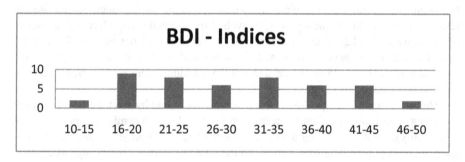

Fig. 1. The distribution of BDI indices of the depressed people in the database

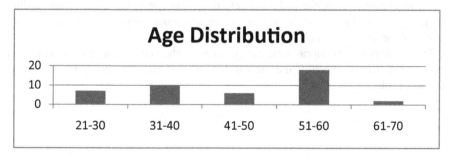

Fig. 2. The distribution of age of the depressed people in the database

Segmentation and Labelling. In order to measure the acoustic-phonetic parameters, first the speech needs to be segmented into phoneme level. For the segmentation we used an automatic language independent segmentation which was developed by our laboratory [19]. Every 10 ms unit was classified into language independent acoustic classes. The classification was made using SVM (Support Vector Machine) on the basis of the spectrum of the speech. Then the final phoneme level segmentation was made using the text of the speech and the acoustic classes. For the higher precision manual correction was made on the automatic segmentation.

Both databases were preprocessed and was segmented and labeled. The speech was segmented into phoneme level and every phoneme was labeled by SAMPA charset. Every plosive was segmented into two parts: the voiced/unvoiced section, and the burst section.

3   Selection of Acoustic-Phonetic Parameters

For the selection of the acoustic-phonetic parameters we used the Seasonal Affective Database. First we segmented the speech into phoneme level, just as was discussed in the Sect. 3.3.

We examined the acoustic-phonetic parameters in two groups according to their segmental and suprasegmental (prosodic) features.

The segmental features were measured at the middle of the same vowel ("E"). The following segmental features were measured according to the international studies [20–22]: fundamental frequency of 'E' vowels (F0), first and second formant frequency of 'E' vowels (F1, F2), jitter of the vowels 'E', shimmer of the vowels 'E'. For the measurement of formants, fundamental frequency and the spectral values, a Hamming window was used with 25 ms frame size and these features were always evaluated from the middle of each vowel 'E'.

The supra-segmental (prosodic) features were measured by the total length of each recording. The following features were measured: volume dynamics of speech (range of intensity), fundamental frequency dynamics of speech (range of fundamental frequency), ratio of total length of pauses and the total length of the recording, articulation and speech rate.

For the measurement of the intensity and the fundamental frequency a 100 ms frame size window was used. For statistical significance testing, we used two-tailed tests. The analysis was performed separately by gender and it was done on the reading text. Every segmental feature showed significant difference between healthy speech and depressed speech (Table 4).

Two (independent) suprasegmental parameters showed significant difference, in both gender, between healthy speech and depressed speech (Range of F0 and Pause Length) (Table 5).

Table 4. Two-sample T tests results for segmented features

Feature	Gender	Group	Mean	Standard deviation	Significance level
Pitch [Hz]	Women	Depressed	154	24	99.9 %
		Normal	199	33	
	Men	Depressed	101	14	95 %
		Normal	116	19	
F1 [Hz]	Women	Depressed	613	74	99.5 %
		Normal	695	47	
	Men	Depressed	512	56	<90 %
		Normal	531	46	
F2 [Hz]	Women	Depressed	1764	112	99.9 %
		Normal	1955	87	
	Men	Depressed	1565	83	99 %
		Normal	1672	80	
Jitter [%]	Women	Depressed	3.4	4.2	95 %
		Normal	1.9	2.3	
	Men	Depressed	5.3	4.7	95 %
		Normal	1.7	2.7	
Shimmer [%]	Women	Depressed	14.1	8.8	95 %
		Normal	8.6	7.7	
	Men	Depressed	17	9.3	95 %
		Normal	9.6	6.2	

Table 5. Two-sample T tests results for suprasegmental features

Feature	Gender	Group	Mean	Standard deviation	Significance level
Range of F0 [Hz]	Women	Depressed	49.8	12.5	95 %
		Normal	61.4	14.3	
	Men	Depressed	32.5	11.4	97.5 %
		Normal	47.6	17.8	
Pause length [sec]	Women	Depressed	9.2	3.3	95 %
		Normal	6.6	3.1	
	Men	Depressed	12.2	5.6	95 %
		Normal	7.6	4.2	

4 Physiological and Cognitive Status Monitoring of the Crew Members at Concordia

We built an online monitoring system which tracks the acoustic-phonetic parameters of a subject's speech in time and compares the collected data with the reference. The system has two aims. The primary aim is the alert at early stage of cognitive dysfunction, the second on is the suggestions that the system gives about the reason of the vocal disorder. An automatic (semi - automatic) online alerting system is planned which records speech at a given time rate and gives alerts (and predicts). The flow chart of an online alerting system is shown on Fig. 3.

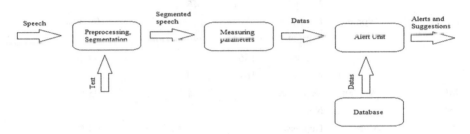

Fig. 3. The online alerting system flowchart

The system has three main parts: Preprocessing and Segmentation unit, the unit that measures the parameters (Measuring Parameters) and the Alert unit. Operation of each part are described in the next Sects. 4.1–4.3).

4.1 Preprocessing and Segmentation

The system gets the recorded speech and preprocesses it. With the help of the transcription of the recorded speech, it makes the automatic phoneme level segmentation, as it was described in Sect. 3. This part of the processing is only semi-automatic, because, for the segmentation the program needs the text of the speech. We plan to make it automatic, then the preprocessing and segmentation would be fully automatic.

4.2 Measuring the Selected Features

The following parameters of speech were selected for monitoring: F0, F1, F2, jitter, shimmer, range of F0, pauses of total length for segmental and supra-segmental parameters. These parameters showed significant differences between healthy and depressed people in case of Seasonal Affective Disorder Database, as it was described in Sect. 4. Furthermore the Voice Onset Time (VOT), average length of the vowels, and the ratio of stationary and transient sections of the speech were measured, which correlate highly with hypoxia [9].

4.3   Alert Unit

The changing of the values of the selected acoustic-phonetic parameters as a function of time was examined. The system compares the actual measured parameters with the parameters of the reference data (measured before the departure in Europe in normal circumstances). For the comparison two-sample T test was used, because these parameters shows normal distribution and the variances are equal. 98 % significance level (1-alpha) was selected. The acoustic phonetic parameters could change from a variety of reasons, not only from the examined two cognitive dysfunctions (hypoxia and SAD). For example alcoholic condition, flew, or sleepiness could cause significant changes of certain parameters together with other body conditions. The separate examination of the hypoxia and SAD is almost impossible. This is the reason that our work includes the examination of the relationship between the acoustic-phonetic parameters and other biological and psychological data of crew members measured in the station, like: Long term medical survey data (LTMS) and O2 saturation data. Unfortunately, we have not received these data until now. Thus we can measure and examine only the acoustic-phonetic parameters in this follow up study. In the following, the parameters of two crew members will be presented: person "A" where a big change was detected (Fig. 4, Table 6), and person "B" where there was no big change in the values of the selected parameters (Fig. 5, Table 7).

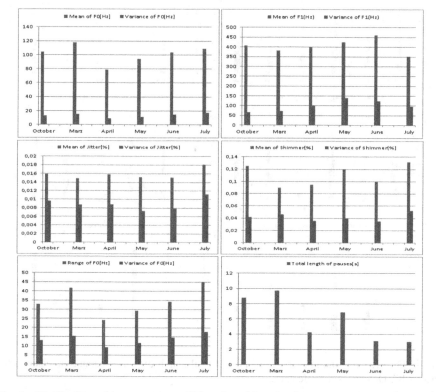

Fig. 4. F0, F1, jitter, shimmer, range of F0 and total length of pauses parameters changes by person "A", October is the reference.

Table 6. The results of the two-sample t tests in case person "A"

		March	April	May	June	July
F0	P_Value	9,27	20,14	7,07	0,34	3,09
	Critical_Value	2,33	2,33	2,33	2,33	2,33
	IsSignficant	**Yes**	**Yes**	**Yes**	No	**Yes**
F1	P_Value	1,41	0,41	0,57	1,97	2,56
	Critical_Value	2,41	2,41	2,41	2,41	2,41
	IsSignficant	No	No	No	No	**Yes**
Jitter	P_Value	0,41	0,04	0,3	0,3	0,74
	Critical_Value	2,4	2,4	2,4	2,4	2,4
	IsSignficant	No	No	No	No	No
F1	P_Value	2,87	2,33	0,41	2,01	0,87
	Critical_Value	2,4	2,4	2,4	2,4	2,4
	IsSignficant	**Yes**	No	No	No	No

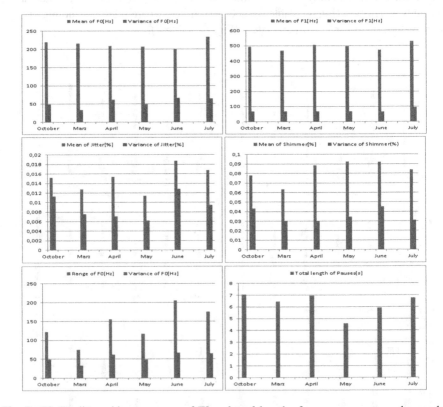

Fig. 5. F0, F1, jitter, shimmer, range of F0 and total length of pauses parameters changes by person "B", October is the reference.

Table 7. The results of the two-sample t tests in case person "A"

		March	April	May	June	July
F0	P_Value	0,56	0,47	0,87	0,08	1,32
	Critical_Value	2,33	2,33	2,33	2,33	2,33
	IsSignficant	No	No	No	No	No
F1	P_Value	0,88	0,06	1,27	0,78	0,52
	Critical_Value	2,41	2,41	2,41	2,41	2,41
	IsSignficant	No	No	No	No	No
Jitter	P_Value	0,81	0,42	0,11	1,14	1,61
	Critical_Value	2,47	2,32	2,47	2,48	2,47
	IsSignficant	No	No	No	No	No
Shimmer	P_Value	1,35	0.96	1,17	1,05	0,52
	Critical_Value	2,41	2,41	2,41	2,41	2,41
	IsSignficant	No	No	No	No	No

Person "A" had some big changes in acoustic phonetic parameters around April compared to the reference, and then values started to move back to the reference.

Person "B" had almost the same values of the selected acoustic phonetic parameters. There was no significant difference at any measured parameter compare to reference. Person B is supposed to have no change of physical and psychological condition.

15 crew members were monitored in such a way, and checked when the differences from the references rich the 98 % significance level. The psychological and biological results of these crew members were measured parallel in the Concordia research station, but we have not received those data until now. Of course we have to examine those data parallel with the speech parameters. The final decision can be taken only on the evaluation of those data too.

For example the linear combination of the differences of the corresponding (appropriate) parallel parameters reaches a threshold, alert can be given. Determination of this threshold is a very difficult task and need further important researches.

5   Conclusions, Future Tasks

The reviewed study shows an online monitoring system which tracks the cognitive condition of a subject by voice. This system was developed to follow up the crew members' psychological condition of Concordia Station and to give alerts if it experiences any vocal disorder which can be a sign of cognitive dysfunction (hypoxia and SAD). We have found some segmental and supra-segmental acoustic-phonetic parameters from continuously read speech that can show significant differences between the speech of depressed people and a healthy reference group. We have found that segmental parameters: fundamental frequency, F1, F2 formants frequencies, jitter, shimmer; and supra-segmental parameters: number of phonemes, speech rate, length of pauses, intensity and fundamental frequency dynamics in the speech of depressed people shows significant changes compared to a healthy reference group.

The examined number of depressed people is small according to the wide range of degree of the depression. But the database is under continuous expansion with more and more recordings with further people, speaking different languages as mother thong. A proper number of sound samples will allow us to perform a full analysis, and thus we can select a complete set of acoustic features that enables more precise conclusions to deduct. The ultimate goal would be to find a clear correlation between the severity of depression and the change of the acoustic-phonetic parameters.

Until now standard read speech (the folk tale) was used for our research, but free speech have also been recorded from the same persons and stored in both type of the databases, in the Seasonal Affective Disorder Database and in Concordia Speech Databases, for further analysis. Finally, the results of this presented study show that on the base of the extended databases the development of a metric that alert crews at early stage of cognitive dysfunction (Automatic detection) is possible, and the research should continue.

Acknowledgments. The authors would like to thank the COALA project: Psychological Status Monitoring by Computerised Analysis of Language phenomena (COALA) (AO-11-Concordia). Moreover this work was partially supported by the European Union and the European Social Fund through project FuturICT.hu (grant no.: TAMOP-4.2.2.C-11/1/KONV-2012-0013) organized by VIKING Zrt, Balatonfüred.

References

1. Goberman, A.M.: Correlation between acoustic speech characteristics and non-speech motor performance in Parkinson disease. Med. Sci. Monit. **11**(3), 109–116 (2005)
2. Goberman, A.M., McMillan, J.: Relative speech timing in Parkinson's disease. Commun. Sci. Disord. **32**, 22–29 (2005)
3. Metter, E.J., Hanson, W.R.: Clinical and acoustical variability in hypokinetic dysarthria. J. Commun. Disord. **19**(5), 347–366 (1986)
4. Doyle, P., Raade, A., Pierre, A., Desai, S.: Fundamental frequency and acoustic variability associated with production of sustained vowels by speakers with hypokinetic dysarthria. J. Med. Speech. Lang. Pathol. **3**, 41–50 (1995)
5. McNeil, M.R., Rosenbeck, J.C., Aronson, A.E.: The Dysarthrias: Physiology, Acoustics, Perception, Management. College Hill Press, San Diego (1984)
6. Logemann, J.A., Fisher, H.B.: Vocal tract control in Parkinson's disease: phonetic feature analysis of misarticulations. J. Speech Hear. Disord. **46**(4), 348–352 (1981)
7. Ackermann, H., Konczak, J., Hertrich, I.: The temporal control of repetitive articulatory movements in Parkinson's disease. Brain Lang. **56**(2), 312–319 (1997)
8. Kent, R.D.: Acoustic Analysis of Speech, 2nd edn. Singular, San Diego (2001)
9. Lieberman, P., Morey, A., Hochstadt, J., Larson, M., Mather, S.: Mount Everest: a space analogue for speech, monitoring of cognitive deficits and stress. Aviat. Space Environ. Med. **76**(6, Section II), 198–207 (2005)
10. Ivry, R.B., Justus, T.C., Middleton, C.: The cerebellum, timing, and language: implications for the study of dyslexia. In: Wolf, M. (ed.) Dyslexia Fluency and the Brain, pp. 198–211. York Press, Timonium (2001)

11. Esposito, A., Bourbakis, N.: The role of timing in speech perception and speech production processes and its effects on language impaired individuals. In: Sixth Symposium on BioInformatics and BioEngineering (BIBE'06), pp. 348–356. IEEE Computer Society (2006)

12. Vicsi, K., Sztahó, D.: Problems of the automatic emotion recognitions in spontaneous speech; an example for the recognition in a dispatcher center. In: Esposito, A., Esposito, A. M., Martone, R., Müller, V.C., Scarpetta, G. (eds.) COST 2010. LNCS, vol. 6456, pp. 331–339. Springer, Heidelberg (2011)

13. Tóth, S.L., Sztahó, D., Vicsi, K.: Speech emotion perception by human and machine. In: Esposito, A., Bourbakis, N.G., Avouris, N., Hatzilygeroudis, I. (eds.) HH and HM Interaction. LNCS (LNAI), vol. 5042, pp. 213–224. Springer, Heidelberg (2008)

14. France, D.J., Shiavi, R.G., Silverman, S., Silverman, M., Wilkes, D.M.: Acoustical properties of speech as indicators of depression and suicidal risk. IEEE Trans. Biomed. Eng. **47**, 829–837 (2000)

15. Cannizzaro, M., Harel, B., Reilly, N.,• Chappell, P., Snyder, P.J.: Voice acoustical measurement of the severity of major depression. Brain Cogn. **56**, 30–35 (2004)

16. Cannizzaro, M., Reilly, N., Mundt, J.C., Snyder, P.J.: Remote capture of human voice acoustical data by telephone: a methods study. Clin. Linguist. Phon. **19**, 649–658 (2005)

17. Garcia-toro, M., Talavera, J.A., Saiz-Ruiz, J., Gonzalez, A.: Prosody impairment in depression measured through acoustic analysis. J. Nerv. Ment. Dis. **188**, 824–829 (2000)

18. Abela, J.R.Z., D'Allesandro, D.U.: Beck's cognitive theory of depression: the diathesis-stress and causal mediation components. Br. J. Clin. Psychol. **41**, 111–128 (2002)

19. Kiss G., Sztahó D., Vicsi K.: Language independent automatic speech segmentation into phoneme-like units on the base of acoustic distinctive features. In: 4th IEEE International Conference on Cognitive Infococommunications - CogInfoCom 2013, Budapest, Hungary, 2–6 Dec 2013, pp. 579-582. IEEE Press, Piscataway (2013). ISBN: 978-1-4799-1-1543-9, IEEE Catalog Number: CFP1326R-PRT

20. Alghowinem, S., Goecke, R., Wagner, M., Epps, J., Breakspear, M., Parker, G.: Detecting depression – a comparison between spontaneous and read speech. In: 38th International Conference on Acoustics, Speech, and Signal Processing (ICASSP) (2013)

21. Helfer, B.S., Quatieri, T.F., Williamson, J.R., Mehta, D.D., Horwitz, R., Yu, B.: Classification of depression state based on articulatory precision. In: 14th Annual Conference of the International Speech Communication Association (2013)

22. Mundt, J.C., Snyder, P.J., Cannizzaro, M.S., Chappie, K., Geralts, D.S.: Voice acoustic measures of depression severity and treatment response collected via interactive voice response (IVR) technology. J. Neurolinguist. **20**, 50–64 (2007)

Automatic Phonetic Transcription in Two Steps: Forced Alignment and Burst Detection

Barbara Schuppler[(⊠)], Sebastian Grill, André Menrath,
and Juan A. Morales-Cordovilla

Signal Processing and Speech Communication Laboratory,
Graz University of Technology, Graz, Austria
b.schuppler@tugraz.at

Abstract. In the last decade, there was a growing interest in conversational speech in the fields of human and automatic speech recognition. Whereas for the varieties spoken in Germany, both resources and tools are numerous, for Austrian German only recently the first corpus of read and conversational speech was collected. In the current paper, we present automatic methods to phonetically transcribe and segment (read and) conversational Austrian German. For this purpose, we developed an automatic two-step transcription procedure: In the first step, broad phonetic transcriptions are created by means of a forced alignment and a lexicon with multiple pronunciation variants per word. In the second step, plosives are annotated on the sub-phonemic level: an automatic burst detector automatically determines whether a burst exists and where it is located. Our preliminary results show that the forced alignment based approach reaches accuracies in the range of what has been reported for the inter-transcriber agreement for conversational speech. Furthermore, our burst detector outperforms previous tools with accuracies between 98 % and 74 % for the different conditions in read speech, and between 82 % and 52 % for conversational speech.

Keywords: Speech transcription · Austrian German · Conversational speech · Automatic burst detection · Forced alignment

1 Introduction

In the last decade, there was a growing interest in spontaneous and conversational speech in the fields of human and automatic speech recognition. Therefore, large conversational speech corpora have been created for many languages (e.g., for English [17], for Japanese [13], for Dutch [5], and for French [28]). For conversational German, large speech resources are limited to the varieties spoken in Germany (e.g., [7,12,31]). For the varieties of Austria, only recently the first corpus of conversational speech was recorded (i.e., Graz corpus of Read and Conversational Speech (*GRASS*) [27]). In order to make the GRASS corpus accessible for speech technology as well as linguistic and phonetic research,

© Springer International Publishing Switzerland 2014
L. Besacier et al. (Eds.): SLSP 2014, LNAI 8791, pp. 132–143, 2014.
DOI: 10.1007/978-3-319-11397-5_10

it needs to be segmented and transcribed phonetically. The aim of the current paper is to present a transcription tool for read and conversational German. The tool is operating in two subsequent steps. First, a broad phonetic transcription is created by means of a forced alignment (i.e., with a HMM-based approach). Second, a non-stochastic MATLAB tool annotates whether plosives are realized with a burst and, in case of an existing burst, where it is positioned. The resulting transcriptions are exported to PRAAT TextGrid format [3].

1.1   STEP 1: Broad Phonetic Transcription

Traditionally, phonetic transcriptions are produced manually by one or more transcribers. Since this approach is time consuming, methods have been developed to create broad phonetic transcriptions with the help of an ASR system (e.g., [2,4,9]). The accuracy of these systems has steadily increased and the agreement between automatic and manual transcriptions for some systems already is in the range of the agreements reported for human transcribers (e.g., [26]). Furthermore, automatically created transcriptions have successfully been used for phonetic investigations concerning pronunciation variation (English: [34]; German: [1]; French: [2] and Dutch: [26]).

There are different methods for creating broad transcriptions automatically. For instance, free and constrained phone recognition have been reported to work well for read speech but not for spontaneous telephone dialogues [30]. Since we aim at using the transcription tool for the casual conversations, which are part of the GRASS corpus, we did not follow this approach.

A method which does not make use of a phone recognizer based on Hidden Markov models, has been presented by Leitner et al. [11]. Their example-based approach is non-parametric and uses methods from template-based speech recognition. This tool has been trained on isolated words read by male trained Austrian speakers. Even though this tool reaches a high accuracy on carefully pronounced speech, it does not capture the variation found in spontaneous Austrian German.

Another method for creating broad phonetic transcriptions automatically is *forced alignment* (e.g., [2,4]). For instance, the tool MAUS (Munich Automatic Segmentation) is a forced-alignment based tool which is available for German (among other languages) [20]. It works as follows: The orthographic transcription and the speech files of an utterance are uploaded to an online-tool. Then, a canonical transcription is created for each word with the Balloon tool [19]. Then, possible pronunciation variants are created based on phonological rules. Finally, an HMM based ASR system chooses the most probable pronunciation variant for each word and places the segment boundaries. We have tested this tool for our Austrian German data of the GRASS corpus and we have observed a good accuracy of the segmentation for the read speech component. For the conversational speech, however, the MAUS tool did not cover well typical characteristics of Austrian German pronunciation. For instance, MAUS annotated the alveolar fricative, which in Austrian German is typically pronounced voiceless, as voiced. Furthermore, words which tend to be reduced in spontaneous speech were not transcribed correctly. For example, the highly frequent word *ich* 'I', which in Austria is typically pronounced as [iː], was transcribed in its canonical form /iç/. To conclude, none of the existing transcription tools fulfilled our requirements. Therefore, we decided to develop a HMM based ASR system in forced alignment mode to transcribe Austrian German. The main difference of our approach to the MAUS tool (described above) is that our method to creates an Austrian German pronunciation dictionary with several variants per word type (see Sect. 3.2). Most importantly, MAUS does not provide a sub-phonemic annotation of plosives, which is the task of Step 2.

1.2   STEP 2: Sub-phonemic Annotation of Plosives

Figure 1 shows three examples for different realizations of /t/ in conversational Dutch. The example of the left panel is the canonical realization of /t/, which consists of a voiceless complete closure followed by a strong burst and a subsequent release friction. The example on the right panel shows a realization of /t/ where all characteristic properties of a plosive are absent. [25] showed that 80.4 % of /t/s in conversational Dutch are realized somewhere in between these two extremes (e.g., example in the middle panel). Recently, numerous studies investigated the different realizations of plosives in spontaneous speech and the conditions for their occurrence (for English (e.g., [18]), for Dutch (e.g., [22]), for French (e.g., [29]) and for German (e.g., [10,35]). In these studies, sub-phonemic annotations of the plosives were created manually for a relatively small set of tokens. At the same time, they used high level statistical modeling techniques to estimate which are the predictors for the variation observed. In this paper, we propose a method to create such annotations automatically, which allows to enlarge the amount of data available for future phonetic investigations.

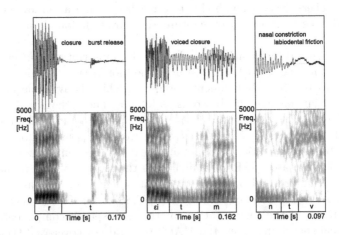

Fig. 1. Three realizations of /t/ in spontaneous Dutch. Left panel: canonical /t/. Middle panel: reduced /t/. Right panel: absent /t/ [23]

2   Speech Material

The speech material transcribed with the developed tool is the *Graz corpus of Read And Spontaneous Speech (GRASS)* [27]. For each of the 38 speakers (male and female), this corpus contains 62 phonetically balanced sentences, 20 (read and spontaneous) commands elicited with pictures and one hour of conversation (approximately 1200 utterances per speaker). All conversations were between family members or friends and the speakers were relaxed and talked completely freely about everyday topics (in the absence of an experimenter). Therefore, the style of the conversational speech is informal and casual. The speakers are gender balanced, with a similar average age per group. They were born in one of the eastern provinces of Austria and they all were living in Graz at the time of the recordings.

Since the corpus was collected with speech technology applications in mind, it fulfills the requirements for automatic processing (e.g., [26]): the recordings took place in a soundproof studio with both head-mounted and large-membrane microphones at 48 kHz. The relative position of the speakers and the according directivity of the microphones was adjusted to optimize the SNR in the presence of overlapping speech. On average over all conversations, the resulting SNR was 46.4 dB [27].

Since for a forced alignment, the orthographic transcription is needed as input (see Sect. 3), the quality and consistency of the orthographic transcriptions is especially relevant. For instance, Gubian et al. [6] reported that mistakes on the orthographic level can not be compensated on the overlying transcription layers, the contrary is the case. The orthographic transcriptions of the *GRASS* corpus were created having also such further (semi-) automatic transcription layers in mind: Speakers were transcribed on separate tiers with speech stretches of less than six seconds. Transcriptions contain information of overlapping speech, hesitations, disfluencies and other vocal and non-vocal noises [27].

3   STEP 1: Creation of a Broad Phonetic Transcription

As motivated in Sect. 1, we used a forced alignment to create broad phonetic transcriptions automatically. For this purpose, we used the HTK speech recognition toolkit [33]. A forced alignment needs the following input: (1) the acoustic signal, (2) the orthographic transcriptions, (3) acoustic phone models and (4) a lexicon containing pronunciation variants for each word. With this input, the alignment system determines the most likely pronunciation variant for each word appearing in the transcription of an utterance and places the corresponding segment boundaries. Finally, we exported the output of HTK to the PRAAT TextGrid format [3].

3.1   Monophone Acoustic Models

The 35 (34 phones + silence) acoustic models were continuous density 3-state monophone acoustic models with 5 Gaussians per state. The models have been trained on 5000 utterances from 50 German speakers of the BAS read speech corpus [21]. The acoustic parameterization was as follows: 16 kHz sampling frequency, frame shift and length of 10 and 32 ms, 1024 frequency bins, 26 mel channels and 13 cepstral coefficients with cepstral mean normalization. After adding delta and delta-delta features, each final MFCC vector had 39 components (see also [24]).

3.2   Pronunciation Dictionary

The only existing pronunciation dictionary is the Austrian Phonetic Database [15]. It is based on isolated words produced by a trained speaker and thus does not cover the pronunciation variation found in the conversational speech of the *GRASS* corpus. In the following, we describe how we created a pronunciation dictionary for Austrian German, with several pronunciation variants per word type.

First, for each word a canonical pronunciation (German standard) was created with the Balloon tool [19], which makes use of a set of 49 SAMPA phoneme symbols providing syllabic and morphological boundaries, as well as primary and secondary stress. This tool is also used by the MAUS transcription system [20]. Whereas in MAUS the output is not corrected manually, we corrected the resulting canonical transcriptions. Errors mainly concerned proper names, foreign words and compounds, especially regarding the syllable boundaries and primary and secondary stress marks. The correction of the syllable boundaries and stress marks is especially important since the automatic creation of pronunciation variants is based on rules which are specific for certain syllabic structures (e.g., deletion of /r/ in coda position) and certain stress patterns (e.g., schwa deletion in unstressed syllables).

Subsequently, we applied a set of 32 rules to the canonical pronunciations. These rules can be divided into three groups. The first group is formed by those rules covering co-articulation, assimilation and reduction rules which are also typical for spontaneous German spoken by speakers from Germany. These rules include those mentioned by Wesenick [32] and by Schiel [20]. Secondly, we applied rules formulated on the basis of literature on standard Austrian German. Several of these rules have earlier been used for a text-to-speech engine for Austrian German [16]. The majority of these rules, however, have been formulated on the basis of phonetic studies and have not yet been used in speech technology (e.g., [14]). These rules include the deletion and lenition of plosives in all word positions. For a detailed description of each rule and their frequencies see [24].

Finally, variants were created manually for the 150 most frequent words and for certain verbs which tend to have typical Austrian realizations which cannot be easily derived from the citation form (e.g., *möchte* 'would like to': citation form /m'ɘxtə/ as /m'ɛxatn/.

Table 1. Discrepancy between automatically created and manually corrected broad phonetic transcriptions in absolute number of phones and in % of deletions, insertions and substitutions

	Read	Commands	Conversational
Total # Phones	1826	429	10836
Deletions	0.4 % [8]	0.0 % [0]	1.3 % [133]
Insertions	1.7 % [31]	0.2 % [1]	2.1 % [228]
Substitutions	16.8 % [307]	17.0 % [73]	15.1 % [1637]
Total discrepancy	18.9 % [346]	17.2 % [74]	18.4 % [1998]

3.3   Validation

In order to validate the created broad phonetic transcriptions, a phonetically trained transcriber corrected the labels of the created transcriptions of part of the *GRASS* material. Then the number of substitutions, insertions and deletions was calculated (for all phones but the silence segments). Table 1 shows the discrepancies between the components of the corpus. Overall, there was a 18.5 % discrepancy between the phone labels of the forced alignment and the manually corrected ones. This was mainly due to substitutions, with only a small number of insertions and deletions. These deviations between automatic and manual transcriptions are in the range of earlier reported inter-annotator discrepancies on manual transcriptions (21.2 % for spontaneous speech [9]). Furthermore, the accuracy of our system lies within the range of other automatic transcription systems. For instance, Cucchiarini and Binnenpoorte [4] reported a discrepancy of 24.3 % for spontaneous speech.

4   STEP 2: Automatic Sub-phonemic Annotation of Plosives

The following section describes the components of a burst-detector which is used to annotate plosives at the sub-phonemic level. The detector determines whether the plosive contains a closure and a burst and in case of a burst, it determines its position. Similar as in [8], the detector uses the power and its derivative with respect to time as principal source of information. We, however, developed a more elaborate decision stage.

4.1   Preprocessing

In a first step, the signal is Fourier transformed, high pass filtered and subsequently the power densities for each sample are accumulated to a power curve. Then, the signal passes an envelope generator that interpolates all local maxima. To suppress erroneous behavior, the interpolation stage discards all envelope points previous to the first or after the last detected maximum (Fig. 2).

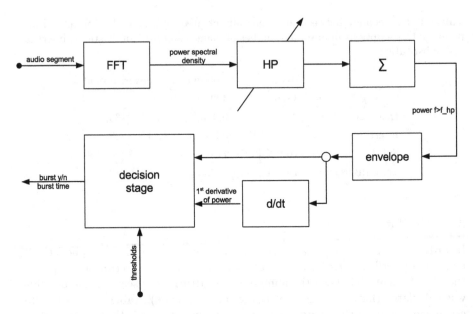

Fig. 2. Block diagram of the complete algorithm: FFT: Fouriertransform/Spectral analysis; HP: High Pass; $\sum$: Generation of power curve from power spectral densities; d/dt: Discrete Derivative

If insufficient supporting points are found to generate an envelope, the detection is categorically aborted and the result is set to contain no burst, as the majority of signals which result in such a condition do not contain any significant spikes in power that would hint at a burst event. We tested the positive impact of this feature on detection performance (see Sect. 4.3). Finally, the discrete derivative of the resulting envelope is calculated. The resulting two signals are then passed to the decision stage.

4.2 Decision Stage

First, the maximum of the derivative is compared to a plosive dependent threshold (i.e., different for /p/, /t/, /k/, etc.). If the threshold is not exceeded, the decision process is aborted with the decision that no burst is present.

Second, thresholds for power as well as for the derivative are obtained by taking the maximum value of each signal multiplied by a manual parameter set. These parameters are chosen individually for each plosive. The algorithm then starts at the maximum power and proceeds backwards along the time axis until both power and its derivative fall below their respective threshold. If the values do not fall below the threshold, the decision is that no burst is present. If these two conditions are met, the process is aborted and a burst is detected.

Finally, if a burst was detected, a plosive dependent offset is added to the sample at which the algorithm stopped to obtain the burst time. The reason for this offset stems from the usage of an envelope in the power signal, which shifts the onset of the burst forward, as well as there being an optimization problem between overall burst detection and temporal precision. If parameters are optimized to obtain optimal burst detection, the temporal precision suffers and vice versa. We found that an offset was an easy method to avoid this optimization conflict.

4.3   Sub-phonemic Annotation of Plosives in Read Speech

For evaluating the accuracy of the burst detector in read speech, we used a subset of the German *Kiel Corpus* [7]. The subset contains German read speech of the same text spoken by nine male and seven female speakers. The corpus comes with detailed manually created phonetic transcriptions, also at the sub-phonemic level of plosives, which made it an ideal reference to validate the burst detector. In total, we used 1579 bursts for the validation of our automatically created sub-phonemic annotation of the plosives.

We evaluated both the decision of the detector (is there a burst?, yes or no) and the position of the plosive (distance from manual burst in ms). For the analysis, we calculated the deviations from the manual transcription separately for the different plosives (/p/,/t/,/k/,/b/,/d/,/g/) and we grouped them in terms of position within the word (word initial, word medial and word final). For each of these combinations the following measures were calculated to estimate the accuracy of the burst detector:

- P_1: A burst was detected and it was present in the manual transcription.
- P_2: No burst was detected and it was absent in the manual transcription.
- P: Detector decision is correct.
- Δb: Arithmetic mean of the temporal error between the detected and the manually labeled burst position (in numbers of samples).

Table 2 shows the results separately for the different plosives as well as the overall result. The numbers in the squared brackets represent the number of occurrences of the respective case. In 161 cases spanning all plosives in all possible positions, no burst was detected because insufficient supporting points were found to generate a hull curve (also see Sect. 4.1). This detector decision was correct in 98 % of the cases.

Overall plosive categories, the decision of whether a burst was present was correct in 93 % of the cases, with a maximum of 98 % for initial, voiceless plosives and a minimum of 74 % for word-final /k/. These values are much better than previously reported. [8], for instance, reached a similarly high maximum of 97 % agreement for the presence of bursts in word-initial position, but only 47 % agreement for the absence of bursts in word-medial position.

4.4   Sub-phonemic Annotation of Plosives in Austrian German

In order to evaluate the accuracy of the burst detector on Austrian German, we extracted 2071 word tokens containing a plosive from the *GRASS* corpus.

Table 2. Automatic annotation of bursts in plosives in the Kiel Corpus of Read Speech. Percentages P_1 - P_3: P_1: Burst detected and it was present in the manual transcription. P_2: No burst detected and it was absent in the manual transcription. P: Detector decision is correct. Δb: Temporal error (in numbers of samples, the sampling frequency was 44 kHz)

	Total	/p/	/t/	/k/	/b/	/d/	/g/
Overall							
P	0.93 [1579]	0.96 [50]	0.96 [673]	0.92 [119]	0.92 [112]	0.89 [494]	0.87 [131]
P₁	0.91 [822]	0.92 [12]	0.92 [263]	0.91 [93]	0.97 [35]	0.89 [351]	0.88 [68]
P₂	0.95 [757]	0.97 [38]	0.98 [410]	0.92 [26]	0.90 [77]	0.91 [143]	0.86 [63]
Δb	43 [744]	76 [11]	50 [243]	44 [85]	50 [34]	37 [311]	35 [60]
Initial							
P	0.92 [528]	-	0.98 [48]	1.00 [16]	0.94 [64]	0.92 [352]	0.88 [48]
P₁	0.93 [407]	-	-	1.00 [16]	1.00 [33]	0.93 [313]	0.89 [45]
P₂	0.90 [121]	-	0.98 [48]	-	0.87 [31]	0.85 [39]	0.67 [3]
Δb	39 [379]	-	-	25 [16]	51 [33]	38 [290]	37 [40]
Medial							
P	0.92 [801]	0.97 [32]	0.96 [424]	0.95 [80]	0.90 [48]	0.82 [134]	0.87 [83]
P₁	0.89 [338]	1.00 [6]	0.94 [208]	0.95 [61]	0.50 [2]	0.55 [38]	0.87 [23]
P₂	0.94 [463]	0.96 [26]	0.98 [216]	0.95 [19]	0.91 [46]	0.93 [96]	0.87 [60]
Δb	46 [302]	26 [6]	48 [196]	50 [58]	38 [1]	27 [21]	32 [20]
Final							
P	0.94 [250]	0.94 [18]	0.96 [201]	0.74 [23]	-	1.00 [8]	-
P₁	0.82 [77]	0.83 [6]	0.85 [55]	0.69 [16]	-	-	-
P₂	0.99 [173]	1.00 [12]	0.99 [146]	0.86 [7]	-	1.00 [8]	-
Δb	58 [63]	135 [5]	54 [47]	39 [11]	-	-	-

For these tokens, the bursts in the plosives were annotated manually by a trained transcriber. The results for the different plosives are shown in Table 3. Since basically all tools work better for read than for spontaneous speech, it could be expected that also our burst detector did not achieve as high accuracies for the material from the *GRASS* corpus as for the carefully pronounced speech from the Kiel Corpus. Nevertheless, the tool reached a maximum accuracy of 82 % for /g/ and a minimum for /b/ of 52 %. These values are still within the range of what [8] (max. 97 %, min. 47 %), and [23] (average 63 %) reported for spontaneous American English.

One explanation for the lower accuracy reached for detection of bursts in /b/ might be that /b/ is frequently realized as voiced labiodental fricative in spontaneous Austrian German. Another reason might be the different recording

Table 3. Automatic annotation of bursts in plosives in the GRASS corpus.
Percentages P_1 - P_3: P_1: Burst detected and it was present in the manual transcription. P_2: No burst detected and it was absent in the manual transcription. P: Detector decision is correct. Δb: Temporal error (in numbers of samples; the sampling frequency was 16 kHz)

	/p/	/t/	/k/	/b/	/d/	/g/
P	0.59 [144]	0.68 [917]	0.81 [198]	0.52 [158]	0.67 [466]	0.82 [188]
P_1	0.39 [95]	0.60 [676]	0.84 [176]	0.26 [102]	0.63 [355]	0.82 [137]
P_2	0.98 [49]	0.90 [241]	0.64 [22]	0.98 [56]	0.82 [111]	0.82 [51]
Δb	4 [37]	13 [403]	9 [147]	5 [27]	10 [222]	9 [112]

conditions of the two corpora. In future work, we will develop automatic methods to optimize the parameterization for the different plosives, specifically for conversational speech.

5   Conclusions

In the current paper, we presented automatic methods to phonetically transcribe and segment the recently collected *GRASS* corpus, which is the first corpus of read and conversational Austrian German [27]. For this purpose, we developed a two-step procedure: In the first step, broad phonetic transcriptions were created by means of a forced alignment and a lexicon with multiple pronunciation variants per word. In order to create pronunciation variants typical for Austrian German, we applied 32 rules to the canonical pronunciations of the words. In a second step, all plosives were annotated on the sub-phonemic level: a burst detector automatically determined whether a burst existed in a plosive and where it was located. After this step, both the broad phonetic transcription and the sub-phonemic plosive annotation are exported in form of a PRAAT TextGrid.

The quality of both steps was evaluated separately by comparison with manually created transcriptions. We found that the forced alignment based approach reached accuracies in the range of what has been reported for inter-transcriber agreement for conversational speech. Furthermore, our burst detector outperformed previous tools with accuracies between 98 % and 74 % for the different conditions in read speech, and between 82 % and 52 % for conversational speech.

In future work, we will tune the parameters of the burst detector for the conversational speech. Then, we will use the created annotations to model which are the predictors for plosive reduction in read and conversational Austrian German in comparison to the varieties spoken in Germany. These plosive-reduction models will not only inform linguists interested in conversational speech, but they will also be incorporated into the pronunciation model of an ASR system for Austrian German.

Acknowledgements. The work by Barbara Schuppler was supported by a Hertha-Firnberg grant from the FWF (Austrian Science Fund). The work of Juan A. Morales-Cordovilla was funded by the European project DIRHA (FP7-ICT-2011-7-288121).

References

1. Adda-Decker, M., Lamel, L.: Modeling reduced pronunciations in German. Phonus 5, Institute of Phonetics, University of the Saarland, pp. 129–143 (2000)
2. Adda-Decker, M., Snoeren, N.D.: Quantifying temporal speech reduction in French using forced speech alignment. J. Phonetics **39**, 261–270 (2011)
3. Boersma, P.: Praat, a system for doing phonetics by computer. Glot Int. **5**(9/10), 314–345 (2001). http://www.praat.org (last viewed 25-3-2014)
4. Cucchiarini, C., Binnenpoorte, D.: Validation and improvement of automatic phonetic transcriptions. In: Proceedings of ISCLP, Denver, USA, pp. 313–316 (2002)
5. Ernestus, M.: Voice assimilation and segment reduction in casual Dutch. A corpus-based study of the phonology-phonetics interface. Ph.D. thesis, LOT, Vrije Universiteit Amsterdam, The Netherlands (2000)
6. Gubian, M., Schuppler, B., van Doremalen, J., Sanders, E., Boves, L.: Novelty detection as a tool for automatic detection of orthographic transcription errors. In: Proceedings of the 13-th International Conference on Speech and Computer SPECOM-2009, pp. 509–514 (2009)
7. IPDS: CD-ROM: The Kiel Corpus of Spontaneous Speech, vol i-vol iii. Corpus description available at http://www.ipds.uni-kiel.de/forschung/kielcorpus.de.html (1997) (last viewed 25/11/2012)
8. Khasanova, A., Cole, J., Hasegawa-Johnson, M.: Assessing reliability of automatic burst location. In: Proceedings of Interspeech (2009)
9. Kipp, A., Wesenick, M., Schiel, F.: Pronunciation modeling applied to automatic segmentation of spontaneous speech. In: Proceedings of Eurospeech, pp. 1023–1026 (1997)
10. Kuzla, C., Ernestus, M.: Prosodic conditioning of phonetic detail in German plosives. J. Phonetics **39**, 143–155 (2011)
11. Leitner, C., Schickbichler, M., Petrik, S.: Example-based automatic phonetic transcription. In: Proceedings of the Seventh Conference on International Language Resources and Evaluation (LREC'10), pp. 3278–3284 (2010)
12. Lücking, A., Bergman, K., Hahn, F., Kopp, S., Rieser, H.: The Bielefeld speech and gesture alignment corpus (SaGA). In: Proceedings of LREC 2010 Workshop: Multimodal Corpora: Advances in Capturing, Coding and Analyzing Multimodality, pp. 92–98 (2010)
13. Makimoto, S., Kashioka, H., Nick, C.: Tagging structure and relationships in a Japanese natural dialogue corpus, In: Proceedings of Interspeech, pp. 912–917 (2007)
14. Moosmüller, S.: The process of monophthongization in Austria (reading material and spontaneous speech). In: Papers and Studies in Contrastive Linguistics, pp. 9–25 (1998)
15. Muhr, R.: Österreichisches Aussprachewörterbuch – Österreichische Aussprache-datenbank. Peter Lang Verlag, Frankfurt/M., Wien u.a. 525 S. mit DVD (2007)
16. Neubarth, F., Pucher, M., Kranzler, C.: Modeling Austrian dialect varieties for TTS. In: Proceedings of Interspeech, pp. 1877–1880 (2008)

17. Pitt, M.A., Johnson, K., Hume, E., Kiesling, S., Raymond, W.D.: The Buckeye corpus of conversational speech: labeling conventions and a test of transcriber reliability. Speech Commun. **45**, 89–95 (2005)
18. Raymond, W.D., Dautricourt, R., Hume, E.: Word-internal /t, d/ deletion in spontaneous speech: modeling the effects of extra-linguistic, lexical and phonological factors. Lang. Var. Change **18**, 55–97 (2006)
19. Reichel, U.D.: PermA and Balloon: tools for string alignment and text processing. In: Proceedings of Interspeech 2012, pp. 346 (2012)
20. Schiel, F.: Automatic phonetic transcription of non-prompted speech. In: Proceedings ICPhS 1999, pp. 607–610 (1999)
21. Schiel, F., Baumann, A.: Phondat1, corpus version 3.4. Internal report, Bavarian Archive for Speech Signals (BAS) (2006). http://www.bas.uni-muenchen.de/bas/basformatseng.html
22. Schuppler, B., van Dommelen, W., Koreman, J., Ernestus, M.: How linguistic and probabilistic properties of a word affect the realization of its final /t/: studies at the phonemic and sub-phonemic level. J. Phonetics **40**, 595–607 (2012)
23. Schuppler, B.: automatic analysis of acoustic reduction in spontaneous speech. Ph.D. thesis, Radboud University Nijmegen, The Netherlands (2011)
24. Schuppler, B., Adda-Decker, M., Morales-Cordovilla, J.A.: Pronunciation variation in read and conversational Austrian German. In: Interspeech'14 (2014, accepted for publication)
25. Schuppler, B., van Dommelen, W., Koreman, J., Ernestus, M.: Word-final [t]-deletion: an analysis on the segmental and sub-segmental level. In: Proceedings of Interspeech, pp. 2275–2278 (2009)
26. Schuppler, B., Ernestus, M., Scharenborg, O., Boves, L.: Acoustic reduction in conversational Dutch: a quantitative analysis based on automatically generated segmental transcriptions. J. Phonetics **39**, 96–109 (2011)
27. Schuppler, B., Hagmüller, M., Morales-Cordovilla, J.A., Pessentheiner, H.: GRASS: the Graz corpus of read and spontaneous speech. In: Proceedings of LREC'14, pp. 1465–1470 (2014)
28. Torreira, F., Adda-Decker, M., Ernestus, M.: The Nijmegen corpus of casual French. Speech Commun. **52**(3), 201–212 (2010)
29. Torreira, F., Ernestus, M.: Probabilistic effects on French [t] duration. In: Proceedings of Interspeech, pp. 448–451 (2009)
30. Van Bael, C.: Validation, automatic generation and use of broad phonetic transcriptions. Ph.D. thesis, Radboud Universiteit Nijmegen, Nijmegen (2007)
31. Weilhammer, K., Reichel, U., Schiel, F.: Multi-tier annotations in the Verbmobil corpus. In: Proceedings of LREC, pp. 912–917 (2002)
32. Wesenick, M.B.: Automatic generation of German pronunciation variants. In: Proceedings of the ICSLP, pp. 125–128 (1996)
33. Young, S., Evermann, G., Kershaw, D., Moore, G., Odell, J., Ollason, D., Povey, D., Valtchev, V., Woodland, P.: The HTK book (v. 3.2). Technical report, Cambridge University. Engineering Department (2002)
34. Yuan, J., Liberman, M.: Investigating /l/ variation in English through forced alignment. In: Proceedings of Interspeech, pp. 2215–2218 (2009)
35. Zimmerer, F., Scharinger, M., Reetz, H.: When BEAT becomes HOUSE: factors of word final /t/-deletion in German. J. Phonetics **39**, 143–155 (2011)

Machine Learning Methods

Supervised Classification Using Balanced Training

Mian Du, Matthew Pierce, Lidia Pivovarova$^{(\boxtimes)}$, and Roman Yangarber

Department of Computer Science, University of Helsinki, Helsinki, Finland
`lidia.pivovarova@cs.helsinki.fi`

Abstract. We examine supervised learning for multi-class, multi-label text classification. We are interested in exploring classification in a real-world setting, where the distribution of labels may change dynamically over time. First, we compare the performance of an array of binary classifiers trained on the label distribution found in the original corpus against classifiers trained on *balanced* data, where we try to make the label distribution as nearly uniform as possible. We discuss the performance trade-offs between balanced vs. unbalanced training, and highlight the advantages of balancing the training set. Second, we compare the performance of two classifiers, Naive Bayes and SVM, with several feature-selection methods, using balanced training. We combine a Named-Entity-based rote classifier with the statistical classifiers to obtain better performance than either method alone.

Keywords: Text categorisation · Information extraction

1 Introduction

In much research on supervised classification it is traditional to assume not only that the test data has the same distribution of labels as the training data, but also that the classifier will be applied in the future to data drawn from the same distribution. However, this is not always the case: the label distribution may change over time, even within the same news stream. For example, it is unlikely that the distribution of industry-sector labels in the RCV1 corpus, which was collected over 15 years ago, is similar to that in the current Reuters news-wire. Furthermore, a single set of classifiers may be required to label data from multiple sources, such as a variety of news feeds.

We present PULS, a framework for Information Extraction (IE) from text, designed for decision support in various domains and scenarios, including business [10]. PULS works with a large business corpus, currently consisting of over 1.5 M news articles. Articles are collected daily from multiple sources, therefore, one of our goals is to build classifiers that are not biased toward the particular distribution of labels in a given training set. Rather than using all available documents from a training set, we experiment with smaller subsets of balanced data. We use a balancing procedure, suitable for the multi-label setting.

© Springer International Publishing Switzerland 2014
L. Besacier et al. (Eds.): SLSP 2014, LNAI 8791, pp. 147–158, 2014.
DOI: 10.1007/978-3-319-11397-5_11

Using a collection of test sets, with different label distributions, we demonstrate that classifiers trained on balanced data perform better, on average, than classifiers trained using the original distribution of labels in the corpus.

We compare several classification methods, including Naive Bayes (NB) and Support Vector Machine (SVM), with two well-known feature selection methods, Information Gain (IG) and Bi-Normal Separation (BNS). We also combine supervised classification with a "baseline" Rote classifier, which uses knowledge collected from the corpus via IE.

2   Related Work

There are two principal approaches to adapt methods for single-label classification to the multi-label task: *problem transformation* and *algorithm adaptation* [20]. In problem transformation, multi-label classification is converted into a series of single-label classification sub-tasks, while algorithm adaptation is an extension of single-label methods to handle the multi-label data directly. One common method for problem transformation, which we adopt in our work, is *cross-training* [1]: a single *binary* classifier is trained for each label, using instances having the given label as positive examples, and all remaining instances as negative.

Text datasets are typically "naturally skewed," [15], since topics differ both in frequency and importance, depending on where the data originates; additional skew may be introduced by annotator bias. Such imbalance poses a challenge for categorization, especially when the classes have a high degree of overlap [16]. This problem can be tackled on the *data* level or the *algorithmic* level [13]. The data-level approach is based on various re-sampling techniques [2]. Some re-sampling techniques applied to the text classification task are described in [4,6,18]. Two approaches to re-sampling are *oversampling*, i.e., adding more instances of the minor classes into the training set, and *under-sampling*, i.e., removing instances of the major classes from the training set [11]. Over- and under-sampling can be either *random* or *focused* (i.e., *informed*). We follow the random under-sampling approach, which means that documents in the training set are randomly selected from each class.

A commonly used *data representation* for text categorization is the "bag of words" model, which ignores any document structure and assumes that words occur independently [12]. This model can be extended by using n-grams [5,23]. We use the bag-of-words model with a combination of unigrams and bigrams. Information Extraction (IE) can be used to obtain additional features for classification [9,10]. We use company names extracted from the text by PULS named-entity recognition system, to build a baseline, Rote classifier (Sect. 5).

Text data is characterised by a very large number of distinct word types, which can exceed the number of training documents by an order of magnitude [7]. Thus *dimensionality reduction* becomes a key step in most text classification approaches. This aims not only to accelerate processing but also to improve categorization performance [12,19] through avoidance of over-fitting [15]. Reduction can be done either by *selection* of highly-relevant features or by *grouping* (i.e., clustering) features [12]. In this paper we use feature selection which is based on comparing the discriminative power of a given word, relative to all other words in the feature set. Comparative studies of various feature selection methods can be found in, e.g., [7,22].

3   Data

We focus on supervised-learning techniques to classify news articles into industry sectors. Although we are primarily interested in our own news collection, all experiments we present here are conducted on the publicly available Reuters corpus (RCV1),[1] to allow meaningful comparison and to assure replicability. RCV1 contains 800,000 news stories published by Reuters between 1996–1997. Documents are labeled using 103 *Topic* labels, 350 *Industry* labels and 296 *Region* codes; the labels are organised hierarchically. In this paper we use a subset of 200 industry sectors.[2]

Although RCV1 is a popular dataset, relatively few papers use its sector classification, and not all of them are directly comparable with our study. To the best of our knowledge there are four papers directly comparable to our work in that they use a large number of sector labels and report micro- and/or macro-averaged F-measures: [3,14,17,24]. In Table 5 (in the Results section) we present a detailed comparison between their results on RCV1 industry labels and ours.

We use the raw text data from RCV1.[3] We only use documents that have sector labels, 351,810 in total. These documents were manually classified into 350 industry sectors. There are seven- and five-digit industry codes; seven-digit codes are children of the corresponding five-digit codes: e.g., *Fruit Growing* (I0100206), *Vegetable Growing* (I0100216) and *Soya Growing* (I0100223) are all children of *Horticulture* (I01002).

This sector classification has some inconsistencies, as observed by others, e.g., [14]. We map all seven-digit codes to their corresponding parent codes, and merge labels that have the same name but different code.[4] After this pre-processing, 245 distinct sector labels remain.

4   Array of Balanced Binary Classifiers

As mentioned in Sect. 2, we split the multi-label classification task into many binary classification sub-tasks, carried out by an array of statistical classifiers, one trained for each individual sector. All classifiers in the array use exactly the same training set, where all documents labeled with a given sector are used as

---

[1] http://about.reuters.com/researchandstandards/corpus/

[2] Henceforth we use the terms *label*, *class* and *(industry) sector* interchangeably.

[3] The commonly-used pre-processed data from [14] is not suitable, for two reasons: (a) we need plain text as input for IE, and (b) the preprocessed dataset contains only unigrams, while we use a combination of unigrams and bigrams as features.

[4] For example, we merge I64000 and I65000, both called *Retail Distribution*.

positive instances for that sector's classifier, while all remaining training documents are used as negative instances. We experiment with two supervised-learning algorithms: Naive Bayes and Support Vector Machines (SVM). We use implementations from the open-source WEKA toolkit [8].

4.1   Text Representation

Each training and test document is represented using bag-of-words features from the text. We use only nouns, adjectives, and verbs in our feature set, and apply simple filters to remove all stop-words, proper names, locations, dates, and common verbs such as "have" and "do". We also generate bigrams that consist of these three parts of speech. When indexing documents after feature selection, we use a unigram as a feature only if it appears *outside of any bigram features* extracted from that document. For example if the phrase "power plant" appears in a document we will consider "power" or "plant" as independent features, only if they also appear elsewhere in the document (and not in another extracted bigram). This allows us to resolve ambiguity to some extent; for example, we can more easily distinguish documents containing the feature "SIM card," which may be relevant for *Telecommunications*, from "credit card," which is relevant for *Commercial Banking*.

In total, 77,636 training instances (documents) have 49,262 unique features; each binary classifier has 49,262 features. We use two standard feature-selection methods—we select the top 500 features, as ranked by Information Gain (IG) [22], and Bi-Normal Separation (BNS) [7]. We then try different learning algorithms and feature selection methods to find the combination with the best performance.

4.2   Training and Test Data Pools

If a particular sector S_1 is dominant in the training set, the negative features for other classifiers could become dominated by features drawn from S_1, which may hurt performance on some other sector, S^*, since it won't learn negative features from other, "minor" sectors (those having fewer documents in the corpus). If S_1 is also over-represented in the test set, we run the risk of over-fitting. For these reasons we try to keep the training data as balanced as possible across sectors, and ensure that the test set will contain a sufficient number of instances for every binary classifier in the array.

In cross-training (defined in Sect. 2) we use a *single* pool of training instances and a *single* pool of test instances; recall that documents may have multiple labels. In creating a *balanced* training pool, we aim to provide each of the 245 binary classifiers a sufficient number of examples in both pools. Ranking the sectors by size, from 1 to N, we begin collecting data into the pools from the sector,

S_N, that has the smallest number of instances in the corpus.[5] We randomly select up to 600 documents labeled with S_N, and split them into two subsets: 3/4 for the training pool and 1/4 for test. If there are not enough documents (< 600) for S_N, all available instances are collected, with the same training/test proportion. In this way we try to guarantee some data will be available for testing, even for the smallest sectors.

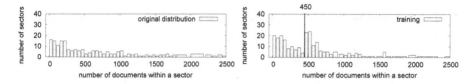

Fig. 1. Document distribution among sectors in the training pool (right): aiming for approximately 450 documents per sector; distribution in the original corpus (left).

Table 1. Number of *positive* instances in the training pool, for the most frequent sectors

Sector	Instances	Sector	Instances
Diversified Holding Companies	3644	Electricity Production	1986
Commercial Banking	3153	Agriculture	1980
Petroleum and Natural Gas	2628	Computer Systems and Software	1805
Telecommunications	2145	Air Transport	1754
Metal Ore Extraction	2099	Passenger Cars	1713

We then move on to the second smallest sector, S_{N-1}, and repeat the collection process, except now we first check how many documents labeled with S_{N-1} are *already present* in the training and test pools—which may happen due to multiple labeling (label overlap). The number of documents collected for S_{N-1} at this step is reduced by the number already collected. The collection process continues in this manner for all sectors. Collection may be skipped for a sector if it already has more than 450 documents in the training pool (this happens for sectors with high label overlap). As stated, it is also possible that some sectors will have fewer documents for training, based on total availability. These are inherent limitations of the skew in the original corpus, and cannot be avoided.

The resulting set, called the "balanced training data pool" has 77,636 documents. It is still skewed, as seen in Fig. 1, on the right, though much more balanced than the initial distribution, shown on the left. As can be inferred from the Figure, between 50 and 60 sectors contain fewer than 150 instances each. Since a lower amount of data makes it difficult to obtain reliable results, we use

---

[5] Otherwise we cannot guarantee that each sector will have a sufficient number of instances in the training and test pools. For example, if we collect the training and testing data in random order and happen to start with the largest sectors, then by the time we come to the smallest sectors all of its data may already be included in the training pool (due to multiple labeling of documents), leaving none for testing.

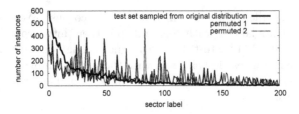

Fig. 2. Label distributions of an original test set, and permuted test-sets (2 of 50 shown).

Table 2. Sector distribution for company "Apple"

Sector	Freq	Prob
Computer Systems and Software	549	0.61
Electronic Active Components	61	0.07
Datacommunications and Networking	36	0.04
Telecommunications	19	0.02
Electrical and Electronic Engineering	13	0.01

only the 200 largest sectors in our experiments, which cover approximately 99 % of the original corpus.

Table 1 shows the most frequent sectors in the balanced training pool. We can see, e.g., that although we only collected 450 positive training instances for *Diversified Holding Companies*, it still receives 3644 positive instances in the pool, most of which were picked up when collecting data for other sectors.

For comparison, (Sect. 7.2), we use an *unbalanced* training pool, which is simply half of the corpus.

All data *outside* the balanced and unbalanced training pools—called the "test pool"—are available for the construction of test sets. From the test pool, we generate 10 samples of 10,000 documents each, using the original distribution in the corpus. We use one of these samples as a held-out *development* set for parameter tuning (Sect. 4.3), and the remaining nine as test sets.

To simulate the effect of changing trends in news streams, we generate 50 additional datasets. To build these sets, we calculate the individual proportions of the sectors in the original distribution, then assign these proportions to 50 *random permutations* of the sector labels. We then attempt to sample 10,000 documents from the testing pool according to the new, permuted distributions. Each set among these 50 has its own label distribution, different from both the original and from each other. The distribution of labels in these random test sets will appear "naturally skewed," since it mimics the original shape.

Three example test sets are shown in Fig. 2, one "original," and two "permuted." The permuted distributions are still somewhat biased toward the largest classes in the original corpus. This is expected because some larger classes (such as *Diversified Holding Companies*) still have a high degree of overlap, and because the smallest sectors may not have enough data to dominate the permuted distribution. However, the distributions of the permuted test sets look substantially different from the original distribution and contain significantly more instances from small- and medium-sized sectors. We use the original and permuted test sets in our comparison of balanced and unbalanced training (Sect. 7.2).

4.3 Classification

The SVM classifiers output a binary decision for every document. For Naive Bayes, the output for each sector is a confidence score between 0.01 and 1; thus a decision threshold is required to make a classification. We learn the best threshold over a range of thresholds (in increments of 0.01), using a held-out *development* set (one of the test sets, described in Sect. 4.2). We then evaluate on the remaining test sets using the learned threshold.

5 Rote Classifier

The Rote classifier labels documents based on the company–sector relationships present in the RCV1 corpus. PULS finds mentions of companies in the corpus, using a named-entity (NE) recognition module. It distinguishes company names from other proper names in the text, e.g., persons and locations. NE also merges together variants of the same name, for example, "Apple," "Apple Inc.," "Apple Computer, Inc.," etc. For each company we collect all sector labels from all documents where it is mentioned; sectors co-occurring with a company fewer than 3 times are discarded. For example, Table 2 shows the top sectors that co-occur with "Apple."; it shows the frequency (the co-occurrence count of the company with the sector), and the proportion, which is the normalized count.

For every document, the Rote classifier returns a sector associated with the companies found in the text if the proportion for this sector is higher than a certain threshold; the threshold is chosen from the range 0.01 to 1, using the development set. If the same sector co-occurs with more than one company found in the text, we apply the highest proportion.

6 Combined Classifiers

We experiment with several methods of combining the Rote classifier, described in Sect. 5, with the balanced probabilistic classifiers, described in Sect. 4, to see whether the combination can produce better *overall* prediction of the sector labels. One method of combining is a simple two-stage process: for each document, we first try to identify sectors using the Rote classifier; if that does not return any sectors, we then attempt to classify using the statistical classifiers. We also experiment with the reverse order of these classification stages. The motivation for this method is to give the overall system a "second chance" at classification, in the hope that together the two methods may overcome their respective shortcomings. Another method of combining classifiers is to return the *union* of the results of the two classifiers—rote and probabilistic. Again, We learn the optimal threshold for each classifier in the combination using the development set.

7   Experiments and Results

7.1   Evaluation Measures

Common measures in text classification are precision, recall, and F-measure. For a given class c, these are calculated as:

$$Rec_c = \frac{TP_c}{TP_c + FN_c} \qquad Prec_c = \frac{TP_c}{TP_c + FP_c} \qquad F1_c = \frac{2 \times Rec \times Prec}{Rec + Prec}$$

where TP_c, TN_c, FP_c and FN_c are the number of true positive, true negative, false positive, and false negative classified instances for the class, respectively; $|c|$ is the number of documents in the test pool labeled with this class.

In evaluating multi-label classification, *macro-averaging* and *micro-averaging* are commonly reported [21]. In micro-average evaluation, first the numbers of true- and false-positives, and true- and false-negatives are counted for all instances in the test set, and then the standard measures, e.g., recall or precision, are calculated using these numbers:

$$Rec_\mu = \frac{\Sigma_{i \in S} TP_i}{\Sigma_{i \in S}(TP_i + FN_i)} \qquad Prec_\mu = \frac{\Sigma_{i \in S} TP_i}{\Sigma_{i \in S}(TP_i + FP_i)} \qquad \mu\text{-}F1 = \frac{2 \times Rec_\mu \times Prec_\mu}{Rec_\mu + Prec_\mu}$$

where S is the set of all classes. In the macro-average evaluation scheme, the measures are calculated for each class *separately* first, and then these are averaged across all classes:

$$Rec_M = \frac{\Sigma_{i \in S} Rec_i}{|S|} \qquad Prec_M = \frac{\Sigma_{i \in S} Prec_i}{|S|} \qquad M\text{-}F1 = \frac{\Sigma_{i \in S} F1_c}{|S|}$$

We report both evaluation schemes, although we focus more on the macro-average scores, as explained below, since they are less dependent on the particular distribution of labels in the corpus. Henceforth we denote the macro-averaged F-measure by M-F1, and micro-averaged F-measure by μ-F1.

7.2   Balanced vs. Unbalanced Training

To justify the use of balanced training data in building our classifiers, we compare two sets of classifiers, built using two distinct training pools: one balanced, under-sampled training set and one unbalanced training set, comprised of half the total data, selected at random. All data outside these training pools are available for the construction of test sets. As described in Sect. 4.2, we generate 10 "original" test sets that preserve the original label distribution, and 50 "permuted" test sets with label distributions that are meant to simulate the effect of changing

Table 3. Results for SVM + IG classifiers trained on balanced vs. unbalanced training sets, applied to originally-distributed and permuted test sets

10 originally distributed testsets				50 permuted testsets			
training	Rec	Pre	F1	training	Rec	Pre	F1
	M-average				*M-average*		
balanced	**31.8**±1.3	59.1±1.1	**37.1**±1.1	balanced	**32.6**±0.9	70.9±1.3	**41.8**±0.9
unbalanced	24.3±0.9	**73.6**±1.3	31.8±0.9	unbalanced	23.5±0.9	**74.0**±1.5	31.4±0.8
	μ-average				*μ-average*		
balanced	30.4±0.4	72.6±0.6	42.9±0.5	balanced	**34.4**±0.4	**78.6**±1.4	**47.8**±0.2
unbalanced	**36.8**±0.6	**79.5**±0.5	**50.3**±0.6	unbalanced	29.8±1.8	76.9±1.4	43.0± *2.1*

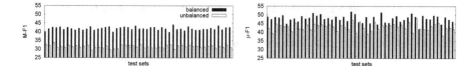

Fig. 3. F-measure obtained by SVM+IG classifiers trained on balanced vs. unbalanced data, for all *permuted* test sets.

trends in news streams, over time or due to shifts in emphasis toward new sectors in a particular source.

The averaged results obtained on both original and permuted test sets are presented in Table 3. To save space we present only the SVM + IG results, since results for all classifiers follow the same pattern: classifiers trained on the original distribution have higher μ-F1 on originally distributed test sets, but lower on the permuted test sets; the classifiers trained on the balanced training set yield higher M-F1 on *all test sets*, both original and permuted.

A comparison of balanced and unbalanced training is presented in Fig. 3, where we plot macro- and micro-averaged F-measure obtained by classifiers trained on balanced vs. unbalanced data for each *permuted* test set. As can be seen from the plot in the left figure, the classifier trained on balanced data has significantly and consistently higher M-F1: for each test set M-F1 is over 30 % higher for the balanced classifiers.

As seen from the right plot, in the majority of cases, the classifier trained on balanced data also yields higher μ-F1 than the classifier trained on unbalanced data, although the difference between two classifiers has somewhat higher variance (also seen from Table 3, standard deviation scores). Thus the M-F1 appears to be more stable for both classifiers. This suggests that focusing on macro-averaged results is more appropriate for real-world news classification tasks.

7.3   Comparison of Classifiers and Feature Selection Methods

Results obtained by all classifiers are shown in Table 4; we present only results obtained with *balanced* training data, since they are consistently higher—in terms of M-F1—than results obtained using unbalanced training.

Table 4. Results from all classifiers and feature selection methods, averaged across 9 test sets randomly sampled from original distribution; single classifiers on top, combined classifiers on bottom. For each classifier, the best threshold is trained on one random, originally-distributed development set; $\rightarrow$ and $\cup$ denote, respectively, two-stage and union combining methods, described in Sect. 6.

Classifier	M-average			μ-average		
	Rec	Pre	F1	Rec	Pre	F1
NB + IG	31.3 ± 0.9	21.9 ± 0.6	19.7 ± 0.6	31.5 ± 0.5	22.4 ± 0.6	26.2 ± 0.5
NB + BNS	34.2 ± 1.1	16.6 ± 0.6	15.8 ± 0.5	33.1 ± 0.7	13.4 ± 0.4	19.0 ± 0.5
SVM + IG	31.9 ± 1.3	59.2 ± 1.1	37.1 ± 1.2	30.5 ± 0.4	$\mathbf{72.7 \pm 0.6}$	42.9 ± 0.4
SVM + BNS	32.7 ± 0.9	55.2 ± 1.0	36.2 ± 0.7	30.1 ± 0.5	70.8 ± 0.6	42.2 ± 0.5
Rote	$\mathbf{35.0 \pm 0.8}$	$\mathbf{67.6 \pm 1.0}$	43.8 ± 0.8	$\mathbf{42.4 \pm 0.6}$	64.2 ± 0.4	$\mathbf{51.1 \pm 0.5}$
Rote $\rightarrow$ NB+BNS	51.5 ± 0.9	33.6 ± 0.4	36.1 ± 0.4	57.6 ± 0.6	39.1 ± 0.4	46.6 ± 0.4
NB + BNS $\rightarrow$ Rote	49.7 ± 1.0	24.0 ± 0.2	26.9 ± 0.3	53.3 ± 0.4	23.7 ± 0.3	32.8 ± 0.3
Rote $\cup$ NB+BNS	$\mathbf{59.2 \pm 0.9}$	25.4 ± 0.3	30.7 ± 0.3	$\mathbf{64.3 \pm 0.5}$	26.2 ± 0.3	37.2 ± 0.3
Rote $\rightarrow$ NB+IG	51.8 ± 0.9	39.8 ± 0.6	41.5 ± 0.6	59.1 ± 0.5	47.3 ± 0.4	52.5 ± 0.4
NB + IG $\rightarrow$ Rote	48.7 ± 1.0	31.5 ± 0.5	33.4 ± 0.4	53.0 ± 0.5	36.3 ± 0.3	43.1 ± 0.3
Rote $\cup$ NB+IG	57.2 ± 0.9	32.7 ± 0.4	37.3 ± 0.4	63.2 ± 0.5	38.1 ± 0.3	47.5 ± 0.4
Rote $\rightarrow$ SVM+BNS	48.2 ± 1.0	67.5 ± 1.0	54.7 ± 0.9	53.7 ± 0.5	70.1 ± 0.3	60.8 ± 0.4
SVM + BNS $\rightarrow$ Rote	48.0 ± 1.1	63.0 ± 1.0	52.6 ± 1.0	50.2 ± 0.4	70.8 ± 0.4	58.7 ± 0.4
Rote $\cup$ SVM + BNS	54.0 ± 0.9	62.0 ± 0.8	56.1 ± 0.8	58.5 ± 0.4	68.2 ± 0.3	63.0 ± 0.3
Rote $\rightarrow$ SVM + IG	46.2 ± 1.0	$\mathbf{73.7 \pm 0.8}$	55.1 ± 0.8	52.5 ± 0.5	$\mathbf{75.9 \pm 0.4}$	62.0 ± 0.4
SVM + IG $\rightarrow$ Rote	47.0 ± 1.2	67.7 ± 0.9	53.7 ± 1.1	49.9 ± 0.3	73.9 ± 0.3	59.6 ± 0.3
Rote $\cup$ SVM + IG	52.2 ± 1.1	66.3 ± 0.8	$\mathbf{56.9 \pm 0.9}$	57.7 ± 0.4	71.1 ± 0.3	$\mathbf{63.7 \pm 0.4}$

Table 5. Classification results on RCV1 industry sectors, compared with state of the art.

Reference	Algorithm	M-F1	μ-F1
[14]	SVM	29.7	51.3
[24]	SVM	30.1	52.0
[17]	Naive Bayes	-	70.5
[3]	Bloom Filters	47.8	**72.4**
Our best results	Rote $\rightarrow$ SVM + IG	**56.9**	63.7

As seen from the table, the SVM classifier yields higher performance than NB, independently of the feature selection method used. IG performs better than BNS with both Naive Bayes and SVM.

The baseline Rote classifier yields the highest F-measure among single classifiers; combining Rote with SVM + IG yields the best combined performance. The M-F1 obtained by this two-stage classifier is higher than the best previously reported results, as shown in Table 5. It also can be seen from the table that the difference between M-F1 and μ-F1 for our classifiers is smaller than that reported in prior work. This supports the claim that our classifiers are less sensitive to changes in label distribution (due to the balancing of the training), which is one of our main objectives.

The μ-F1 in our experiments is lower than the best μ-F1 reported in the literature on RCV1. This is likely due to the fact that both [3,17] try to model inter-dependencies among the labels in the corpus. This is not done in [14] or [24]. We plan to investigate this further in future work; however, our results suggest that balancing the training data improves the classifier performance overall, regardless of the method used.

8   Conclusion

We have described an approach using supervised learning for labeling business-news documents with multiple industry sectors. We treat the multi-class, multi-label problem as a set of binary sub-tasks, with one binary classifier for each sector. We attempt to create robust classifiers, suitable for real-world text classification (rather than improving performance on a given static corpus), by balancing the training data given to each classifier. Our results suggest that, compared to classifiers trained on labels drawn from the original corpus distribution, the balanced training helps improve the scores—M-F1 in particular—when classifying data drawn from different distributions of labels.

We explore several combinations of learning algorithms and feature selection methods, and evaluate them using a large number of manually-labeled documents. Combining a named-entity-based Rote classifier with the set of balanced classifiers, into a two-stage classifier, yields better results than either classifier alone. Additionally, this method improves on the best M-F1 previously reported, while using the same amount of training data for the Rote classifier, and considerably less for the statistical classifiers.

References

1. Boutell, M.R., Luo, J., Shen, X., Brown, C.M.: Learning multi-label scene classification. Pattern Recogn. **37**(9), 1757–1771 (2004)
2. Chawla, N.V., Bowyer, K.W., Hall, L.O., Kegelmeyer, W.P.: Smote: synthetic minority over-sampling technique. J. Artif. Intell. Res. **16**(1), 321–357 (2002)
3. Cisse, M.M., Usunier, N., Arti, T., Gallinari, P.: Robust Bloom filters for large multilabel classification tasks. In: Advances in Neural Information Processing Systems, pp. 1851–1859 (2013)
4. Dendamrongvit, S., Kubat, M.: Undersampling approach for imbalanced training sets and induction from multi-label text-categorization domains. In: Theeramunkong, T., Nattee, C., Adeodato, P.J.L., Chawla, N., Christen, P., Lenca, P., Poon, J., Williams, G. (eds.) New Frontiers in Applied Data Mining. LNCS, vol. 5669, pp. 40–52. Springer, Heidelberg (2010)
5. Dhondt, E., Verberne, S., Weber, N., Koster, C., Boves, L.: Using skipgrams and pos-based feature selection for patent classification. Comput. Linguist. Neth. **2**, 52–70 (2012)

6. Erenel, Z., Altınçay, H.: Improving the precision-recall trade-off in undersampling-based binary text categorization using unanimity rule. Neural Comput. Appl. **22**(1), 83–100 (2013)
7. Forman, G.: An extensive empirical study of feature selection metrics for text classification. J. Mach. Learn. Res. **3**, 1289–1305 (2003)
8. Hall, M., Frank, E., Holmes, G., Pfahringer, B., Reutemann, P., Witten, I.H.: The WEKA data mining software: an update. ACM SIGKDD Explor. Newsl. **11**(1), 10–18 (2009)
9. Huang, R., Riloff, E.: Classifying message board posts with an extracted lexicon of patient attributes. In: Proceedings of the 2013 Conference on Empirical Methods in Natural Language Processing, pp. 1557–1562 (2013)
10. Huttunen, S., Vihavainen, A., Du, M., Yangarber, R.: Predicting relevance of event extraction for the end user. In: Poibeau, T., Saggion, H., Piskorski, J., Yangarber, R. (eds.) Multi-source, Multilingual Information Extraction and Summarization. Theory and Applications of Natural Language Processing, pp. 163–176. Springer, Berlin (2012)
11. Japkowicz, N., Stephen, S.: The class imbalance problem: a systematic study. Intell. Data Anal. **6**(5), 429–449 (2002)
12. Koller, D., Sahami, M.: Hierarchically classifying documents using very few words. Technical report 1997–75, Stanford InfoLab, February 1997
13. Kotsiantis, S., Kanellopoulos, D., Pintelas, P., et al.: Handling imbalanced datasets: a review. GESTS Int. Trans. Comput. Sci. Eng. **30**(1), 25–36 (2006)
14. Lewis, D.D., Yang, Y., Rose, T.G., Li, F.: RCV1: a new benchmark collection for text categorization research. J. Mach. Learn. Res. **5**, 361–397 (2004)
15. Liu, Y., Loh, H.T., Sun, A.: Imbalanced text classification: a term weighting approach. Expert Syst. Appl. **36**(1), 690–701 (2009)
16. Prati, R.C., Batista, G.E.A.P.A., Monard, M.C.: Class imbalances *versus* class overlapping: an analysis of a learning system behavior. In: Monroy, R., Arroyo-Figueroa, G., Sucar, L.E., Sossa, H. (eds.) MICAI 2004. LNCS (LNAI), vol. 2972, pp. 312–321. Springer, Heidelberg (2004)
17. Puurula, A.: Scalable text classification with sparse generative modeling. In: Anthony, P., Ishizuka, M., Lukose, D. (eds.) PRICAI 2012. LNCS, vol. 7458, pp. 458–469. Springer, Heidelberg (2012)
18. Stamatatos, E.: Author identification: using text sampling to handle the class imbalance problem. Inf. Process. Manage. **44**(2), 790–799 (2008)
19. Tikk, D., Biró, G.: Experiments with multi-label text classifier on the Reuters collection. In: Proceedings of the International Conference on Computational Cybernetics (ICCC 03), pp. 33–38 (2003)
20. Tsoumakas, G., Katakis, I.: Multi-label classification: an overview. Int. J. Data Warehouse. Min. (IJDWM) **3**(3), 1–13 (2007)
21. Yang, Y.: An evaluation of statistical approaches to text categorization. Inf. Retrieval **1**(1–2), 69–90 (1999)
22. Yang, Y., Pedersen, J.O.: A comparative study on feature selection in text categorization. In: ICML, vol. 97, pp. 412–420 (1997)
23. Zhang, W., Yoshida, T., Tang, X.: A comparative study of TF*IDF, LSI and multi-words for text classification. Expert Syst. Appl. **38**(3), 2758–2765 (2011)
24. Zhuang, D., Zhang, B., Yang, Q., Yan, J., Chen, Z., Chen, Y.: Efficient text classification by weighted proximal SVM. In: Fifth IEEE International Conference on Data Mining (2005)

Exploring Multidimensional Continuous Feature Space to Extract Relevant Words

Márius Šajgalík[(⊠)], Michal Barla, and Mária Bieliková

Faculty of Informatics and Information Technologies,
Slovak University of Technology in Bratislava, Ilkovičova 2,
842 16 Bratislava, Slovakia
{marius.sajgalik,michal.barla,maria.bielikova}
@stuba.sk

Abstract. With growing amounts of text data the descriptive metadata become more crucial in efficient processing of it. One kind of such metadata are keywords, which we can encounter e.g. in everyday browsing of webpages. Such metadata can be of benefit in various scenarios, such as web search or content-based recommendation. We research keyword extraction problem from the perspective of vector space and present a novel method to extract relevant words from an article, where we represent each word and phrase of the article as a vector of its latent features. We evaluate our method within text categorisation problem using a well-known 20-newsgroups dataset and achieve state-of-the-art results.

1 Introduction

The distributional hypothesis that features of a word can be learned based on its context was formulated sixty years ago by Harris [5]. In the area of natural language processing (NLP), the context of a word is commonly represented by neighbouring words that surround it (i.e. both the preceding and succeeding words). Since Harris, many NLP researchers built on his hypothesis and developed methods for learning both syntactic and semantic features of words. Hinton [6] was one of the pioneers who represented words as vectors in continuous feature space. Collobert and Weston [2] showed that such feature vectors can be used to significantly improve and simplify many NLP applications.

With advent of deep learning, more efficient methods for training feature vectors of words have been discovered. This allows us to use larger sets of training data, leading to higher quality of learned vectors, which in turn open possibilities for more practical applications of those vectors.

The advantage of such distributed representation of words, which maps words onto points in hyperspace is that we can employ standard vector operations to achieve interesting tasks e.g., to measure similarity between pairs of words. We can also calculate what words are the most similar by finding the closest vectors, or a vector that encodes a relationship between a pair of words, e.g., a vector transforming singular

L. Besacier et al. (Eds.): SLSP 2014, LNAI 8791, pp. 159–170, 2014.
DOI: 10.1007/978-3-319-11397-5_12

form into plural one, etc. Mikolov et al. showed that with such word vectors, we can encode many semantic and also syntactic relations [14].

One of the open problems in the current NLP is the computation of meaning of complex structures. The solvability of this problem is unanimously presupposed by notion of language compositionality, which is regarded as one of the defining properties of human language. One of the definitions of language compositionality, as almost an uncontroversial notion among linguists, can be found in [3]:

"(...) the meaning of (that) complex expression is fully determined by the meanings of its component expressions plus the syntactic mode of organization of those component expressions."

There are various approaches to solve the problem of compositionality in distributional semantics. The standard approach to model the composed meaning of a phrase is the vector addition [13]. There are also other approaches to model composition in semantic spaces that include vector point-wise multiplication [15], or even more complex operations like tensor product [4]. For purpose of this paper, we utilise vector addition to compose meaning of a phrase as it is reported to give very good results and to hold multiple relations between vectors [14] on the corpus we use in our experiments. We do not experiment with other methods of composition, but focus our experiments on applying various weighting schemes to analyse and compare their performance.

In this paper, we leverage feature vectors of words to extract relevant words from documents and evaluate such representation in text categorization problem. Such representation has a big potential in NLP and we can only anticipate that in a few years it will gradually supersede all those manually crafted taxonomies, ontologies, thesauri and various dictionaries, which are often rather incomplete (like most of artificial data). Moreover, there is no means of measuring similarity directly between pairs of words in such hand-crafted data. Most relations are just qualitative and described by their type (e.g. approach in [1] reveals only a relation type, but it cannot determine the relation quantitatively) and thus, all existing methods for measuring (semantic) similarity are limited to achieving only rather imprecise results.

2 Related Work

In the past, we were studying a problem of extracting key-concepts from documents. Our approach described in [17], uses key-concepts (instead of features) to classify documents. We evaluated the proposed method using 20-newsgroups dataset, giving classification accuracy 41.48 % using naïve Bayes classifier. However, 20-newsgroups dataset is one of the most commonly used in text categorisation problem and most of the researchers use micro-average F1 score to evaluate their classification performance. If we calculate micro-averaged F1 score for results in [17], it yields 58.63 % using naïve Bayes classifier and 55.85 % using kNN classifier.

There are many other researchers, who used the same dataset to evaluate and compare their work. The most important and relevant are those that study novel weighting schemes. Authors of [8] propose TF-RF for relevant term extraction, which can be viewed as an improvement of well-known TF-IDF weighting. The problem of TF-IDF is that it treats each word equally across different categories. The novelty of TF-RF weighting is that it discriminates the relevance of words based on the frequency differences in different categories. Evaluation using 20-newsgroups dataset gives micro-averaged F1 score 69.13 % using kNN and 80.81 % using SVM classifier.

Approach in [19] is based on two concepts – category frequency and inverse category frequency. Authors propose ICF-based weighting scheme that is reported to give better results than TF-RF using Reuters-21578 dataset, which has many categories (52). However, using 20-newsgroups dataset, they report not as good performance as TF-RF. Our explanation is that 20-newsgroups dataset has fewer categories and thus, the discriminating power of their "icf assumption" is weaker.

There are also approaches that do not focus on weighting of words, but try to employ various more sophisticated methods. In [9] we can find an approach using Discriminative Restricted Boltzmann Machines that achieves 76.2 % micro-averaged F1 score using 20-newsgroups dataset and in [10] authors propose an error-correcting output coding method, which using naïve Bayes classifier achieves 81.84 %.

3   Data Pre-processing

We processed each article with Stanford CoreNLP [16] to transform raw text into sequence of words labelled with part-of-speech tags. Approach in [18] inspired us to build a finite automaton (Fig. 1) that accepts candidate phrases for further processing. We assume that extracted candidate phrases describe the content of given article and have a potential to contribute with good features to topical representation, which would help us to extract better words. Using these patterns of labelled words, we got rid of stopwords and retained only valid candidate phrases that could possibly influence the process of choosing the most relevant words that would be most distinctive for classifying the article into its correct category.

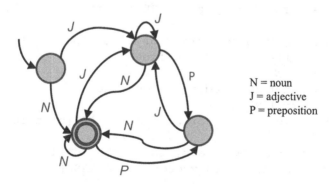

N = noun
J = adjective
P = preposition

Fig. 1. Finite automaton for generating candidate phrases.

Also, we tried choosing only noun phrases obtained by chunking, however, it skipped noun phrases prefixed by adjectives, which sometimes significantly influence the meaning of a phrase. We also tried choosing only the words that are nouns, adjectives, verbs or adverbs, but it gave us worse results than considering candidate phrases accepted by this automaton.

We simulate the understanding of word semantics by using distributed word representation [13], which represents each word as vector of latent features. We use pre-trained feature vectors[1] trained on part of Google News dataset (about 100 billion words) using neural network and negative sampling algorithm. The model contains 300-dimensional vectors for 3 million words and phrases. The vectors of phrases were obtained using a simple data-driven approach described in [13].

4 Transforming Phrases into Vectors

After we have obtained the list of noun phrases for each article, we transform each phrase into the corresponding feature vector. The model contains not just words, but also many multi-word phrases that have a specific meaning different from what we would get if we just summed up the vectors of words in those phrases. Let us call both the words and the phrases present in the model simply as unit phrases. Since there is no direct mapping in the model for every possible phrase, we try to assemble the longer phrases by concatenating the most probable unit phrases that are present in the model. We use a simple dynamic programming technique to minimise number of concatenated unit phrases and thus, prefer the longer unit phrases present in the model. In case of ambiguity we choose the sequence of unit phrases that are more similar to each other. Here follows the description of our algorithm:

```
Input: string PHRASE of length LEN
Output: feature vector VECTOR

Initialise array of integers NO_UNITS of size (LEN + 1),
set all elements to 0
Initialise array of floats SCORE of size (LEN + 1), set
all elements to 0
Initialise array of vectors V of size LEN+1, set V[0] to
zero vector, the rest is undefined
For 0 <= a <LEN do
  If V[a] is not defined then continue
  For a < b <= LEN do
    If there is not end of word at PHRASE[b] then contin-
ue
    If substring PHRASE[a:b] is in the model
      Let V1 be the word vector for PHRASE[a:b]
```

---
[1] Available online at - https://drive.google.com/file/d/0B7XkCwpI5KDYNlNUTTlSS21pQmM/edit?usp=sharing (last accessed on May 14, 2014).

```
    If V[a] is undefined
      Let SCORE1 = 0
    Else
      Let SCORE1 be similarity of V1 and V[a]
    If V[b] is undefined or
    NO_UNITS[a] + 1 < NO_UNITS[b] or
    (NO_UNITS[a] + 1 = NO_UNITS[b] and
    SCORE[a] + SCORE1 < SCORE[b])
        NO_UNITS[b] = NO_UNITS[a] + 1
        V[b] = V[a] + V1
        SCORE[b] = SCORE[a] + SCORE1
 If no substring PHRASE[a:b] was found in the model
   Let b be the ending position of nearest word in
   PHRASE[a:]
   NO_UNITS[b] = NO_UNITS[a] + 1
   V[b] = V[a]
   SCORE[b] = SCORE[a]
Let VECTOR = V[LEN]
```

We use NO_UNITS array to track the minimal number of unit phrases that comprise longer phrases. Thus, we prefer unit phrases with multiple words, since their learned feature vector has higher quality than if we just summed up feature vectors of their individual words. In case of ambiguity, we use SCORE array to track the similarity of the composing unit phrases. Thus, we prefer more common expressions. The array V serves as an intermediate storage of computed vectors, which is used to retrieve the final result. The step 4.3 is a handler for unknown words (tokens) present in the phrase, which simply skips to the next available word. This mostly handles the common cases of various punctuation characters present in the phrase, which are not included in the model we used.

5 Searching for Relevant Words with t-SNE Visualisation

Since in the corpus we use each word is a real-valued vector, we can map each word to a point in 300 dimensional feature space. We used t-SNE [11] (t-Distributed Stochastic Neighbour Embedding), which is a technique for dimensionality reduction particularly well suited for visualisation of high-dimensional datasets, and visualised word vectors to analyse and better understand what exactly is going on with these vectors. A priori, we state the following hypothesis:

"If a word in the article is more relevant (i.e. if it is a keyword), it means that it is more relevant to the discussed topic. On the other hand, if a word is not relevant to the discussed topic, it will diverge randomly in the vector space. If divergence of non-relevant words was not random, those words would form another topic in the article."

There are multiple possibilities of what exactly to visualise. First, we can visualise all the words and phrases extracted from the article by Stanford CoreNLP parser. However, that results in a big incomprehensible mess of words. Alternatively, we can reduce the set to only the candidate phrases as described in Sect. 3. Although it is much better, we cannot really see any natural word clusters as could be expected a priori. Although similar words are grouped together, there are not just a topical word clusters relevant to the article and thus, we cannot infer easily which of those words should be chosen as keywords. An explanation for why the sole candidate phrases are not sufficient, is that since we do not use any frequency statistics, we cannot infer which words are more or less common than others and thus approximate the relevance.

To cope with that we could visualise the whole corpus along the extracted words and phrases to simulate general language understanding. However, this proves to be not usable and mostly impractical. Although there exists the fastest t-SNE implementation so far called Barnes-Hut-SNE that is specifically designed for big dataset and utmost performance, we did not succeed to process the whole corpus of word vectors due to its huge size. We analysed the source code of t-SNE implementation and applied multiple optimisations, which improved the overall performance even more, however, it did not suffice. We identified the bottleneck of the computation, which was the search of k nearest neighbours. We found out that in case of our corpus of word vectors, the simple linear search is faster than the vantage-point tree used in Barnes-Hut-SNE. This is probably due to bigger ratio between number of dimensions (300) and number of words (3000000) and division by vantage-point does not help to speed up the algorithm in practice. However, computation of k nearest neighbours for each word in vocabulary would require roughly at least $(3*10^6)^2*300 = 2.7*10^{15}$ operations, which is impractically too much even with parallelisation over multiple processor cores.

Finally, we can use a golden mean of above approaches and enrich the feature vectors of candidate phrases in the article with feature vectors of k nearest neighbours present in the corpus. For each such unit phrase that we get, we increase its relevance relative to its cosine similarity to the query noun phrase. Thus, we sum up the total relevance for each unit phrase and are able to visualise the top 100 keywords.

Initially, we optimised our method to extract relevant words from articles on Washington Post website. The reason was that we intended to create a dataset of articles annotated with keywords. Although it turned out to be not suitable for our needs, it helped us to tune our method and not to overfit on the evaluation dataset used in this paper. We can see the t-SNE visualisation of top 100 most relevant words for a random article[2] produced by our method in Fig. 2.

We can see that visualisation of top 100 most relevant words forms several clusters. Most of these clusters hide at least one of the keywords. The red words are those keywords that are present in the keyword list on the webpage of the article. The blue ones are keywords that we would also manually and subjectively choose as the most relevant for this article and we can clearly see their relevance just from looking at the headline of the article. The absence of these words in the original list of keywords is probably due to SEO (search engine optimisation), since SEO probably focuses just on a few words and their variants.

---

[2] Available online at - http://www.washingtonpost.com/world/national-security/cybersecurity-poll-americans-divided-over-government-requirements-on-companies/2012/06/06/gJQAmWqnJV_story.html (last accessed on May 14, 2014).

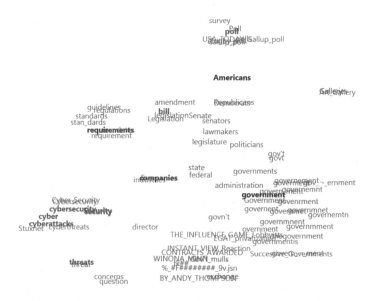

Fig. 2. Top 100 most relevant words computed by our method and visualised by t-SNE. Red words are keywords from website and blue words are keywords that we would also manually and subjectively choose as the most relevant for this article (Color figure online).

However, what is more significant for our method is that there are many words in used corpus, which represent only a noise (mostly typos). Thus, we reduced the original corpus to only 200 000 most frequent words. We calculated frequency of each word in corpus by processing whole Google N-gram dataset.[3] This cleaned up the corpus and also sped up further processing, which would be intractable in some cases. We can see that many words representing words with typos disappeared. We can see the result of this second alteration in Fig. 3.

As we can see, most of the keywords remained in the top, which is a positive result. However, there are still some stopwords, which represent determiners, prepositions, names and some letters that probably emerged from name initials present in the training data. Although the finite automaton described in Sect. 3 ignores such stopwords present in text, after the selection of k nearest neighbours, there still emerge such stopwords. Therefore, we employed TF-IDF statistics to filter out these stopwords. We also removed all words shorter than 3 letters. As we can see in Fig. 4, it didn't clean out all stopwords (names) as discussed in previous paragraph, but still it removed a good portion of them (there are only 4 stopwords, which are all in one small cluster). However, it might be speculative if names are really stopwords, since there are some persons mentioned in the article as well. We can see that most words are more relevant than in the previous case, since they are more coherent and focused on given topic.

Based on analysis in this section, we finally used TF-RF weighting instead of TF-IDF, since it has been reported to give better results in text categorisation task [8].

---

[3] Available online at - http://storage.googleapis.com/books/ngrams/books/datasetsv2.html (last accessed on May 14, 2014).

Fig. 3. Top 100 most relevant words computed by our method using top 200 k word vector corpus and visualised by t-SNE.

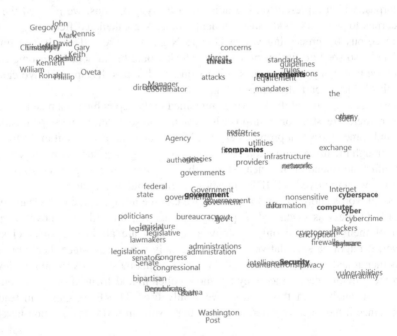

Fig. 4. Top 100 most relevant words computed by our method with TF-IDF weighting using top 200 k word vector corpus and visualised by t-SNE.

6 Evaluation by Text Categorisation

We evaluated proposed method in text categorisation problem. We used 20-news-groups dataset, which consists of 18 846 news documents divided into 20 categories. We used the version "by date" divided into training and testing sets, consisting of 11 314 and 7 532 documents respectively.

For each document, we computed one feature vector as a normalised sum of vectors of top relevant words output from our method. Thus, each document was expressed by 300 features. We tried different number of top words to create the feature vector of a document and also multiple common classifiers, which seemed most reasonable to us.

We tried using k-NN classifier, however, it didn't yield very good results (only below 80 %). We used linear discriminant analysis presuming that each category can be expressed as a linear combination of features in a vector and thus not treating all features as equal could yield better results. Truly, we achieved better F1 score 82.85 % for top 1200 words. We also applied linear SVM classifier, which we found having better performance in other researchers' work [8]. Our expectation was fulfilled by SVM yielding state-of-the-art performance of 84.5 % micro-averaged F1 score (compared to approaches from Sect. 2). We can see summarisation of achieved results in Fig. 5.

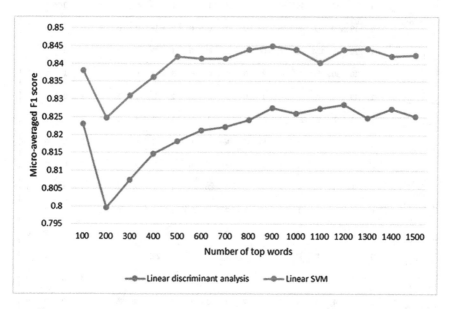

Fig. 5. Micro-averaged F1 score using different number of top words and different classifiers.

In results of SVM classification for sum of top 900 words (Table 1), we can see that the vast majority of the erroneous predictions fall into categories that are very similar to the predicted ones. We can see that categories like rec.sport.* and talk.politics.mideast that are rather different from the rest, were classified with very high precision. On the other hand, we can see that wrong predictions from categories comp.* mostly retained within these, since they are very similar. Also, misclassifications in categories about politics remain roughly within the respective categories. We can see that classification performance for articles in talk.religion.misc category is very poor, although majority of misclassification are classified into very similar categories – alt.atheism and soc. religion.christian, which are also about religion. Our explanation is that although we form better thematic representation by using higher number of top words to compute feature vector of a document, we are limited in differentiation of articles about miscellaneous religion topics, since they probably include also topics discussed in other two more specific categories about religion.

Table 1. Classification results using linear SVM and sum of top 900 word vectors used as a feature vector for document.

	alt.atheism	comp.graphics	comp.os.ms-windows.misc	comp.sys.ibm.pc.hardware	comp.sys.mac.hardware	comp.windows.x	misc.forsale	rec.autos	rec.motorcycles	rec.sport.baseball	rec.sport.hockey	sci.crypt	sci.electronics	sci.med	sci.space	soc.religion.christian	talk.politics.guns	talk.politics.mideast	talk.politics.misc	talk.religion.misc
alt.atheism	183	1	3	1	2	1	0	1	0	2	0	2	5	4	3	20	8	10	9	57
comp.graphics	4	273	34	18	14	58	9	1	3	1	0	1	18	16	4	1	2	0	2	1
comp.os.ms-windows.misc	4	19	211	33	30	49	3	0	0	0	0	2	16	2	0	4	0	0	1	2
comp.sys.ibm.pc.hardware	1	19	35	194	42	7	12	0	0	0	0	1	22	0	0	0	0	0	0	0
comp.sys.mac.hardware	1	10	30	56	202	6	8	0	0	0	0	3	14	0	2	1	0	1	1	1
comp.windows.x	1	19	41	10	7	230	1	0	1	1	0	2	7	1	1	0	1	0	1	1
misc.forsale	3	11	9	17	27	7	314	1	0	2	0	3	9	4	2	1	3	0	3	2
rec.autos	0	3	3	1	2	0	15	357	36	0	0	1	10	3	1	0	2	0	0	3
rec.motorcycles	10	1	3	2	2	2	5	4	321	1	0	2	5	2	1	1	3	0	3	4
rec.sport.baseball	3	3	0	2	3	2	5	0	5	376	6	1	1	2	1	0	1	4	2	0
rec.sport.hockey	0	2	7	1	1	1	1	0	2	10	390	2	3	1	0	1	2	0	0	2
sci.crypt	3	7	6	4	6	4	0	0	2	0	0	321	23	0	1	0	8	3	5	0
sci.electronics	1	12	2	49	32	8	9	15	7	1	1	18	245	5	3	1	2	1	1	1
sci.med	7	1	2	0	1	2	2	1	0	1	0	3	6	338	6	2	4	0	5	8
sci.space	11	1	4	4	5	3	0	2	1	0	0	1	2	2	360	0	2	0	7	9
soc.religion.christian	48	1	0	0	1	1	1	0	0	1	0	0	0	2	0	356	2	1	2	93
talk.politics.guns	8	0	0	0	3	0	3	10	7	1	0	7	2	3	1	2	298	2	90	34
talk.politics.mideast	8	4	1	0	0	1	1	0	0	0	0	3	0	1	0	0	2	344	3	1
talk.politics.misc	14	1	2	0	4	1	0	3	7	0	1	15	2	6	6	0	20	8	170	5
talk.religion.misc	9	1	1	0	1	12	1	1	6	0	1	8	3	4	2	8	4	2	5	27

7 Conclusions

We can see in this paper that word vectors can significantly help in the task of relevant words extraction. We have presented a working method of extracting relevant words based on the feature vectors of words. We evaluated our method in text categorisation and succeeded to achieve state-of-the-art results, which we consider the main contribution of our work.

The method we have proposed in this paper has wide application. Basically, relevant words can be used as a metadata for compact description of a document. Further, such metadata can be used for indexing purposes and can be utilise to ease search in document collections. Perhaps the most active area is web, since it contains ever growing huge amounts of web documents that need to be organised. We have shown that extracted relevant words are good for categorisation, but that also implies that the underlying representation captures the semantics of documents quite accurately.

On web, the application is twofold. In the first case, it can be used on the "Wild Web" to categorise web pages by their topic, e.g. in personalised search [7], or more broadly to assist in regular keyword-based web search [12]. Alternatively, it can be utilised also in some specific domains, not just to help categorise new documents or to facilitate the search using keywords where we are aware of the topical structure of the domain, but even to create the topical structure of the domain from scratch, based solely on the documents in a collection and thus let the data speak itself.

In our future research, we plan to focus on more thorough analysis of vector representation of words for the purpose of text classification, where we have found a great open space for further research. We would also like to research utilisation of vector representation in other tasks, which till now, used manually crafted semantic representations like taxonomies.

Acknowledgement. This work was partially supported by grants No. VG1/0675/11, APVV-0208-10 and it is the partial result of the Research and Development Operational Programme project "University Science Park of STU Bratislava", ITMS 26240220084, co-funded by the European Regional Development Fund.

References

1. Barla, M., Bieliková, M.: On deriving tagsonomies: keyword relations coming from crowd. In: Nguyen, N.T., Kowalczyk, R., Chen, S.-M. (eds.) ICCCI 2009. LNCS, vol. 5796, pp. 309–320. Springer, Heidelberg (2009)
2. Collobert, R., Weston, J.: A unified architecture for natural language processing: deep neural networks with multitask learning. In: Proceedings of the 25th International Conference on Machine Learning, pp. 160–167. ACM (2008)
3. Fara, D.G., Russell, G.: The Routledge Companion to Philosophy of Language, p. 92. Routledge, New York (2013). ISBN: 978-0-203-20696-6
4. Giesbrecht, E.: In search of semantic compositionality in vector spaces. In: Rudolph, S., Dau, F., Kuznetsov, S.O. (eds.) ICCS 2009. LNCS, vol. 5662, pp. 173–184. Springer, Heidelberg (2009)

5. Harris, Z.S.: Distributional structure. Word **10**(23), 146–162 (1954)
6. Hinton, G.E., McClelland, J.L., Rumelhart, D.E.: Distributed representations. In: Rumelhart, D.E., McClelland, J.L. (eds.) Parallel Distributed Processing: Explorations in the Microstructure of Cognition, vol. 1: Foundations, pp. 77–109. MIT Press, Cambridge (1986)
7. Kramár, T., Barla, M., Bieliková, M.: Personalizing search using socially enhanced interest model, built from the stream of user's activity. J. Web Eng. **12**(1–2), 65–92 (2013)
8. Lan, M., Tan, C., Low, H.: Proposing a new term weighting scheme for text categorization. In: Proceedings of the 21st National Conference on Artificial Intelligence, vol. 1, pp. 763–768. AAAI Press (2008)
9. Larochelle, H., Bengio, Y.: Classification using discriminative restricted Boltzmann machines. In: Proceedings of the 25th International Conference on Machine Learning, pp. 536–543. ACM (2008)
10. Li, B., Vogel, C.: Improving multiclass text classification with error-correcting output coding and sub-class partitions. In: Farzindar, A., Kešelj, V. (eds.) Canadian AI 2010. LNCS, vol. 6085, pp. 4–15. Springer, Heidelberg (2010)
11. Van der Maaten, L.J.P., Hinton, G.E.: Visualizing high-dimensional data using t-SNE. J. Mach. Learn. Res. **9**, 2579–2605 (2008)
12. Martinský, L., Návrat, P.: Query formulation improved by suggestions resulting from intermediate web search results. Comput. Inf. Syst. J. **16**(1), 56–73 (2012)
13. Mikolov, T., Sutskever, I., Chen, K., Corrado, G., Dean, J.: Distributed representations of words and phrases and their compositionality. In: Advances in Neural Information Processing Systems 26, pp. 3111–3119. Curran Associates (2013)
14. Mikolov, T., Yih, W., Zweig, G.: Linguistic regularities in continuous space word representations. In: Proceedings of NAACL HLT, pp. 746–751. ACL (2013)
15. Mitchell, J., Lapata, M.: Vector-based models of semantic composition. In: Proceedings of the 46th Annual Meeting of the ACL, pp. 236–244. ACL (2008)
16. Bauer, J., Socher, R., Manning, C.D., Ng, A.Y.: Parsing with compositional vector grammars. In: Proceedings of the 51st Annual Meeting of the Association for Computational Linguistics, pp. 455–465. ACL (2013)
17. Šajgalík, M., Barla, M., Bieliková, M.: From ambiguous words to key-concept extraction. In: Proceedings of the 24th International Workshop on Database and Expert Systems Applications, pp. 63–67. IEEE (2013)
18. Vu, T., Aw, A.T., Zhang, M.: Term extraction through unithood and termhood unification. In: Proceedings of the Third International Joint Conference on NLP, pp. 631–636. ACL (2004)
19. Wang, D., Zhang, H.: Inverse-category-frequency based supervised term weighting scheme for text categorization (2010). arXiv preprint arXiv:1012.2609

Lazy and Eager Relational Learning Using Graph-Kernels

Mathias Verbeke[1]([✉]), Vincent Van Asch[2], Walter Daelemans[2], and Luc De Raedt[1]

[1] Department of Computer Science, KU Leuven, Celestijnenlaan 200A, 3001 Heverlee, Belgium
{mathias.verbeke,luc.deraedt}@cs.kuleuven.be
[2] Department of Linguistics, Universiteit Antwerpen, Lange Winkelstraat 40-42, 2000 Antwerpen, Belgium
{vincent.vanasch,walter.daelemans}@uantwerpen.be

Abstract. Machine learning systems can be distinguished along two dimensions. The first is concerned with whether they deal with a feature based (propositional) or a relational representation; the second with the use of eager or lazy learning techniques. The advantage of relational learning is that it can capture structural information. We compare several machine learning techniques along these two dimensions on a binary sentence classification task (hedge cue detection). In particular, we use SVMs for eager learning, and kNN for lazy learning. Furthermore, we employ kLog, a kernel-based statistical relational learning framework as the relational framework. Within this framework we also contribute a novel lazy relational learning system. Our experiments show that relational learners are particularly good at handling long sentences, because of long distance dependencies.

1 Introduction

Solutions to NLP problems require one to take into account both structural information and domain knowledge. Learning systems have essentially two ways of dealing with such information. First, several methods encode the relational information using a set of (automatically or manually) derived features [20]. This effectively propositionalizes the data after which standard machine learning algorithms apply. The drawback of these techniques is that propositionalization results in information loss. On the other hand, (statistical) relational [17] and graph-based learners use the structural information directly, but are often more complex to use. Today there is a growing interest in the use of such statistical relational learning approaches in NLP and several successes have been reported.

Learning techniques can also be distinguished along another dimension that indicates whether they are *eager* or *lazy*. Eager techniques (such as SVMs) compute a concise model from data, while lazy (or memory-based) learners (MBL) simply store the data and use (a variant) of the famous kNN algorithm to classify unseen data. Today, eager methods are much more popular than memory-based

© Springer International Publishing Switzerland 2014
L. Besacier et al. (Eds.): SLSP 2014, LNAI 8791, pp. 171–184, 2014.
DOI: 10.1007/978-3-319-11397-5_13

ones. Nevertheless, it has been argued [10] that MBL is particularly suited for NLP, since language data contains in addition to regularities, also a lot of sub-regularities and productive exceptions. Consequently, *lazy* learning may identify these subregularities and exceptions, while *eager* learning often discards them as noise. Thus, lazy learning also has some advantages. It has proven to be successful in a wide range tasks in computational linguistics (e.g., for learning of syntactic and semantic dependencies [21]).

The key contributions of this paper are that we evaluate the performance of learning systems on an NLP task (hedge cue detection) alongst these two dimensions: *propositionalized* versus *relational* representations, and *eager* versus *lazy* learning. As part of this investigation we also contribute a novel lazy relational learning technique.

A wide range of *eager* learners have been developed and tailored for NLP tasks. Support vector machines (SVM) [7] are one of the most prominent methods, and will be used here as a representative eager learning method. For *lazy* learning we shall employ a kNN framework.

Our relational framework shall be based on kLog [15], a kernel-based statistical relational learning framework, that uses *graphicalization*; a technique that transforms the relational data into a graph-based representation and employs graph-kernels afterwards. The graphicalization does not result in information loss, and is easily understandable. Furthermore, kLog offers a declarative specification of the domain that supports the use of domain knowledge. Our novel lazy relational memory-based learner is based on the kLog representation, and it employs the kLog kernel to define its similarity measure in a memory-based setting.

Thanks to its focus on relations between abstract objects, graph-based relational learning offers the possibility to model a problem on different levels simultaneously, and provides the user with the possibility to represent the problem at the right level of abstraction. For example, sentence classification can be carried out using instances on the token level, without having to resort to a two-step system in which the first step consists of labeling the tokens and the second step is an aggregation step to reach a prediction on the sentence level. Attributes on a higher level, e.g., sentences, can be predicted on the basis of lower level subgraphs, e.g., sequences of tokens, taking into account the relations in the latter, e.g., the dependency tree. This approach has already proven successful for several tasks in NLP [19,30] and computer vision [1].

Our analysis is performed on a partition of the CoNLL-2010 Shared Task dataset on hedge cue detection, a binary sentence classification task. Sentence classification is particularly interesting for our evaluation as long sentences tend to contain complex dependencies that can be represented with relational representations.

The paper is organized as follows: Sect. 2 gives an overview of related work. In Sect. 3 we review kernel-based relational learning with kLog and introduce a new relational memory-based learner. An empirical analysis on the different aspects of the declarative, relational representation is given in Sect. 4, and discusses the advantages in more depth. Finally, Sect. 5 concludes.

2   Related Work

Since our analysis comprises a comparison of lazy versus eager learning, both in the relational and propositional setting, the discussion of related approaches is structured along these lines. Since the evaluation uses a dataset from the CoNLL-2010 shared task, state-of-the-art approaches for this problem will be discussed as well.

A wide range of *statistical relational learning* (SRL) systems exist [17]. In principle, many of these are useful to solve problems in computational linguistics. The most popular formalism, Markov Logic, has already been used for tasks such as coreference resolution [24]. Most SRL systems are based on a combination of learning and inference techniques from probabilistic graphical models.

The technique that we propose is based on kLog[1] [15], a declarative language for kernel-based logical and relational learning with graphs. kLog has two distinguishing features when compared to Markov Logic. First, it employs kernel-based methods grounded in statistical learning theory. Second, it employs Prolog for defining and using background knowledge. As Prolog is a programming language, this is more flexible than the formalism used by Markov Logic. Furthermore, kLogNLP [28] offers a natural language processing module for kLog that enriches kLog with NLP-specific preprocessors, and enables the use of existing libraries and toolkits within this powerful declarative machine learning framework.

A number of approaches have combined relational and instance-based learning. RIBL [12] is a *relational instance-based learning* algorithm that combines memory-based learning with statistical relational learning. It was extended by Horváth et al. [18] to support representations of lists and terms. Armengol and Plaza [2] introduced Laud; a distance measure that can be used to estimate similarity among relational cases, with Shaud [3] as an improvement that is able to take into account the complete structure provided by the feature terms. Ramon [25] proposes a set of methods to perform IBL using a relational representation, and extends distances and prototypes to more complex objects. To the best of the authors' knowledge, these lazy relational learners have not been applied to natural language processing tasks.

In this paper, the classification problem of the *CoNLL-2010 Shared Task*, hedge cue detection, is tackled in an in-domain, closed manner. Hedge cues are words that indicate speculative language. The goal is identifying if a sentence contains such type of words, thus distinguishing factual from uncertain sentences. Since this task involves analyzing the language beyond its propositional meaning, in addition to the lexico-syntactic features of individual words, also the context of the individual words in the sentence or document plays a more important role. This motivates the use of the graph-based relational representation.

During the shared task, Georgescul [16] obtained the best score for the closed task of in-domain Wikipedia hedge cue detection with a macro-averaged F1-score

---

[1] http://klog.dinfo.unifi.it/

of 75.13 % [2]. This score was obtained despite the fact that the system does not use any intricate feature architecture. Each hedge cue of the training set is taken as a feature and prediction occurs directly at the sentence level. For generalization, the set of hedge cues is extended with n-gram subsets of the cues. In a sense, this system resembles a bag-of-words approach. Interestingly, Georgescul [16] also reports the scores for a simple, but effective baseline algorithm: if a test sentence contains any of the hedge cues occurring in the training corpus, the sentence is labeled as UNCERTAIN. This baseline system obtains a macro-averaged F1-score of 69 %. During the shared task and for the Wikipedia data, only the top 3 is able to do better than baseline on the UNCERTAIN class.

3   Kernel-Based Relational Learning with Graphs

kLog [15] is a logical and relational language for kernel-based learning, that is embedded in Prolog, and builds upon and links together concepts from database theory, logic programming and learning from *interpretations* (i.e., each interpretation is a set of tuples that are true in the example, and can be seen as a small relational database). It is based on a technique called *graphicalization* that transforms relational representations into graph based ones and derives features from a grounded entity/relationship diagram using graph kernels. This leads to an extended high-dimensional feature space on which a statistical learning algorithm can be applied. The general workflow is illustrated in Fig. 1 and will be explained in the following paragraphs.

3.1   Declarative Domain Specification and Feature Construction

Since kLog is rooted in database theory, the modeling of the problem domain is done using an entity-relationship (ER) model [6]. It gives an abstract representation of the interpretations. An example ER model for the hedge cue detection task is given in Fig. 3. This ER model is coded declaratively in kLog using an

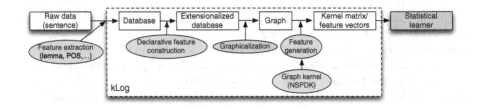

Fig. 1. General kLog workflow.

---

[2] This equals an F1-score on the UNCERTAIN class of 60.17 %, but in this paper we prefer reporting the macro-averaged F1-score because it takes the performance on both class labels into account.

extension of the logic programming language Prolog[3]. Every entity or relationship is declared with the keyword `signature`, which can either be *extensional* or *intensional*. Extensional signatures represent information that is readily available from the input data. An example is the dependency relationship (`depRel`) between two word (`w`) entities, where each relation has its type (`depType`) as a property.

```
signature dependency(word1::w, word2::w, dep_rel::property).
```

On top of these extensional signatures, intensional ones can be defined. In contrast to extensional signatures, intensional signatures introduce novel relations using a mechanism resembling deductive databases. For this type of signatures, due to the declarative nature, no additional preprocessing is required. This type of signatures is mostly used to add domain knowledge about the task at hand. For the hedge cue detection task, the following features provide meaningful additional knowledge [29].

```
signature cw(cw_id::self, lemma::property, pos::property).
cw(CW, L, P) :- w(W,L,P,_,1,_), atomic_concat(c,W,CW).
signature leftof(cw_id::cw, lemma::property, pos::property).
leftof(CW,L,P) :- cw(W,_,_), atomic_concat(c,W,CW),
                  next(W1,W), w(W1,L,P,_,_,_).
```

`cw` retains only the words, together with their respective lemma and pos-tag, that appear in a predefined list of hedge cues that was compiled from the training data. `leftOf`, and a similarly defined predicate `rightOf`, also take the two surrounding words of a `cw` in the sentence into account.

3.2   Relational Feature Generation

Subsequently, these interpretations are *graphicalized*, i.e., transformed into graphs. This can be interpreted as unfolding the ER-diagram over the data. We will now extract features from these graphs using a feature generation technique that is based on Neighborhood Subgraph Pairwise Distance Kernel (NSPDK) [8], a particular type of graph kernel. Informally the idea of this kernel is to decompose a graph into small neighborhood subgraphs of increasing radii $r < r_{max}$. Then, all pairs of such subgraphs whose roots are at a distance not greater than $d < d_{max}$ are considered as individual features. The kernel notion is finally given as the fraction of features in common between two graphs. For the sake of completeness we briefly report the formal definitions.

For a given graph $G = (V, E)$, and an integer $r \geq 0$, let $N_r^v(G)$ denote the subgraph of G rooted in v[4] and induced[5] by the set of vertices $V_r^v \doteq \{x \in V : d(x,v) \leq r\}$, where $d(x,v)$ is the shortest-path distance between x and v. A neighborhood $N_r^v(G)$ is therefore a topological *ball* with center v and radius r.

---

[3] The full relational model and data are available at http://people.cs.kuleuven.be/~mathias.verbeke/klogmbl.html.

[4] A graph is *rooted* when we distinguish one of its vertices as *root*.

[5] In a graph G, the *induced-subgraph* on a set of vertices $W = \{w_1, \ldots, w_k\}$ is a graph that has W as vertex set and contains every edge of G whose endpoints are in W.

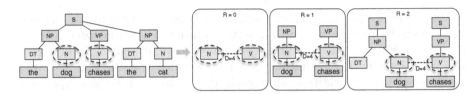

Fig. 2. Illustration of the NSPDK feature concept. Left: two root nodes at distance 4 highlighted; Right: (a subset of) the resulting features for radius 0; radius 1; radius 2.

Formally the relation is defined in terms of neighborhood subgraphs as $R_{r,d} = \{(N_r^v(G), N_r^u(G), G) : d(u,v) = d\}$, that is, a relation $R_{r,d}$ that identifies pairs of neighborhoods of radius r whose roots are exactly at distance d. Finally:

$$\kappa_{r,d}(G, G') = \sum_{\substack{A,B \in R_{r,d}^{-1}(G) \\ A',B' \in R_{r,d}^{-1}(G')}} \mathbf{1}_{A \cong A'} \cdot \mathbf{1}_{B \cong B'} \tag{1}$$

where $R_{r,d}^{-1}(G)$ indicates the multiset of all pairs of neighborhoods of radius r with roots at distance d that exist in G, and where $\mathbf{1}$ denotes the indicator function and $\cong$ the isomorphism between graphs.

The NSPDK graph kernel is illustrated in Fig. 2 for a distance of 4 between two roots of the neighborhood subgraphs and varying radii. In this toy example, the graph kernel takes a graphicalized sentence parse tree as input, and outputs the subgraphs on the right as (a subset of the) resulting features. This yields a high-dimensional feature space that is much richer than most of the other direct propositionalization approaches, as the relations are explicitly encoded. The result is a propositional learning setting, which enables the use of this set of features in any statistical learner.

3.3 Relational Memory-Based Learning

In order to construct a relational memory-based learner, the relational information constructed with kLog and MBL are combined, using the NSPDK graph kernel as a relational distance measure. The similarities between the instances are readily available from the kernel matrix (also known as the *Gram matrix*), which is calculated by the graph kernel, and thus can be exploited efficiently. A kernel K can be easily transformed into a distance *metric*, using $d_K(x, y) = \sqrt{K(x, x) - 2K(x, y) + K(y, y)}$. This will be referred to as *kLog-MBL*. We both employed a regular kNN setup [14], referred to as *kLog-MBL (NW)*, as well as a distance-weighted variant [11], referred to as *kLog-MBL (W)*. In the latter, a neighbor that is close to an unclassified observation is weighted more heavily than the evidence of another neighbor, which is at a greater distance from the unclassified observation.

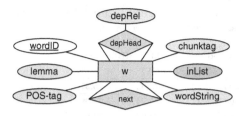

Fig. 3. ER-diagram modeling the hedge cue detection task. Attributes are represented as oval nodes.

Table 1. Number of sentences per class in the training, downsampled training and test partitions (CoNLL-ST 2010, Wikipedia dataset).

Wikipedia	Train	TrainDS	Test
CERTAIN	8,627	2,484	7,400
UNCERTAIN	2,484	2,484	2,234
Total	11,111	4,968	9,634

4   Advantages of the Relational Representation

In order to illustrate the different steps and the respective advantages of the relational representation, we will evaluate our approach on the CoNLL 2010 Shared Task dataset (Sect. 4.1). In Sect. 4.2, we will describe the details of the baseline and benchmarks that were used in the comparison, before turning to an in-depth discussion of the characteristics of the relational representation.

4.1   Dataset

The dataset under consideration consists of sentences from Wikipedia articles that were manually annotated [13]. Due to the collaborative nature of Wikipedia, sentences in this dataset show a very diverse structure. A sentence is considered UNCERTAIN if it contains at least one hedge cue, which is referred to as a *weasel* in the context of Wikipedia. The number of instances per class in the training and test partitions are listed in Table 1.

As can be seen from the table, the data is unbalanced. This can lead to different issues for machine learning algorithms [27]. For MBL, the majority class tends to have more examples in the k-neighbor set, due to which a test instance thus tends to be assigned the majority class label. As a result, the majority class tends to have high classification accuracy, in contrast to a low classification accuracy for the minority class, which affects the total performance and partly obfuscates the influence of the distance measure [26].

Since the goal of this paper is to show the influence of the relational representation and distance measure, we want to reduce the influence of the imbalancedness of the dataset. Several approaches have been proposed to deal with this (i.e., adjusting misclassification costs, learning from the minority class, adjusting the weights of the examples, etc.). One of the two most commonly used techniques to deal with this problem is sampling [5], where the training dataset is resized to compensate for the imbalancedness. We created a downsampled version of the training sets. This was done in terms of the negative examples (the CERTAIN sentences), i.e., we sampled as many negative examples as there are positive examples. We will refer to this dataset as *TrainDS*.

Table 2. General evaluation of the different systems on the CoNLL-2010 ST Wikipedia corpus with downsampled training set. All scores are macro-averaged.

	Baseline	TiMBL (5%)	TiMBL (30%)	SVM	kLog-MBL (NW)	kLog-MBL (W)	kLog-SVM
Precision	11.59	65.65	69.45	73.03	73.02	73.02	75.37
Recall	50.00	67.83	76.76	79.00	75.24	75.24	73.99
F1-score	18.82	50.92	69.34	74.59	73.96	73.96	74.63

4.2 Baseline and Benchmarks

In this paper, we want to examine the behavior of a system that uses a relational, graph-based representation to classify the sentence as a whole (i.e., using the graphicalization process) and contrast it with lazy and eager learning systems that do not use this extra step. For this reason, several baselines and benchmarks are included in the result table (Table 2).

The first, simple baseline is a system that labels all sentences with the UNCERTAIN class. This enables us to compare against a baseline where no information about the observations is used.

The first group of benchmarks consists of systems that operate without relational information. These systems typically use a two-step approach; first the individual words in the sentence are classified, whereafter the target label for the sentence is determined based on the number of tokens that are labeled as hedge cues. This requires an extra parameter to threshold this number of individual tokens from which the sentence label is derived (i.e., if more than $X\%$ of the token-level instances are marked as being a hedge cue, the sentence is marked as UNCERTAIN). To optimize this parameter, the training set was split in a reduced training set and a validation set (70/30% split). The influence of this parameter is discussed in more detail in Sect. 4.5.

The Tilburg Memory-based Learner[6] (TiMBL), a software package implementing several memory-based learning algorithms, among which IB1-IG, an implementation of k-nearest neighbor classification with feature weighting suitable for symbolic feature spaces, and IGTree, a decision-tree approximation of IB1-IG. We will use it in the same setup and with the same feature set as Morante et al. [22]. They used a 5% percentage threshold for sentences, however, our optimization procedure yielded better results with a 30% threshold. In the result table, these variants are referred to as *TiMBL (5%)* and *TiMBL (30%)*.

In order to parameterize TiMBL for the word classification, we used paramsearch[7] [4], a wrapped progressive sampling approach for algorithmic parameter optimization for TiMBL. The IB1 algorithm was chosen as optimal setting, which is the standard MBL algorithm in TiMBL.

---

[6] http://ilk.uvt.nl/timbl/

[7] http://ilk.uvt.nl/paramsearch/

The same is done with SVMs[8]. In a first step the (non-graphicalized) representation of the) data is converted into binary feature vectors[9]. Subsequently, the SVMs are optimized in terms of the cost parameter C using a grid search with 10-fold cross-validation on the reduced training set. Hereafter the percentage threshold was optimized on the validation set. The SVM without relational information is referred to as *SVM* in the result tables.

We contrasted these systems with a lazy and eager learning approach that use the graph-based relational representation from kLog. For *kLog-SVM*, we used an SVM as statistical learner at the end of the kLog workflow. The results were obtained using the model and graph kernel hyperparameter settings from our previous work [29], for which the cost parameter of the SVM was optimized using cross-validation on the downsampled training set.

The second pair of relational systems use memory-based learning, as discussed in Sect. 3.3. The value of k was optimized using the reduced training and validation set as discussed above.

4.3 Performance

Table 2 contains the macro-averaged F1-scores of these seven systems. Looking at the tables, one may conclude that all systems perform better than the UNCERTAIN baseline.

The systems are best compared in a pairwise manner. A first interesting observation is that the memory-based learners that use a relational representation (i.e., *kLog-MBL*), perform significantly better than those that use a relational approach, i.e., the *TiMBL* systems. The weighted and unweighted variants of *kLog-MBL* score equally well.

For the SVM setups, *SVM* and *kLog-SVM* score equally well. At first sight, the relational representation does not seem to add much to the performance of the learner. However, when comparing the relational approach on the full (unbalanced) dataset to an *SVM* using the propositional representation, *kLog-SVM* performs significantly better (Table 3). Furthermore, when comparing the results of the regular *SVM* on the balanced and full dataset, a decrease in performance is observed, which is not present for the *kLog-SVM* setup. This indicates that the relational representation increases the generalization power of the learner. We also compared the *kLog-SVM* setup to the best scores obtained during the CoNLL 2010 Shared Task, *viz.* Georgescul [16], in which *kLog-SVM* is able to outperform the state-of-the-art system. This can be attributed to the relational representation, which offers the possibility to model the sentence as a whole and perform the classification in a single step (i.e., avoiding the need for a two step approach where first token-based classification is performed followed by a thresholding step to obtain the sentence-level classification). We will study this effect in more detail in Sect. 4.5. In addition, the relational representation

---

[8] We used the implementation from scikit-learn [23].

[9] The exact implementation is available at http://www.cnts.ua.ac.be/~vincent/scripts/binarize.py.

Table 3. Comparison of kLog-SVM with the state-of-the-art results and an SVM using the propositional representation on the full dataset. All scores are macro-averaged.

	Georgescul	SVM	kLog-SVM
Precision	79.29	81.59	77.27
Recall	72.80	68.96	74.16
F1-score	75.13	72.17	75.48

Table 4. Relational feature ranking.

#	Feature	Score	Triplet	Score
1	CW	27.20	cw-next-cw	49.79
2	RightOf	6.69	cw-dh-cw	49.73
3	LeftOf	4.30	cw-dh-word	37.29
4	Next	4.02	cw-next-word	31.79
5	DH	3.42	word-dh-word	12.62
6	WString	−2.35	word-next-word	−5.76
6	InList	−2.35		
6	Chunk	−2.35		
6	Lemma	−2.35		
6	PoS	−2.35		

is able to model the relations between the words in the sentence explicitly. The graph kernel thus seems to provide a good way to translate the context of the words in a sentence.

4.4 Relational Regularization and Feature Ranking

The importance of the relational features can now be estimated using kLog's relational regularization and feature ranking methods [9], which lift regularization and feature selection to a relational level. The techniques use the relational structure and topology of the domain. Based on a notion of locality, relevant features in the ER-model are tied together. It enables to get deeper insights into the relative importance of the elements in the ER model of the domain. As the added declarative features (CW, LeftOf and RightOf) showed a clear improvement in the results, it is to be expected that these relational features are the main discriminative predicates, while the propositional lexico-syntactic features should be less informative. When measured for kLog-SVM on the full dataset, the results in Table 4 are obtained. In addition, the method also enables to rank the importance of predicate triplets. The right hand side of Table 4 lists the ranking of the possible triplets of predicates of type entity - relationship - entity. Pairs of consecutive hedged words and hedged words that are linked by a dependency relation are clearly very informative relational features.

4.5 Level of Abstraction

The graph-based representation has the advantage that attributes on a higher level, e.g., sentences, can be predicted on the basis of lower level subgraphs, e.g., tokens. It furthermore enables taking into account the relations in the latter, e.g., the dependency tree. This leads to a one-step classification, without the need for an additional thresholding parameter to go from the lower-level classification (e.g., the classification of the individual tokens) to the higher level (e.g., the sentences). The goal of this section is to show when sentence-based systems are more fit for the task than token-based systems.

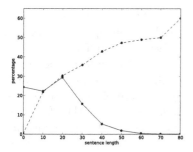

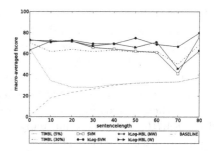

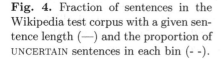

Fig. 4. Fraction of sentences in the Wikipedia test corpus with a given sentence length (—) and the proportion of UNCERTAIN sentences in each bin (- -).

Fig. 5. Macro-averaged F1-score as a function of sentence length, expressed in number of tokens.

The baseline system predicts only one type of class label, namely the minority class. The other systems label sentences with both labels and apart from the observation that one system is more inclined to assign the CERTAIN label than the other, the general scores are not of much help to get more fundamental insights. For this reason, an extra dimension is introduced, namely sentence length. It is an intuitive dimension and other dimensions, like the number of uncertainty cues, are indirectly linked to the sentence length. Figure 5 shows the *evolution* of the macro-averaged F1-score when the sentences to be labeled contain more tokens. To create this figure, the sentences are distributed over 9 bins centered on the multiples of 10. The last bin contains all larger sentences.

Figure 4 shows the fraction of the corpus that is included in each bin (solid line) and the fraction of sentences in each bin that is labeled as UNCERTAIN (dashed line). There are fewer long sentences and long sentences tend to be labeled as UNCERTAIN. As a sanity check, we can look at the behavior of the baseline system in Fig. 5. The observation of an increasing number of UNCERTAIN sentences with increasing sentence length (Fig. 4) is consistent with the increasing F1-score for the baseline system in Fig. 5.

A more interesting observation is the curve of *TiMBL (5 %)*, that quickly joins the baseline curve in Fig. 5. Although this system performs significantly better than the baseline system, it behaves like baseline systems for longer sentences. Because a large fraction of the sentences is short, this undesirable behavior is not readily noticeable when examining the scores of Table 2. Optimizing the threshold can be a solution to this problem. Changing the threshold influences the chances of a sentence being labeled as UNCERTAIN depending on the sentence length. Increasing the threshold leads to a more unequal distribution of the chances over sentence length. As a result, the behavior of the optimized TiMBL system is more stable with varying sentence length (see *TiMBL (30 %)*).

The token-based systems (*SVM* and *TiMBL (30 %)*) behave very similar after optimization of the threshold. Indeed, the *SVM* and optimized *TiMBL* curves follow more or less the same course; a course that is different from the other systems. This indicates that by using a two-step approach, the choice of the classifier is of a lesser importance. Although to a more limited extent, this behaviour is also noticeable for *kLog-MBL* in the case of longer sentences. However, when contrasting the kLog-based systems, the curves of the *kLog-SVM* and *kLog-MBL* systems are mutually divergent for longer sentences, indicating the importance of the classifier.

The importance of the threshold parameter for the propositional, two-step approaches (*TiMBL (5 %)*, *TiMBL (30 %)* and *SVM*) may be an argument to opt for relational systems. Indeed, the threshold is an extra parameter that has to be learned during training and may introduce errors because of its rigidness. It is the same fixed value for all sentences and it weakens the, possibly positive, influence of the classifier. The kLog-based systems do not require such a threshold and are thus able to *dynamically* look for the best prediction on sentence level using the dependencies between the separate tokens.

The claim that dynamically looking for the best prediction on sentence level is better, is based on the observation that, in general, the *kLog*-based systems perform better than their non-relational counterparts. For the dataset under consideration, the *SVM* system performs not significantly different in F1 than the *kLog-SVM* system, but if we look at their behavior in Fig. 5 we see that for almost all sentence lengths *kLog-SVM* performs better. Furthermore, as shown in Sect. 4.3, *kLog-SVM* generalizes better to the unbalanced version of the dataset when compared to *SVM*, and also obtains the most stable predictions across all sentence lengths.

5   Conclusions

We have used the task of hedge cue detection to evaluate several types of machine learning systems along two dimensions. The results show that relational representations are useful, especially for dealing with long sentences and capturing complex dependencies amongst constituents. The relational representation also allows for one-step classification, without the need for an additional thresholding parameter to go from word to sentence level predictions. We have shown that the kLog framework can be used in both an eager SVM type of learner and a lazy or memory-based learning framework. Especially useful for natural language is that its declarative representation offers a flexible experimentation approach and more interpretable results. In future work we will investigate a hybrid approach that combines lazy and eager learning in the relational case.

Acknowledgments. This research is funded by the Research Foundation Flanders (FWO project G.0478.10 - Statistical Relational Learning of Natural Language).

References

1. Antanas, L., Frasconi, P., Costa, F., Tuytelaars, T., De Raedt, L.: A relational kernel-based framework for hierarchical image understanding. In: Gimel'farb, G., Hancock, E., Imiya, A., Kuijper, A., Kudo, M., Omachi, S., Windeatt, T., Yamada, K. (eds.) SSPR & SPR 2012. LNCS, vol. 7626, pp. 171–180. Springer, Heidelberg (2012)
2. Armengol, E., Plaza, E.: Similarity assessment for relational CBR. In: Proceedings of the 4th International Conference on Case-Based Reasoning, pp. 44–58. Springer, London (2001)
3. Armengol, E., Plaza, E.: Relational case-based reasoning for carcinogenic activity prediction. Artif. Intell. Rev. **20**(1–2), 121–141 (2003)
4. van den Bosch, A.: Wrapped progressive sampling search for optimizing learning algorithm parameters. In: Proceedings of the Sixteenth Belgian-Dutch Conference on Artificial Intelligence, pp. 219–226 (2004)
5. Chawla, N.V.: Data mining for imbalanced datasets: an overview. In: Maimon, O., Rokach, L. (eds.) Data Mining and Knowledge Discovery Handbook, pp. 875–886. Springer, New York (2010)
6. Chen, P.P.S.: The entity-relationship model - toward a unified view of data. ACM Trans. Database Syst. **1**(1), 9–36 (1976)
7. Cortes, C., Vapnik, V.: Support-vector networks. MLJ **20**(3), 273–297 (1995)
8. Costa, F., De Grave, K.: Fast neighborhood subgraph pairwise distance kernel. In: Proceedings of the 26th International Conference on Machine Learning, pp. 255–262. Omnipress (2010)
9. Costa, F., Verbeke, M., De Raedt, L.: Relational regularization and feature ranking. In: Proceedings of the 14th SIAM International Conference on Data Mining (2014)
10. Daelemans, W., van den Bosch, A.: Memory-Based Language Processing, 1st edn. Cambridge University Press, New York (2009)
11. Dudani, S.A.: The distance-weighted k-nearest-neighbor rule. IEEE Trans. Syst. Man Cybern. **SMC–6**(4), 325–327 (1976)
12. Emde, W., Wettschereck, D.: Relational instance-based learning. In: ICML, vol. 96, pp. 122–130 (1996)
13. Farkas, R., Vincze, V., Móra, G., Csirik, J., Szarvas, G.: The CoNLL-2010 shared task: learning to detect hedges and their scope in natural language text. In: Proceedings of the Conference on Computational Natural Language Learning – Shared Task, pp. 1–12. ACL, Stroudsburg (2010)
14. Fix Jr., E.: Discriminatory analysis: nonparametric discrimination: consistency properties. Technical report 4, USAF School of Aviation Medicine, TX, USA (1951)
15. Frasconi, P., Costa, F., De Raedt, L., De Grave, K.: kLog: a language for logical and relational learning with kernels. CoRR abs/1205.3981 (2012)
16. Georgescul, M.: A hedgehop over a max-margin framework using hedge cues. In: Proceedings of the Conference on Computational Natural Language Learning, pp. 26–31. ACL, Uppsala (2010)
17. Getoor, L., Taskar, B.: Introduction to Statistical Relational Learning. MIT, Cambridge (2007)
18. Horváth, T., Wrobel, S., Bohnebeck, U.: Relational instance-based learning with lists and terms. MLJ **43**(1–2), 53–80 (2001)
19. Kordjamshidi, P., Frasconi, P., Van Otterlo, M., Moens, M.-F., De Raedt, L.: Relational learning for spatial relation extraction from natural language. In: Muggleton, S.H., Tamaddoni-Nezhad, A., Lisi, F.A. (eds.) ILP 2011. LNCS, vol. 7207, pp. 204–220. Springer, Heidelberg (2012)

20. Kramer, S., Lavrač, N., Flach, P.: Propositionalization approaches to relational data mining. In: Dězeroski, S. (ed.) Relational Data Mining, pp. 262–286. Springer-Verlag New York Inc., New York (2000)
21. Morante, R., Van Asch, V., van den Bosch, A.: Joint memory-based learning of syntactic and semantic dependencies in multiple languages. In: Stevenson, S., Carreras, X., Hajič, J. (eds.) Proceedings of the Thirteenth Conference on Computational Natural Language Learning (CoNLL) - Shared Task. ACL, Boulder (2009)
22. Morante, R., Van Asch, V., Daelemans, W.: Memory-based resolution of in-sentence scopes of hedge cues. In: Proceedings of the Conference on Computational Natural Language Learning – Shared Task, pp. 40–47. ACL, Stroudsburg (2010)
23. Pedregosa, F., Varoquaux, G., Gramfort, A., Michel, V., Thirion, B., Grisel, O., Blondel, M., Prettenhofer, P., Weiss, R., Dubourg, V., Vanderplas, J., Passos, A., Cournapeau, D., Brucher, M., Perrot, M., Duchesnay, E.: Scikit-learn: machine learning in Python. J. Mach. Learn. Res. **12**, 2825–2830 (2011)
24. Poon, H., Domingos, P.: Joint unsupervised coreference resolution with Markov Logic. In: Proceedings of the Conference on Empirical Methods in Natural Language Processing, pp. 650–659. ACL (2008)
25. Ramon, J.: Conceptuele clustering en instance based leren in eerste orde logica. Ph.D. thesis, Informatics Section, Dept. of Computer Science (2002)
26. Tan, S.: Neighbor-weighted k-nearest neighbor for unbalanced text corpus. Expert Syst. Appl. **28**(4), 667–671 (2005)
27. Van Hulse, J., Khoshgoftaar, T.M., Napolitano, A.: Experimental perspectives on learning from imbalanced data. In: Proceedings of the 24th International Conference on Machine learning, pp. 935–942. ACM, New York (2007)
28. Verbeke, M., Frasconi, P., De Grave, K., Costa, F., De Raedt, L.: kLogNLP: graph kernel-based relational learning of natural language. In: Bontcheva, K., Jingbo, Z. (eds.) Proceedings of the 52nd Annual Meeting of the Association for Computational Linguistics: System Demonstrations, Annual Meeting of the Association for Computational Lingusitics (ACL), Baltimore, Maryland, USA, 22–27 June 2014, pp. 85–90. Association for Computational Linguistics (2014)
29. Verbeke, M., Frasconi, P., Van Asch, V., Morante, R., Daelemans, W., De Raedt, L.: Kernel-based logical and relational learning with kLog for hedge cue detection. In: Muggleton, S.H., Tamaddoni-Nezhad, A., Lisi, F.A. (eds.) ILP 2011. LNCS, vol. 7207, pp. 347–357. Springer, Heidelberg (2012)
30. Verbeke, M., Van Asch, V., Morante, R., Frasconi, P., Daelemans, W., De Raedt, L.: A statistical relational learning approach to identifying evidence based medicine categories. In: Proceedings of the 2012 Joint Conference on Empirical Methods in Natural Language Processing and Computational Natural Language Learning, Joint Conference on Empirical Methods in Natural Language Processing and Computational Natural Language Learning (EMNLP - CoNLL), Jeju Island, Korea, 13–14 July 2012 (2012)

Linear Co-occurrence Rate Networks (L-CRNs) for Sequence Labeling

Zhemin Zhu$^{(\boxtimes)}$, Djoerd Hiemstra, and Peter Apers

Electrical Engineering, Mathematics and Computer Science, University of Twente,
Drienerlolaan 5, 7500 AE Enschede, The Netherlands
{z.zhu,d.hiemstra,p.m.g.apers}@utwente.nl

Abstract. Sequence labeling has wide applications in natural language processing and speech processing. Popular sequence labeling models suffer from some known problems. Hidden Markov models (HMMs) are generative models and they cannot encode transition features; Conditional Markov models (CMMs) suffer from the label bias problem; And training of conditional random fields (CRFs) can be expensive. In this paper, we propose Linear Co-occurrence Rate Networks (L-CRNs) for sequence labeling which avoid the mentioned problems with existing models. The factors of L-CRNs can be locally normalized and trained separately, which leads to a simple and efficient training method. Experimental results on real-world natural language processing data sets show that L-CRNs reduce the training time by orders of magnitudes while achieve very competitive results to CRFs.

Keywords: Sequence labeling · Co-occurrence rate · HMMs · CRFs

1 Introduction

Sequence labeling is a sub-task of structured prediction. A wide range of fundamental applications in natural language processing and speech processing can be formulated as sequence labeling models, such as named entity recognition, part-of-speech tagging and speech recognition. A common nature of these applications is that these applications desire a sequence of labels as output rather than a single label. This makes sequence labeling stand out from the typical supervised classification tasks which normally predict a single label as output. Here we give a simplified example of named entity recognition (NER) to illustrate the typical scenario of sequence labeling. Given a sentence, which consists of a sequence of words, NER systems assign each word of the sentence a label. These labels indicate the types of named entities, such as location (LOC), person (PER), organization (ORG), or out of any named entity (O).

$[\text{Jimmy}]_\text{PER}$ $[\text{de}]_\text{PER}$ $[\text{Graff}]_\text{PER}$ $[\text{is}]_\text{O}$ $[\text{a}]_\text{O}$ $[\text{member}]_\text{O}$ $[\text{of}]_\text{O}$ $[\text{the}]_\text{O}$ $[\text{Dutch}]_\text{ORG}$ $[\text{National}]_\text{ORG}$ $[\text{Research}]_\text{ORG}$ $[\text{School}]_\text{ORG}$ $[\text{for}]_\text{ORG}$ $[\text{Knowledge}]_\text{ORG}$ $[\text{Systems}]_\text{ORG}$.

Our C++ implementation of L-CRNs and the datasets used in this paper can be found at https://github.com/zheminzhu/Co-occurrence-Rate-Networks.

© Springer International Publishing Switzerland 2014
L. Besacier et al. (Eds.): SLSP 2014, LNAI 8791, pp. 185–196, 2014.
DOI: 10.1007/978-3-319-11397-5_14

The words in a sentence are observations. From this example, we can see that intuitively two kinds of information can affect the prediction of the label at current position:

1. Label dependence. Adjacent labels can affect prediction of the current label. For example, if the adjacent labels are ORG, the current label is more likely to be ORG.
2. Observation evidence. The current word observed can affect the current label. For example, the word Dutch is more likely to be ORG than the word is.

Accordingly, a sequence labeling model should do the following three tasks well.

- Task 1. Modeling label dependence.
- Task 2. Modeling observation evidence.
- Task 3. Combining these two parts to obtain results.

Task 3 has been paid less attention. The two parts should be given relative weights properly when we combine them. As we will discuss, failure in doing this will lead to a subtle problem called the label bias problem [11], in which label dependence is given too much weight and observation evidence is underestimated or even ignored.

Due to its wide applications, sequence labeling has been heavily studied for a long history. There exist a rich set of popular models for sequence labeling, such as hidden Markov models (HMMs) [14], conditional Markov models (CMMs) [13] and conditional random fields (CRFs) [11].[1] The general idea under all of these models is factorization. That is to decompose a high-dimensional joint probability into a product of small factors based on some conditional independence assumptions. A model is characterized by its factorization. Hence we can see the pros and cons of a model from its factorization.

1.1   Hidden Markov Models (HMMs)

Figure 1 shows a first order HMM. $S = [s_1, s_2, \ldots, s_n]$ is the label sequence and $O = [o_1, o_2, \ldots, o_n]$ is the observation sequence. In the NER example, S is the sequence of NER labels and O is the sequence of words. HMMs are directed

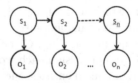

Fig. 1. Hidden Markov models

---

[1] Another popular model is structured (structural) SVM [1] which essentially applies factorization to kernels. Due to its lack of a direct probabilistic interpretation, we leave it for future work.

and generative models. Hence HMMs can also be considered as a special Bayesian network [6]. HMMs factorize a joint probability as follows:

$$p(S, O) \approx p(s_1) \prod_{i=1}^{n} p(o_i|s_i) \prod_{j=1}^{n-1} p(s_{j+1}|s_j). \tag{1}$$

The factors of HMM are probabilities which can be locally normalized. There are two known drawbacks with HMMs [4] which can be observed from Eq. 1. The first drawback is the label transition probabilities $p(s_{j+1}|s_j)$ in HMMs are not conditioned by observations. That is, HMMs use the universal transition probabilities $p(s_{j+1}|s_j)$ without respect to observations. Hence we cannot use observation evidence to help predicting label transition probabilities. Transition features extracted from observation evidence contain valuable information. The second drawback is called mismatch problem. In training stage, HMMs optimize a joint probability $p(S, O)$. But in decoding stage, we search for a sequence of labels which maximizes a conditional probability $p(S|O)$. Klein et al. [9] show that the mismatch problem can reduce accuracy.

To avoid the mismatch problem, we need to directly factorize the conditional probability $p(S|O)$. And in order to encode the transition features, we can set observation evidence to conditions of the transition factors. Conditional Markov models just implement these ideas.

1.2   Conditional Markov Models (CMMs)

Figure 2 shows a CMM. Maximum entropy markov models (MEMMs) [13] are typical CMMs which train the model using a maximum entropy framework, which was later shown to be equivalent to maximum likelihood estimation. CMMs are discriminative models which factorize a conditional probability:

$$p(S|O) = p(s_1|O) \prod_{i=1}^{n-1} p(s_{i+1}|s_i, O). \tag{2}$$

CMMs avoid the mismatch problem of HMMs because they directly factorize $P(S|O)$. And probabilities $p(s_{i+1}|s_i, O)$ predicting the next label are conditioned by previous label s_i together with the observation O. In this way, the transition features can be encoded into CMMs. Hence the first drawback of HMMs is avoided. But this causes a new problem. By putting the previous label s_i

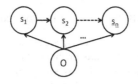

Fig. 2. Conditional Markov models

and observations O together in the condition leads to the label bias problem (LBP). Intuitively, this is because the label dependence (given by s_i) and observation evidence (given by O) are mixed together in one factor. One of them may dominate the factor when its distribution is of low entropy, while the other is underestimated or even ignored. An extreme case is when s_i has only one possible out-going transition s_{i+1}[2], then $p(s_{i+1}|s_i, O)$ is always equal to 1 no matter what O is. That is the observation evidence O is ignored and the label dependence dominates the results. Hence CMMs do not perform the Task 3 perfectly. See [11,12,19] for more examples and discussions.

To avoid the label bias problem, we need to guarantee that the observation evidence can always be used in prediction[3]. This can be done by decoupling the label dependence and observation evidence into different factors, such that none of them can dominate the other. Conditional random fields implement this.

1.3   Conditional Random Fields (CRFs)

Figure 3 shows a linear-chain conditional random field. CRFs [11] are discriminative and undirected graphical models. The factorization for undirected models is based on the Hemmersley-Clifford Theorem [7] which implies a linear-chain CRF can be factorized as follows:

$$p(S|O) = \frac{1}{Z_O} \prod_{i=1}^{n-1} \psi(s_i, s_{i+1}, O) \prod_{j=1}^{n} \phi(s_j, O),$$

Z_O is a global normalization constant, also called partition function, which ensures $\sum_S p(S|O) = 1$. ψ and ϕ are non-negative factors defined over pairwise and unary cliques. The factors of local models, such as HMMs and CMMs, are probabilities. These factors can be locally normalized. By contrast, CRFs are globally normalized models. The factors of CRFs, ψ and ϕ, have no probabilistic interpretations[4] and cannot be locally normalized.

CRFs model the conditional probability $P(S|O)$. Hence they avoid the mismatch problem of HMMs. Also the bigram factors $\psi(s_i, s_{i+1}, O)$ modeling label

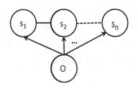

Fig. 3. Conditional random fields

---

[2] In this extreme case, the entropy of $p(s_{i+1}|s_i)$ is the lowest: 0.

[3] HMMs do not suffer from the label bias problem, because the factors $p(o_i|s_i)$ in Eq. 1 guarantee that the observation evidence is always used.

[4] Sometimes they are intuitively explained as the compatibility of the nodes in cliques. But the notion compatibility has no mathematical definition.

dependence include the observations. Hence the transition features can be encoded into CRFs. Furthermore, CRFs decouple the label dependence (modeled by $\psi(s_i, s_{i+1}, O)$) and observation evidence (modeled by $\phi(s_j, O)$) into different factors. This guarantees that none of them can dominate the other. Obviously, unigram factors $\phi(s_j, O)$ guarantee that O is always used for prediction. Therefore the label bias problem is avoided. Nevertheless, training of CRFs can be very expensive [4, 16]. This is because we need to re-calculate the global partition function Z_O for each instance in each optimization iteration.

In this paper, we propose a model called Linear Co-occurrence Rate Networks (L-CRN) for sequence labeling. L-CRNs avoid the problems mentioned above. More specifically, L-CRNs model a conditional probability. Hence they avoid the mismatch problem of HMMs. The label dependence is modeled by the quantity called Co-occurrence Rate (CR), which is conditioned by observations. In this way, transition features can be easily encoded into L-CRNs. Furthermore, in the factorization of L-CRNs, the label dependence and observation evidence are decoupled into difference factors. Thus none of them can dominate the other. The label bias problem is naturally avoided. Finally, L-CRNs are local models. The factors of L-CRNs can be locally normalized and trained separately. This leads to a very efficient maximum likelihood training method. Experiments on real-world datasets show that L-CRNs reduce the training time by orders of magnitudes and achieve very competitive, even slightly better, results to CRFs.

The rest of this paper is organized as follows. In Sect. 2, we present the co-occurrence rate networks and show that this model avoids the problems mentioned above. Section 3 describes the details of learning and decoding. Experiments are reported in Sect. 4. Conclusions follow in the last section.

2   Linear Co-occurrence Rate Networks (L-CRN)

Firstly, we define a quantity which is called Co-occurrence Rate (CR) as follows:

$$\mathrm{CR}(X_1; X_2; \ldots; X_n) := \frac{p(X_1, \ldots, X_n)}{p(X_1) \ldots p(X_n)}.$$

For convenience, CR with a single variable is defined to be 1. Intuitively, if $\mathrm{CR} > 1$, the events are *attractive*; If $\mathrm{CR} = 1$, the events are *independent*; And if $\mathrm{CR} < 1$, the events are *repulsive*. CR is the exponential function of pointwise mutual information [3], and also related to Copulas [21]. Furthermore, we distinguish the following two notations:

$$\mathrm{CR}(X_1; X_2; X_3) := \frac{p(X_1, X_2, X_3)}{p(X_1)p(X_2)p(X_3)}, \quad \mathrm{CR}(X_1 X_2; X_3) := \frac{p(X_1, X_2, X_3)}{p(X_1, X_2)p(X_3)}.$$

The first one is the CR between three variables. By contrast, the second one is the CR between a joint variable $(X_1 X_2)$ and a single variable (X_3). More comprehensive description of CR can be found in [18, 20]. The factorization of L-CRN consists of steps:

1. Decouple the conditional probability into two parts:

$$p(s_1, \ldots, s_n | O) = CR(s_1; s_2; \ldots; s_n | O) \prod_{i=1}^{n} p(s_i | O),$$

where
$$CR(s_1; s_2; \ldots; s_n | O) := \frac{p(s_1, \ldots, s_n | O)}{\prod_{i=1}^{n} p(s_i | O)}.$$

This step seems trivial. But this is the key to avoid the label bias problem. The conditional probability is decoupled into two parts: $CR(s_1; s_2; \ldots; s_n | O)$ models label dependence and $\prod_{i=1}^{n} p(s_i | O)$ models observation evidence. So none of them can dominate the other. $\prod_{i=1}^{n} p(s_i | O)$ guarantees that the observation O is always used for prediction. Hence the label bias problem is naturally avoided. We show this experimentally in [19].

2. Further factorize the joint CR into a product of smaller CRs according to Theorems 1 and 2. See Sect. 6.2 for their proofs.

Theorem 1 (Partition Operation). $CR(X_1; \ldots; X_j; X_{j+1}; \ldots; X_n) = CR(X_1; ..; X_j)CR(X_{j+1}; ..; X_n)CR(X_1..X_j; X_{j+1}..X_n)$

Theorem 2 (Reduce Operation). *If $X \perp\!\!\!\perp Y \mid Z$, then $CR(X; YZ) = CR(X; Z)$.*

$X \perp\!\!\!\perp Y \mid Z$ means X is independent of Y conditioned by Z. Putting two steps together, the factorization of L-CRN is obtained as follow:

$$p(s_1, s_2, \ldots, s_n \mid O) = CR(s_1; \ldots; s_n \mid O) \prod_{i=1}^{n} p(s_i \mid O)$$

$$= CR(s_1|O)CR(s_2; \ldots; s_n|O)CR(s_1; s_2 \ldots s_n|O) \prod_{i=1}^{n} p(s_i|O)$$

$$= CR(s_2; \ldots; s_n \mid O)CR(s_1; s_2 \mid O) \prod_{i=1}^{n} p(s_i \mid O)$$

$$\ldots$$

$$= \prod_{j=1}^{n-1} CR(s_j; s_{j+1} \mid O) \prod_{i=1}^{n} p(s_i \mid O).$$

The second equation is obtained by partitioning s_1 out. We obtain the third equation from the second by $CR(s_1 \mid O) = 1$ and applying the reduce operation to the factor $CR(s_1; s_2 \ldots s_n | O)$ since $s_1 \perp\!\!\!\perp s_3 \ldots s_n | s_2$. By repeating this process, we can get the final factorization. Hence we obtain a L-CRN factorization on a chain graph as follows:

$$p(s_1, s_2, \ldots, s_n | O) = \prod_{j=1}^{n-1} CR(s_j; s_{j+1} \mid O) \prod_{i=1}^{n} p(s_i \mid O). \tag{3}$$

From this factorization, we can see the following facts. L-CRNs model a conditional probability. Hence they avoid the mismatch problem of HMMs. The label dependence of L-CRNs is modeled by $\prod_{j=1}^{n-1} CR(s_j; s_{j+1}|O)$, which is conditioned by observations. Thus transition features can be easily encoded into L-CRNs. Furthermore, $\prod_{i=1}^{n} p(s_i \mid O)$ guarantee the observation is always used for prediction. Therefore the label bias problem is avoided. Finally, L-CRNs are local models. The factors of L-CRNs can be locally normalized and hence separately trained. This leads to a very simple and efficient maximum likelihood training as described in the next section. [2] shows local models can outperform globally normalized models on some NLP tasks. CR factorization can be extended to arbitrary graphs (Sect. 6.2 of [18]).

3 Learning and Decoding

Since the factors of L-CRNs can be normalized locally and trained separately, the learning of L-CRNs becomes very simple and efficient. It is no more than training a set of regression models for each factor in training stage, and combining them together to find a maximum sequence probability in decoding stage. Accroding to Eq. 3, there are two kinds of factors to be trained: unigram factors $p(s|O)$ and bigram factors $CR(s; s'|O)$. We describe the details as follows. See Sect. 6.1 for the justification of this training method. In fact, this is the maximum likelihood estimation of Eq. 3.

3.1 Learning Unigram Factor $p(s|O)$

For the factors $p(s|O)$ in Eq. 3, where s is a label and O is an observation. As described in Sect. 6.1, its MLE is just the relative frequency $\hat{p}(s|O) = \frac{\#(s,O)}{\sum_s \#(s,O)}$, where $\#(s, O)$ is the number of times (s, O) appears in the training dataset. This relative frequency can be easily obtained from the training dataset by counting. Let $g_1(O), g_2(O), \ldots, g_n(O)$ be feature functions of O, which are called unigram features with respect to the unigram label s. For each label s in the label space, we train a regression model ϕ_s:

$$\hat{p}(s|O) = \phi_s : (g_1(O), g_2(O), \ldots, g_n(O)) \mapsto \frac{\#(s, O)}{\sum_s \#(s, O)}.$$

In decoding, for an observation O, we use $\phi_s(g_1(O), \ldots, g_n(O))$ as the estimation of $p(s|O)$. If $g_1(O), g_2(O), \ldots, g_n(O)$ has been seen in the training dataset, we just use $\frac{\#(s,O)}{\sum_s \#(s,O)}$ as the estimation of $p(s|O)$. Because this is the MLE of $p(s|O)$ (see Sect. 6.1). Otherwise, ϕ_s is used.

3.2   Learning Bigram Factor $CR(s; s'|O)$

Similarly, we train regression models $\psi_{s,s'}$ separately for each bigram label s, s' for predicting :

$$\hat{CR}(s; s'|O) = \psi_{s,s'} : (h_1(O), h_2(O), \ldots, h_m(O))$$
$$\mapsto \frac{\#(s, s', O)}{\sum_{s,s'} \#(s, s', O)} \bigg/ \frac{\#(s, O)}{\sum_s \#(s, O)} \bigg/ \frac{\#(s', O)}{\sum_{s'} \#(s', O)}.$$

$h_1(O), h_2(O), \ldots, h_m(O)$ are bigram features extracted from O. Similarly, in decoding, $\psi_{s,s'}(h_1(O), h_2(O), \ldots, h_m(O))$ are used as the prediction of $CR(s; s'|O)$. If $h_1(O), h_2(O), \ldots, h_m(O)$ has been observed in the training dataset, we directly use the empirical value. Otherwise, we use $\psi_{s,s'}$.

We use the traditional Viterbi algorithm for selecting the label sequence with maximum probability in decoding stage.

3.3   Support Vector Regression

There exist a rich set of regression models which may be used for modeling ϕ_s and $\psi_{s,s'}$. In this paper, we adopt the support vector regression (SVR) [15] for modeling ϕ_s and $\psi_{s,s'}$ discussed above. SVR is linear in the high dimensional transformed space and tolerant to low error points with small residuals. Such tolerance seems to fit the natural language processing applications well, in which the input and final output are normally categorical. In text classification, large-margin methods achieve very good results [8]. And there are good implementations of SVR which can handle very large number of instances and features efficiently. These reasons lead us to prefer SVR. In future, we will try other regression models. To avoid endowing unwanted metric and order structures to a single categorical variable, we use the dummy coding as the representation of categorical input variables.

4   Experiments

In this section, we compare L-CRN with CRFs[5]. We adopt CRF++ version 0.58 [10] as the implementation for CRFs and LIBLINEAR version 1.94 [5] for linear SVR in L-CRNs. We set the configurations of LIBLINEAR as L2-regularization L2-loss support vector regression (solving dual). For a fair comparison, we always use a single thread for training[6]. We apply CRFs and L-CRNs to an important natural language processing application: named entity recognition (NER).

---

[5] [11,12] show superiority of CRFs over other models. Hence it is reasonable to compare with CRFs.
[6] L-CRNs can be easily parallellized. Obviously, each regression model can be trained parallely with others.

4.1   Named Entity Recognition

The English part of the CoNLL-2003 NER dataset[7] [17] is used for our NER experiment. There are three data files in this dataset: ner.train, ner.testa and ner.testb. The first one is designed for training and the last two are used for testing. The size of the label space is 8. These three files include 14987, 3466, 3684 sentences and 204567, 51578, 46666 words respectively. We use the same orthographic features as those used by [11]: "whether a spelling begins with a number or upper case letter, whether it contains a hyphen, and whether it ends in one of the following suffixes: −ing, −ogy, −ed, −s, −ly, −ion, −tion, −ity, −ies". Additionally, we use the chunk tags and POS tags provided together with the CoNLL dataset.

Table 1 gives the time taken by CRF and L-CRN. We can see L-CRN reduces the training time significantly.

Tables 2 and 3 show the quality metrics achieved by CRF and L-CRN on ner.testa and ner.testb, respectively. The first three columns show the per-word accuracies (%) on all, known and unknown words[8]. On all and known words, L-CRN consistently outperforms CRF slightly. As described in Sect. 3, L-CRN can directly use empirical values for known word prediction. This may be considered as an advantage of L-CRN. On unknown words, CRF performs better on ner.testa, but L-CRN performs slightly better on ner.testb. The last three columns give the precision, recall and F1 metrics. These metrics were evaluated using the standard CoNLL evaluation tool[9]. CRF obtains better results in precision. L-CRN obtains better results in recall and F1.

Table 1. Training time (seconds) on NER

CRF	L-CRN
1,666	112

Table 2. Metrics on ner.testa

	All	Known	Unknown	Precision	Recall	F1
CRF	97.00	98.27	85.42	84.66	82.31	83.47
L-CRN	97.44	98.80	85.05	84.21	84.45	84.33

Table 3. Metrics on ner.testb

	All	Known	Unknown	Precision	Recall	F1
CRF	95.00	97.46	80.32	75.61	74.70	75.15
L-CRN	95.55	98.06	80.62	75.78	76.43	76.10

---

[7] http://www.cnts.ua.ac.be/conll2003/ner/

[8] Known words are the words that appear in the training data. Unknown words are the words that have not been seen in the training data. All words include both.

[9] http://www.cnts.ua.ac.be/conll2000/chunking/conlleval.txt

5   Conclusions

We propose the linear co-occurrence rate networks (L-CRN) for sequence labeling. This model avoids problems of the existing models, such as mismatching problems and the label bias problem. The transition features can be encoded into L-CRN. Furthermore, the factors of L-CRN can be normalized locally and trained independently, which leads to very efficient training. In this paper, we use support vector regression as the regression models of factors in L-CRN. Experimental results show L-CRNs reduce the training time by orders of magnitudes and achieve very competitive results to CRFs on real-world NLP data.

Acknowledgments. We thank SLSP 2014 reviewers for their comments. This work has been supported by the Dutch national program COMMIT/.

6   Appendix

6.1   Closed-Form MLE Training of L-CRN

We maximize the log likelihood of Eq. 3 over the training dataset D with CR and p as parameters:

$$\max. \sum_{(S,O)\in D} [\sum_{i=1}^{n-1} \log \mathrm{CR}(s_i; s_{i+1}|O) + \sum_{j=1}^{n} \log p(s_j|O)]$$

$$s.t. \quad \sum_{s,s'} \mathrm{CR}(s; s'|O)p(s|O)p(s'|O) = 1, \forall s, s'$$

$$\sum_{s} p(s|O) = 1, \forall s$$

$$\mathrm{CR}(s; s'|O) \geq 0, \forall s, s'$$

$$p(s|O) \geq 0, \forall s$$

First we ignore the last two non-negative inequality constraints. Using Lagrange Multiplier, we obtain the unconstrained objective function:

$$\sum_{(S,O)\in D} [\sum_{i=1}^{n-1} \log \mathrm{CR}(s_i; s_{i+1}|O) + \sum_{j=1}^{n} \log p(s_j|O)]+$$

$$\sum_{s,s'} [\lambda_{s,s'}(\sum_{s,s'} \mathrm{CR}(s; s'|O)p(s|O)p(s'|O) - 1)]$$

$$+ \sum_{s} [\lambda_s(\sum_{s} p(s|O) - 1)].$$

Calculate the first derivative for each parameter and set them to zero, we get the closed form MLE for CR and p:

$$\hat{p}(s|O) = \frac{\#(s,O)}{\sum_s \#(s,O)},$$

$$\hat{CR}(s; s'|O) = \frac{\#(s,s',O)}{\sum_{s,s'} \#(s,s',O)} \bigg/ \frac{\#(s,O)}{\sum_s \#(s,O)} \bigg/ \frac{\#(s',O)}{\sum_{s'} \#(s',O)}.$$

That is, the MLE of p and CR are just their relative frequencies in the training dataset. Fortunately the non-negative inequality constraints which were ignored in optimization are automatically met.

6.2   Theorems of Co-occurrence Rate

Proof of Partition Operation

Proof.

$$\text{CR}(X_1; ..; X_j)\text{CR}(X_{j+1}; ..; X_n)\text{CR}(X_1..X_j; X_{j+1}..X_n)$$
$$= \frac{p(X_1, .., X_j)}{p(X_1)..p(X_j)} \frac{p(X_{j+1}, .., X_n)}{p(X_{j+1})..p(X_n)} \frac{p(X_1, .., X_n)}{p(X_1, .., X_j)p(X_{j+1}, .., X_n)}$$
$$= \frac{p(X_1, .., X_n)}{p(X_1)..p(X_n)} = \text{CR}(X_1; ..; X_n).$$

Proof of Reduce Operation

Proof. Since $X \perp\!\!\!\perp Y \mid Z$, we have $p(X, Y|Z) = p(X|Z)p(Y|Z)$, then $p(XYZ) = \frac{p(X,Z)p(Y,Z)}{p(Z)}$. Hence,

$$\text{CR}(X; YZ) = \frac{p(X, Y, Z)}{p(X)p(Y, Z)} = \frac{p(X, Y, Z)}{p(X)p(Y, Z)} = \frac{p(X, Z)}{p(X)p(Z)} = \text{CR}(X; Z).$$

References

1. Altun, Y., Smola, A.J., Hofmann, T.: Exponential families for conditional random fields. In: Proceedings of the 20th Conference on Uncertainty in Artificial Intelligence, UAI '04, pp. 2–9. AUAI Press (2004)
2. Berg-Kirkpatrick, T., Bouchard-Côté, A., DeNero, J., Klein, D.: Painless unsupervised learning with features. In: NAACL, HLT '10, pp. 582–590 (2010)
3. Church, K.W., Hanks, P.: Word association norms, mutual information, and lexicography. Comput. Linguist. **16**(1), 22–29 (1990)
4. Cohn, T.A.: Scaling conditional random fields for natural language processing. Ph.D. thesis, University of Melbourne (2007)
5. Fan, R.E., Chang, K.W., Hsieh, C.J., Wang, X.R., Lin, C.J.: Liblinear: a library for large linear classification. J. Mach. Learn. Res. **9**, 1871–1874 (2008)

6. Ghahramani, Z.: An introduction to hidden Markov models and Bayesian networks. In: Juang, B.H. (ed.) Hidden Markov Models, pp. 9–42. World Scientific Publishing, Adelaide (2002)
7. Hammersley, J.M., Clifford, P.E.: Markov random fields on finite graphs and lattices. Unpublished manuscript (1971)
8. Joachims, T.: Text categorization with support vector machines: learning with many relevant features. In: Proceedings of the 10th European Conference on Machine Learning, ECML '98, pp. 137–142 (1998)
9. Klein, D., Manning, C.D.: Conditional structure versus conditional estimation in NLP models. In: EMNLP '02, pp. 9–16 (2002). http://dx.doi.org/10.3115/1118693. 1118695
10. Kudo, T.: CRF++: yet another CRF toolkit, free software, March 2012. http:// crfpp.googlecode.com/svn/trunk/doc/index.html
11. Lafferty, J.D., McCallum, A., Pereira, F.C.N.: Conditional random fields: probabilistic models for segmenting and labeling sequence data. In: ICML '01, pp. 282–289 (2001)
12. Le-Hong, P., Phan, X.H., Tran, T.T.: On the effect of the label bias problem in part-of-speech tagging. In: The 10th IEEE RIVF International Conference on Computing and Communication Technologies, pp. 103–108 (2013)
13. McCallum, A., Freitag, D., Pereira, F.C.N.: Maximum entropy markov models for information extraction and segmentation. In: ICML '00, pp. 591–598 (2000)
14. Rabiner, L.R.: A tutorial on hidden markov models and selected applications in speech recognition. In: Proceedings of the IEEE, pp. 257–286 (1989)
15. Smola, A.J., Schölkopf, B.: A tutorial on support vector regression. Stat. Comput. **14**(3), 199–222 (2004)
16. Sutton, C., McCallum, A.: An introduction to conditional random fields. Found. Trends Mach. Learn. **4**(4), 267–373 (2012)
17. Tjong Kim Sang, E.F., De Meulder, F.: Introduction to the CoNLL-2003 shared task: language-independent named entity recognition. In: Proceedings of CoNLL-2003, Edmonton, Canada (2003)
18. Zhu, Z., Hiemstra, D., Apers, P.M.G., Wombacher, A.: Separate training for conditional random fields using co-occurrence rate factorization. Technical report TR-CTIT-12-29, Centre for Telematics and Information Technology, University of Twente, Enschede, October 2012
19. Zhu, Z., Hiemstra, D., Apers, P.M.G., Wombacher, A.: Empirical co-occurrence rate networks for sequence labeling. In: ICMLA 2013, Miami Beach, FL, USA, December 2013, pp. 93–98 (2013)
20. Zhu, Z.: Factorizing probabilistic graphical models using cooccurrence rate, August 2010. arXiv:1008.1566v1
21. Zhu, Z., Hiemstra, D., Apers, P., Wombacher, A.: Comparison of local and global undirected graphical models. In: The 22nd European Symposium on Artificial Neural Networks, Computational Intelligence and Machine Learning, pp. 479–484 (2014)

Text Extraction and Categorization

A Comparison of Sequence-Trained Deep Neural Networks and Recurrent Neural Networks Optical Modeling for Handwriting Recognition

Théodore Bluche[1,2]([✉]), Hermann Ney[2,3], and Christopher Kermorvant[1]

[1] A2iA SA, Paris, France
[2] Spoken Language Processing Group, LIMSI CNRS, Orsay, France
[3] Human Language Technology and Pattern Recognition,
RWTH Aachen University, Aachen, Germany
tb@a2ia.com

Abstract. Long Short-Term Memory Recurrent Neural Networks are the current state-of-the-art in handwriting recognition. In speech recognition, Deep Multi-Layer Perceptrons (DeepMLPs) have become the standard acoustic model for Hidden Markov Models (HMMs). Although handwriting and speech recognition systems tend to include similar components and techniques, DeepMLPs are not used as optical model in unconstrained large vocabulary handwriting recognition. In this paper, we compare Bidirectional LSTM-RNNs with DeepMLPs for this task. We carried out experiments on two public databases of multi-line handwritten documents: Rimes and IAM. We show that the proposed hybrid systems yield performance comparable to the state-of-the-art, regardless of the type of features (hand-crafted or pixel values) and the neural network optical model (DeepMLP or RNN).

Keywords: Handwriting recognition · Recurrent Neural Networks · Deep neural networks

1 Introduction

Handwriting recognition is the problem of transforming an image into the text it contains. Unlike Optical Character Recognition (OCR), segmenting each character is difficult, mainly due to the cursive nature of handwriting. One usually prefers to recognize whole words or lines of text, i.e. the sequence of characters, with HMMs or RNNs.

In HMMs, the characters are modeled as sequences of hidden states, associated with an emission probability model. Gaussian Mixture Models (GMMs) is the standard optical model in HMMs. However, in the last decade, emission probability models based on artificial neural networks have (re)gained considerable interest in the community, mainly due to the *deep learning* trend in computer vision and speech recognition. In this latter domain, major improvements have been observed with the introduction of deep neural networks.

© Springer International Publishing Switzerland 2014
L. Besacier et al. (Eds.): SLSP 2014, LNAI 8791, pp. 199–210, 2014.
DOI: 10.1007/978-3-319-11397-5_15

A significant usage of neural network for handwriting recognition should also be noted. The MNIST database of handwritten digits received a lot of attention in computer vision and in the application of deep learning techniques. Convolutional Neural Networks introduced by Le Cun et al. [20] have soon been applied to handwriting recognition problems, and were recently tested on public databases for handwritten word recognition, yielding state-of-the-art results [5].

The state-of-the-art performance on many public handwriting databases is achieved by RNNs. This type of neural network has the ability to use more context than HMMs and to model the whole sequence directly. The best published results on IAM [19], Rimes [19,26] and OpenHaRT [26,31], were achieved by systems involving an RNN component.

In this work, we compare different approaches to optical modeling in handwriting recognition systems. In particular, we studied different kinds of neural networks (DeepMLPs and RNNs), and features (hand-crafted and pixel values).

We report results on the publicly available IAM [22] and Rimes [1] databases. Major improvements have been recently reported on these tasks, mainly due to a better pre-processing of the images, and an open-vocabulary language model [19]. This work shows that similar Word Error Rates (WERs) can be achieved with different kinds of features (hand-crafted geometric and statistical features, and pixel values), and optical models (DeepMLPs and RNNs), and a rather standard pre-processing. We note that for DeepMLPs to be comparable in performance to RNNs, a sequence training criterion, such as state-level Minimum Bayes Risk (sMBR) [18] should be used.

This paper is divided as follows. Section 2 contains a brief litterature review. Section 3 describes our systems. Section 4 presents the experiments carried out and the results obtained. Conclusions are drawn in Sect. 5.

2   Relation to Prior Work

Recurrent Neural Networks, with the Long Short-Term Memory (LSTM) units, are particularly good for handwriting recognition. State-of-the-art systems for many public databases include an RNN component. Kozielski et al. [19] trained a bidirectional LSTM-RNN (BLSTM-RNN) on sequences of feature vectors, and HMM state targets. They then extract features from hidden layer activations to train a standard GMM-HMM, and report the best known results on the IAM database. On the other hand, Graves et al. [14], and more recently [4,26] trained Multi-Dimensional LSTM-RNNs (MDLSTM-RNNs), which operate directly on the raw image, with a Connectionist Temporal Classification (CTC) objective, which allows to train the network directly using the sequence of characters as targets. With the dropout technique, [26] report the best results on both Rimes and OpenHaRT databases.

Multi-layer Perceptrons with one hidden layer were used for optical modeling in hybrid systems by España-Bocquera et al. [11] and Dreuw et al. [10]. Deep Neural Networks (DeepMLPs), were applied to simple handwriting recognition tasks such as isolated character or digits recognition [7,8]. More recently, they were used in combination with HMMs for keywords spotting in handwritten documents [30]. They enjoy considerable research attention since efficient training methods have been proposed. They achieve excellent results in various computer vision tasks (e.g. object recognition), but also in speech recognition, where they replace efficiently the conventional GMMs in HMMs. Their architecture is simple (multi-layer perceptrons), and their depth seems to contribute to better modeling [25] and robustness [9]. It has been shown [28,32] that optimizing training criteria over whole sequences (e.g. sMBR), including the language constraints, leads to improvements compared with a framewise criterion. Similar (global) training of handwriting recognition systems were already proposed in the 90s [20,21]. In this work, we show that the framework of DeepMLP and sequence training used in speech recognition can successfully be applied to handwriting recognition, with very good results on public databases, and compete with RNNs.

3   System Overview

3.1   Image Pre-processing and Feature Extraction

The goal of pre-processing is to remove the undesirable variabilities from images. First, the lines are deskewed [3] and deslanted [6]. Then, the darkest 5 % of pixels are mapped to black and the lightest 70 % are mapped to white, with a linear interpolation in between, to enhance the contrast. We added 20 columns of white pixels to the beginning and end of each line to account for empty context. Most systems require an image with fixed height. We first detect three regions in the image (ascenders, descenders and core region) [33], and scale these regions to three fixed heights.

We built baseline systems using the handcrafted features described in [2], which gave reasonable performance on several public databases [2,23]. We extracted them with a sliding window, scanned left-to-right through the preprocessed text line image. It is defined by two parameters: its width and shift (controlling the overlap between consecutive windows). To fix these parameters, we trained GMM-HMMs using the handcrafted features and different widths and shifts of the sliding window and keep the parameters yielding the best performance on the validation set. The optimal values we found are a width of 3px, a shift of 3px for both databases.

We also carried out experiments on pixel features (for NNs only). They are extracted with a sliding window of width 45px and shift 3px, rescaled to 20×32px. The pixel values are normalized to lie in the interval $[0, 1]$ (1 corresponding to white), producing 640-dimensional feature vectors. No Principal Component Analysis or other decorrelation or dimensionality reduction algorithm were applied.

3.2   Hidden Markov Models

The topology of the HMM is left-to-right: two output transitions per state, one to itself and one to the next state. We tried different number of HMM states in character models (along with different sliding window parameters), and kept the values yielding the best GMM-HMM results on the validation sets. We built 6-state models for IAM and 5-state models for Rimes. We added two 2-state silence HMMs to model optional empty context on the left and right of words.

3.3   Neural Networks

Multi Layer Perceptrons and Deep Neural Networks. Multi Layer Perceptrons (MLPs) are networks organized in several layers, each one fully connected to the next. The input corresponds to an observation vector, optionally concatenated with a small amount of previous and next frames. The output is a prediction of the HMM states. Deep Neural Networks (DeepMLPs) are MLPs with several hidden layers. We first initialize the weights with unsupervised pretraining, consisting in stacking Restricted Boltzmann Machines, trained with contrastive divergence, as explained in [15]. Then, we perform a supervised discriminative training of the whole network. The targets are obtained by forced alignment of the training set with a bootstrapping model. We optimize the cross-entropy criterion with Stochastic Gradient Descent (SGD).

Sequence training of neural networks consists in optimizing the network parameters with a sequence-discriminative criterion rather than using the frame-level cross-entropy criterion. Sequence training is similar to the discriminative training of GMM-HMMs. Among different possibilities, we chose the state-level Minimum Bayes Risk (sMBR) criterion, described in [18], which yields slightly better WER than other sequence criteria on a speech recognition task (Switchboard) [32]. In speech recognition, sequence training results in relative performance gains of 5–10 % for various tasks [29,32].

Recurrent Neural Networks (RNNs). In RNNs, the input to a given recurrent layer are not only the activations of the previous layers, but also its own activations at the previous time step. This characteristic enables them to naturally work with sequential inputs, and to use the past context to make predictions. Long Short-Term Memory (LSTM) units are recurrent neurons, in which a gating mechanism avoids the vanishing gradient problem, appearing in conventional RNNs [14,16], and enables to learn arbitrarily long dependencies. In Bi-Directional LSTM-RNNs (BDLSTM-RNNs), LSTM layers are doubled: the second layer is connected to the "next" time step rather than the previous one. Thus the input sequence is processed in both directions, so past and future context are used to make predictions (see Fig. 1). The information coming from both directions is summed component-wise after the LSTM layers, and the result is an input for a feed-forward layer. This is a generalization of the MDLSTM-RNN architecture described in [14,26] to the case of sequences of feature vectors.

Finally, the Connectionnist Temporal Classification (CTC) paradigm [13] has been used to train the RNNs. With CTC, no prior segmentation of the training data (line images) is required. Therefore, we do not need a bootstrapping procedure involving forced alignments with a previously trained HMM. Instead,

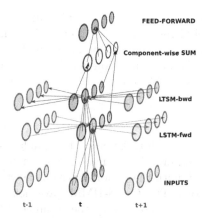

Fig. 1. Bidirectional Recurrent Neural Networks

we can select the target sequence to be the sequence of character in the image annotation, which simplifies the training procedure.

4 Experiments and Results

4.1 Rimes and IAM Databases

The Rimes database [1] consists of images of handwritten paragraphs from simulated French mail. The setup for the ICDAR 2011 competition is a training set of 1,500 images, and an evaluation set of 100 images. We held out the last 149 images from the training set for system validation. We built a 4-gram language model (LM) with modified Kneser-Ney discounting from the training annotations. The vocabulary is made of 12k words. The language model has a perplexity of 18 and out-of-vocabulary (OOV) rate of 2.9 % on the validation set (18 and 2.6 % on the evaluation set).

The IAM database [22] consists of images of handwritten documents. They correspond to English texts extracted from the LOB corpus [17], copied by different writers. The database is split into 747 images for training, 116 for validation, and 336 for evaluation. We used a 3-gram language model limited to the 50k most frequent words from the training set. It was trained on the LOB, Brown and Wellington corpora. The passages of the LOB corpus appearing in the validation and evaluation sets were removed prior to LM training. The resulting model has a perplexity of 298 and OOV rate of 4.3 % on the validation set (329 and 3.7 % on the evaluation set).

4.2 Decoding Method

We used the Kaldi toolkit [27] to decode the sequences of observation vectors (GMMs, DeepMLPs), or the sequences of character predictions (RNNs).

The decoding was done for complete paragraphs rather than lines, to benefit from the language model history across line boundaries. The optical scaling factor, balancing the importance given to the optical model scores and to the language model scores, and the word insertion penalty were tuned on the validation sets. This optimization can yield from 1 to 3 % absolute improvement.

4.3   GMM-HMM

We trained GMM-HMM on both tasks, using the handcrafted features, and the Maximum Likelihood criterion. The number of Gaussians in the mixtures was increased at each iteration until the performance on the validation set decreases for more than 5 iterations. The GMM-HMM have not been discriminatively trained. They were only used to bootstrap the training of DeepMLPs.

4.4   Deep Neural Networks

To train the DeepMLPs, we performed the forced alignments of the training set with the GMM-HMMs, to have a target HMM state for each input observation. We held out 10 % of this dataset for validation and early stopping. Overall, the datasets contain 5,6M examples for Rimes and 3,8M examples for IAM.

DeepMLP on Handcrafted Features. We inverstigated different numbers of hidden layers (1 to 7) in the DeepMLP and different sizes of input context ($\pm\{1,3,5,7,9\}$ frames). The number of hidden nodes in each layer was set to 1,024. The input features were normalized to zero mean and unit variance along each dimension. The networks were pre-trained using 1 epoch of unsupervised training for each layer, followed by a few epochs of supervised training with stochastic gradient descent and a cross-entropy criterion. The training finished when no more improvement was observed on the validation set.

The results are depicted on Fig. 2. The performances of the different networks are similar to each other. It looks like more than one hidden layer is generally better, but the performance gain when we add more layers is not significant. We selected the best architectures based on the performance on the validation sets: 5 hidden layers with 1,024 units and 15 frames of context (central frame ±7) for IAM, 4 hidden layers with 1,024 units and 7 frames of context (central frame ±3) for Rimes. Additionally, training the networks with 5 more epochs of sMBR sequence training allowed to obtain 4 to 6 % relative WER improvement (Table 1).

DeepMLP on Pixels. Instead of adding context frames to the central frames, we extracted the pixels values in a larger sliding window. The means and standard deviations were computed across all dimensions simultaneously, not separately.

For the pixel DeepMLP, we notice a wider difference between one and more hidden layers (Fig. 3) than for DeepMLP on handcrafted features. The justification could be that in hancrafted features DeepMLPs, the inputs are already

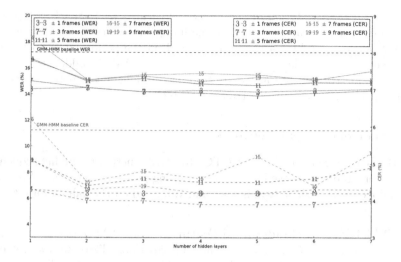

Fig. 2. Effect of depth and size of context (Features DeepMLP, Rimes validation set)

Table 1. Improvement brought by sMBR sequence training (results reported on validation sets)

System	WER - Rimes		CER - Rimes		WER - IAM		CER - IAM	
Features DeepMLP	14.1%		4.0%		12.4%		4.1%	
+ sMBR training	13.5%	(-4.2%)	4.0%	(-0.0%)	11.7%	(-5.6%)	3.9%	(-4.9%)
Pixel DeepMLP	13.6%		3.9%		12.4%		4.4%	
+ sMBR training	13.1%	(-3.7%)	3.8%	(-2.6%)	11.8%	(-4.8%)	4.2%	(-4.5%)

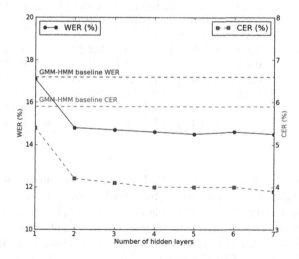

Fig. 3. Effect of increasing the number of hidden layers (Pixel DeepMLP, Rimes validation set)

a higher level representation of the image, while in pixel DeepMLPs, the first layer(s) perform the transformation of the image into a higher level representation. We selected the best architectures based on the performance on the validation sets: 4 hidden layers with 1,024 units for IAM, 7 hidden layers with 1,024 units for Rimes. Again, sMBR training brought a few percents relative improvement over cross-entropy training (Table 1).

4.5   Recurrent Neural Networks

Since the RNNs are trained with a CTC objective function to predict sequences of characters, there is no need for a bootstrapping procedure. All the RNNs have been trained on the whole training set and validated on the validation set.

BDLSTM-RNN on Handcrafted Features. The RNNs naturally takes into account the left and right context to make predictions. Thus, we did not concatenate context feature frames. The input features were normalized to zero mean and unit variance along each dimension. We explored different depths and widths, and also applied the dropout regularization technique in feed-forward layers, as explained in [26].

Table 2. RNNs on handcrafted and pixel features, results for Rimes validation set. uCER stands for unconstrained CER. WER and CER are computed with a lexicon and language model.

Archi. dropout	Handcrafted features				Pixel features			
	CTC cost	uCER	WER	CER	CTC cost	uCER	WER	CER
1x100 -	0.5217	14.5	14.9	4.7	1.201	33.8	24.1	10.3
3x100 -	0.3864	10.6	13.6	4.1	0.4834	12.9	15.1	5.1
5x100 -	0.3516	9.3	14.7	4.3	0.3637	9.8	14.0	4.4
5x200 -	0.3295	8.5	13.5	3.9	0.3724	9.7	15.4	4.9
7x100 -	0.3093	8.0	13.8	4.1	0.3313	8.7	14.5	4.5
7x200 -	0.2969	8.0	14.1	4.1	0.3445	8.9	14.7	5.0
7x200 x	0.2397	5.7	**12.7**	**3.6**	0.2351	6.0	**13.6**	**4.1**
9x100 -	0.2937	7.6	13.2	3.9	0.3229	8.6	14.5	4.5
9x100 x	0.2565	6.0	13.1	3.8	0.2559	6.3	13.8	4.4

We report the results in Table 2. uCER stands for unconstrained CER, and refers to the character error rate when the RNN is used alone to make character predictions, i.e. without lexical and language model. While it seems better to have more than one hidden layer, the biggest improvements were achieved with dropout. The best architectures, selected based on the results on the validation sets, are 7 hidden layers (4 LSTM and 3 feed-forward) of 200 units with dropout for Rimes and IAM.

BDLSTM-RNN on Pixels. For pixel features, the inputs are normalized with the mean and standard deviation of pixel values across all dimensions. We also explored different widths and depths and dropout, and selected the best models based on the validation results. For Rimes and IAM, the best network has 7 hidden layers of 200 units and dropout. The results for different architectures on Rimes database are shown on Table 2. Again, we notice that the effect of having more than one hidden layer is more important for pixel-based models than for models using handcrafted features.

Table 3. Results on IAM database

		Dev.		Eval.	
		WER	CER	WER	CER
g.	GMM-HMM baseline	15.2	6.3	19.6	9.0
df.	Features DeepMLP-5x1024	11.7	3.9	14.7	5.8
dp.	Pixel DeepMLP-4x1024	11.8	4.2	14.7	5.9
rf.	Feature BDLSTM-RNN 7x200 + dropout	11.9	3.9	14.3	5.3
rp.	Pixels BDLSTM-RNN 7x200 + dropout	11.8	4.0	14.8	5.6
	ROVER rf + rp + df + dp	9.7	3.6	**11.9**	**4.9**
	Kozielski et al. [19]	**9.5**	**2.7**	13.3	5.1
	Pham et al. [26]	11.2	3.7	13.6	5.1
	Kozielski et al. [19]	11.9	3.2	-	-

Table 4. Results on Rimes database

		Dev.		Eval.	
		WER	CER	WER	CER
g.	GMM-HMM baseline	17.2	5.9	15.8	6.0
df.	Features DeepMLP-4x1024	13.5	4.0	13.5	4.1
dp.	Pixel DeepMLP-7x1024	13.1	3.8	12.9	3.8
rf.	Feature BDLSTM-RNN 7x200 + dropout	12.7	3.6	12.7	4.0
rp.	Pixels BDLSTM-RNN 7x200 + dropout	13.6	4.1	13.8	4.3
	ROVER rf + rp + df + dp	11.8	3.4	**11.8**	3.7
	Pham et al. [26]	-	-	12.3	**3.3**
	Messina et al. [24]	-	-	13.3	-
	Kozielski et al. [19]	-	-	13.7	4.6

The final results, comparing different models and input features on the one hand, and comparing our proposed systems with other published results on the other hand, are reported on Table 3 (IAM) and Table 4 (Rimes). We see that both handcrafted and pixel features, and both DeepMLPs and RNNs can achieve results that are close to the best reported ones. For DeepMLPs, sequence training seems crucial to attain this performance. Furthermore, we notice that although RNNs have become a standard component of handwriting recognition systems,

DeepMLPs – which have become standard in hybrid speech recognition systems – can perform equally well. Finally, we cannot draw a clear conclusion regarding whether RNNs or DeepMLPs should be preferred, or whether handcrafted features are more suited than pixel values.

Our different optical models and features are also complementary, as shown by their ROVER combination [12], which, to the best of our knowledge constitute the best published results on both databases, outperforming the open-vocabulary approaches proposed in [19,24].

5   Conclusion

In this paper, we shown that state-of-the-art WERs can be achieved with both DeepMLPs - standard method for speech recognition, and RNNs - standard method for handwriting recognition. Even with a pretty simple image preprocessing, the pixel values could replace handcrafted features. Future work may include an evaluation of convolutional neural networks and Multi-Dimensional (MD)LSTM-RNNs for a more comprehensive comparison of neural network optical modeling. An evaluation of a tandem combination (where the neural networks are used to extract features rather than to make predictions) could be carried out. Finally, it would be interesting to evaluate the robustness of the proposed models, i.e. to see how good the results could be when these systems are applied to new databases, not seen during training.

Acknowledgments. The authors would like to thank Michal Kozielsky and his colleagues from RWTH for providing the language model used in IAM experiments. This work was partly achieved as part of the Quaero Program, funded by OSEO, French State agency for innovation and was supported by the French Research Agency under the contract Cognilego ANR 2010-CORD-013.

References

1. Augustin, E., Carré, M., Grosicki, E., Brodin, J.M., Geoffrois, E., Preteux, F.: RIMES evaluation campaign for handwritten mail processing. In: Proceedings of the Workshop on Frontiers in Handwriting Recognition (2006)
2. Bianne, A.L., Menasri, F., Al-Hajj, R., Mokbel, C., Kermorvant, C., Likforman-Sulem, L.: Dynamic and contextual information in HMM modeling for handwriting recognition. IEEE Trans. Pattern Anal. Mach. Intell. **33**(10), 2066–2080 (2011)
3. Bloomberg, D.S., Kopec, G.E., Dasari, L.: Measuring document image skew and orientation. In: IS&T/SPIE's Symposium on Electronic Imaging: Science & Technology, pp. 302–316. International Society for Optics and Photonics (1995)
4. Bluche, T., Louradour, J., Knibbe, M., Moysset, B., Benzeghiba, M., Kermorvant, C.: The A2iA Arabic handwritten text recognition system at the OpenHaRT2013 evaluation. In: 11th IAPR Workshop on Document Analysis Systems (DAS2014), pp. 161–165 (2014)
5. Bluche, T., Ney, H., Kermorvant, C.: Tandem HMM with convolutional neural network for handwritten word recognition. In: 38th International Conference on Acoustics Speech and Signal Processing (ICASSP2013), pp. 2390–2394 (2013)

6. Buse, R., Liu, Z.Q., Caelli, T.: A structural and relational approach to handwritten word recognition. IEEE Trans. Syst. Man Cybern. **27**(5), 847–861 (1997)
7. Ciresan, D.C., Meier, U., Gambardella, L.M., Schmidhuber, J.: Deep, big, simple neural nets for handwritten digit recognition. Neural computation **22**(12), 3207–3220 (2010)
8. Cireşan, D.C., Meier, U., Gambardella, L.M., Schmidhuber, J.: Deep big multilayer perceptrons for digit recognition. In: Montavon, G., Orr, G.B., Müller, K.-R. (eds.) Neural Networks: Tricks of the Trade, 2nd edn. LNCS, vol. 7700, 2nd edn, pp. 581–598. Springer, Heidelberg (2012)
9. Deng, L., Li, J., Huang, J.T., Yao, K., Yu, D., Seide, F., Seltzer, M., Zweig, G., He, X., Williams, J., et al.: Recent advances in deep learning for speech research at microsoft. In: 38th IEEE International Conference on Acoustics, Speech and Signal Processing (ICASSP2013), pp. 8604–8608. IEEE (2013)
10. Dreuw, P., Doetsch, P., Plahl, C., Ney, H.: Hierarchical hybrid MLP/HMM or rather MLP features for a discriminatively trained Gaussian HMM: a comparison for offline handwriting recognition. In: 18th IEEE International Conference on Image Processing (ICIP2011), pp. 3541–3544. IEEE (2011)
11. Espana-Boquera, S., Castro-Bleda, M.J., Gorbe-Moya, J., Zamora-Martinez, F.: Improving offline handwritten text recognition with hybrid HMM/ANN models. IEEE Trans. Pattern Anal. Mach. Intell. **33**(4), 767–779 (2011)
12. Fiscus, J.G.: A post-processing system to yield reduced word error rates: recognizer output voting error reduction (rover). In: IEEE Workshop on Automatic Speech Recognition and Understanding (ASRU1997), pp. 347–354. IEEE (1997)
13. Graves, A., Fernández, S., Gomez, F., Schmidhuber, J.: Connectionist temporal classification: labelling unsegmented sequence data with recurrent neural networks. In: Proceedings of the 23rd International Conference on Machine Learning, pp. 369–376. ACM (2006)
14. Graves, A., Schmidhuber, J.: Offline handwriting recognition with multidimensional recurrent neural networks. In: NIPS, pp. 545–552 (2008)
15. Hinton, G.E., Osindero, S., Teh, Y.W.: A fast learning algorithm for deep belief nets. Neural Comput. **18**(7), 1527–1554 (2006)
16. Hochreiter, S., Schmidhuber, J.: Long short-term memory. Neural Comput. **9**(8), 1735–1780 (1997)
17. Johansson, S.: The LOB corpus of British english texts: presentation and comments. ALLC J. **1**(1), 25–36 (1980)
18. Kingsbury, B.: Lattice-based optimization of sequence classification criteria for neural-network acoustic modeling. In: IEEE International Conference on Acoustics, Speech and Signal Processing (ICASSP 2009), pp. 3761–3764. IEEE (2009)
19. Kozielski, M., Doetsch, P., Ney, H.: Improvements in RWTH's system for off-line handwriting recognition. In: International Conference on Document Analysis and Recognition (ICDAR2013), pp. 935–939 (2013)
20. Le Cun, Y., Bottou, L., Bengio, Y.: Reading checks with multilayer graph transformer networks. In: IEEE International Conference on Acoustics, Speech, and Signal Processing (ICASSP1997), vol. 1, pp. 151–154. IEEE (1997)
21. LeCun, Y., Bottou, L., Bengio, Y., Haffner, P.: Gradient-based learning applied to document recognition. Proc. IEEE **86**(11), 2278–2324 (1998)
22. Marti, U.V., Bunke, H.: The IAM-database: an english sentence database for offline handwriting recognition. Int. J. Doc. Anal. Recogn. **5**(1), 39–46 (2002)

23. Menasri, F., Louradour, J., Bianne-Bernard, A.L., Kermorvant, C.: The A2iA French handwriting recognition system at the Rimes-ICDAR2011 competition. In: IS&T/SPIE Electronic Imaging, pp. 82970–82970. International Society for Optics and Photonics (2012)

24. Messina, R., Kermorvant, C.: Surgenerative finite state transducer n-gram for out-of-vocabulary word recognition. In: 11th IAPR Workshop on Document Analysis Systems (DAS2014), pp. 212–216 (2014)

25. Mohamed, A., Hinton, G., Penn, G.: Understanding how deep belief networks perform acoustic modelling. In: 37th IEEE International Conference on Acoustics, Speech and Signal Processing (ICASSP2012), pp. 4273–4276. IEEE (2012)

26. Pham, V., Bluche, T., Kermorvant, C., Louradour, J.: Dropout improves recurrent neural networks for handwriting recognition. In: 14th International Conference on Frontiers in Handwriting Recognition (ICFHR2014) (2014)

27. Povey, D., Ghoshal, A., Boulianne, G., Burget, L., Glembek, O., Goel, N., Hanne-mann, M., Motlicek, P., Qian, Y., Schwarz, P., et al.: The Kaldi speech recogni-tion toolkit. In: Workshop on Automatic Speech Recognition and Understanding (ASRU2011), pp. 1–4 (2011)

28. Sainath, T.N., Mohamed, A., Kingsbury, B., Ramabhadran, B.: Deep convolutional neural networks for LVCSR. In: 38th IEEE International Conference on Acoustics, Speech and Signal Processing (ICASSP2013), pp. 8614–8618. IEEE (2013)

29. Su, H., Li, G., Yu, D., Seide, F.: Error back propagation for sequence training of context-dependent deep networks for conversational speech transcription. In: 2013 IEEE International Conference on Acoustics, Speech and Signal Processing (ICASSP2013), pp. 6664–6668 (2013)

30. Thomas, S., Chatelain, C., Paquet, T., Heutte, L.: Un modèle neuro markovien profond pour l'extraction de séquences dans des documents manuscrits. Doc. Numérique 16(2), 49–68 (2013)

31. Tong, A., Przybocki, M., Maergner, V., El Abed, H.: NIST 2013 Open Handwriting Recognition and Translation (OpenHaRT'13) evaluation. In: 11th IAPR Workshop on Document Analysis Systems (DAS2014), pp. 81–85 (2014)

32. Veselý, K., Ghoshal, A., Burget, L., Povey, D.: Sequence-discriminative training of deep neural networks. In: 14th Annual Conference of the International Speech Communication Association (INTERSPEECH2013), pp. 2345–2349 (2013)

33. Vinciarelli, A., Luettin, J.: A new normalisation technique for cursive handwritten words. Pattern Recogn. Lett. 22, 1043–1050 (2001)

Probabilistic Anomaly Detection Method for Authorship Verification

Mohamed Amine Boukhaled[(⊠)] and Jean-Gabriel Ganascia

LIP6 (Laboratoire D'Informatique de Paris 6), Université Pierre et Marie Curie
and CNRS (UMR7606), ACASA Team, 4 Place Jussieu,
75252 Paris Cedex 05, France
{mohamed.boukhaled,jean-gabriel.ganascia}@lip6.fr

Abstract. Authorship verification is the task of determining if a given text is
written by a candidate author or not. In this paper, we present a first study on
using an anomaly detection method for the authorship verification task. We have
considered a weakly supervised probabilistic model based on a multivariate
Gaussian distribution. To evaluate the effectiveness of the proposed method, we
conducted experiments on a classic French corpus. Our preliminary results show
that the probabilistic method can achieve a high verification performance that
can reach an F_1 score of 85 %. Thus, this method can be very valuable for
authorship verification.

Keywords: Authorship verification · Anomaly detection · Multivariate
gaussian distribution

1 Introduction

Authorship verification is a special case of the authorship attribution problem. The
authorship attribution problem can be generally formulated as follows: given a set of
candidate authors for whom samples of written text are available, the task is to assign a
text of unknown authorship to one of these candidate authors [17]. This task has been
addressed mainly as a problem of multi-class discrimination, or as a text categorization
task [16]. In the authorship verification problem, though, we are given samples of texts
written by a single author and are asked to assess if a given different text is written by
this author or not [13]. As a categorization problem, modifying the original attribution
problem in this way makes the task of authorship verification significantly more difficult partly because building a characterising model of one author is much harder than
building a distinguishing model between two authors [12].

Authorship verification has two key steps: an indexing step based on style markers
is performed on the text using some natural language processing techniques such as
tagging, parsing, and morphological analysis; then an identification step is applied
using the indexed markers to verify the validity of the authorship. Many style markers
have been used to characterise writing styles, from early studies based on sentence
length and vocabulary richness [19] to more recent and relevant work based on
function words [9, 20], punctuation marks [2], part-of-speech (POS) tags [14], parse

© Springer International Publishing Switzerland 2014
L. Besacier et al. (Eds.): SLSP 2014, LNAI 8791, pp. 211–219, 2014.
DOI: 10.1007/978-3-319-11397-5_16

trees [6] and character-based features [11]. There is an agreement among researchers that function words are the most reliable indicator of authorship [17].

The verification step can be addressed as a one-class problem (written-by-the-author) or as a binary classification problem (written-by-the-author as positive vs not-written-by-the-author as negative). However, both of these formulations of the problem have drawbacks: In the case of binary classification, one should collect a reasonable amount of representative texts of the entire "not-written-by-the-author" class, which is difficult, if not impossible. In the case of one-class classification, one does not take advantage from negative examples that we do not actually lack for them even though they are not representative of the entire class.

In this paper, we address the authorship verification problem as an anomaly detection problem where texts written by the candidate author are seen as normal data while texts not written by that author are seen anomalous data. We propose a probabilistic anomaly detection method that can benefit from negative examples for the authorship verification process.

We first give an overview of the anomaly detection problem in Sect. 2 and then describe our method in Sect. 3. We than experimentally validate the proposed method in Sect. 4 using a classic French corpus. Finally we use this method to settle a literary mystery case.

2 Anomaly Detection

Anomaly detection is a challenging task which consists of identifying patterns in data that do not conform to expected (normal) behaviour. These non-conforming patterns are called anomalies or outliers [3]. Anomaly detection has been successfully used in many applications such as fault detection, radar target detection and hand written digit recognition [15].

This technique has also been used to deal with textual data for various purposes such as detecting novel topics, events, or news stories in a collection of documents or news articles [3]. Anomaly detection is based on the idea that one can never train a classification algorithm on all the possible classes that the system is likely to encounter in real application. Anomaly detection is also suitable for situations in which the class imbalance problem can affect the accuracy of classification (see Fig. 1) [18].

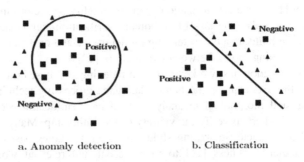

a. Anomaly detection   b. Classification

Fig. 1. The anomaly detection and the classification learning schemas

Many anomaly detection techniques fall under the statistical approach of modelling data based on its statistical properties and using this information to estimate whether a test sample comes from the same distribution or not [15]. Another common method for anomaly detection is the one-class SVM that determines a hyper sphere enclosing the normal data [8]. In this contribution, we describe and use a probabilistic anomaly detection method for authorship verification that straightforwardly follows the definition given above. The method is discussed in the next section.

3 Proposed Method

In our method, we address the authorship verification problem as an anomaly detection problem where texts written by a given author X are seen as normal data, while texts not written by that author X are seen anomalous data. We use a probabilistic anomaly detection method that can benefit from anomalous examples for the authorship verification process based on a multivariate Gaussian modelling. Given the fact that unsupervised anomaly detection approaches often fail to match the required detection rates in many tasks and there exists a need for labelled data to guide the model generation [7], our proposed methods is weakly supervised in the sense that it takes into consideration a small amount of representative anomalous data for the model generation.

The approach to anomalous text detection is to train a multivariate Gaussian distribution model on the style markers extracted from sample of text written by an author X. Every newly arriving text (data instance) that we went to verify as written by X or not is contrasted with the probabilistic model of normality, and a normality probability is computed. The probability describes the likelihood of the new text to have been written by X compared to the average data instances seen during the training. If the probability does not surpass a predefined threshold α, the instance is considered an anomaly and the text is considered not to have been written by the author X. To define the probability threshold, we cross-validate over a data set containing both anomalous and non-anomalous data and we set the threshold to the value that maximizes the authorship verification performance on this data set. The method can be formulated into three steps as follow: Let x_i be a n-dimensional vector representing the text i ($i = 1,..., m$).

1. Train a Multivariate Gaussian distribution model $M(x)$ on the normal data. This is done by estimating the two distribution parameters: the multivariate location μ and the covariance matrix Σ:

$$\mu = \frac{1}{m} \sum_{i=1}^{m} x^{(i)} \tag{1}$$

$$\Sigma = \frac{1}{m} \sum_{i=1}^{m} \left(x^{(i)} - \mu \right) \left(x^{(i)} - \mu \right)^{T} \tag{2}$$

2. Given a new instance x, compute the probability $p(x)$:

$$p(x) = \frac{1}{(2\pi)^{\frac{n}{2}}|\Sigma|^{\frac{1}{2}}} \exp(-\frac{1}{2}(x-\mu)^T \Sigma^{-1}(x-\mu)) \qquad (3)$$

3. Predict the anomaly ($y = 1$) of the instance x given the probability threshold α:

$$y = \begin{cases} 1 \text{ if } p(x) < \alpha \\ 0 \text{ if } p(x) \geq \alpha \end{cases} \qquad (4)$$

The nature of the style markers used as attributes to describe and to get an n-dimensional vector representing the text is very important and determines the applicability of our method. In fact, the nature of these attributes should respect the Gaussian assumption made to train the multivariate Gaussian model. For our experiment, we chose to test this method on two types of style markers separately. Each text in our data set is mapped onto a vector of the frequency of the most frequent function words and a vector of the frequency of POS-tags.

There are two main reasons for using the frequency of function words as attributes. First, because of their high frequency in a written text, function words are very likely to have a Gaussian behaviour (see Fig. 2). Secondary, function words, unlike content words, are difficult to consciously control, thus they are more independent from the topic or the genre of the text [4]. In fact, Koppel and Schler found that all the work of distinguishing the styles of different authors is accomplished with a small set of features containing frequent function words [12]. Based on that information and to get a right balance between the features-set size and the dataset size, we limit our study to the most 30th frequent function words. The part-of-speech-based markers are also shown to be very effective because they partly share the advantages of function words.

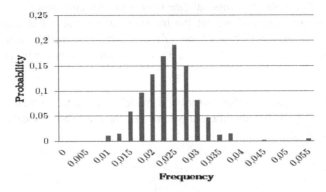

Fig. 2. The probability of frequency of the French function word "de" has a Gaussian behavior

4 Experimental Validation

4.1 Data Set

To test the effectiveness of our method, we used novels written by: Balzac, Dumas and France. This choice was motivated by our special interest in studying the classic French literature of the 19th century, and by the availability of electronic texts from these authors on the project Gutenberg website[1] and in the Gallica electronic library.[2] Our choice of authors was also affected by the fact that we wished to get a challenging problem since these three authors are knows to have relatively comparable syntactic styles. More information about the data set used for the experimentation is summarized in Table 1.

Table 1. Data set used in our experiment

Author name	# of texts
Balzac, Honoré de	126
Dumas, Alexandre	190
France, Anatole	128

For each of the three authors mentioned above, we collected 4 novels, so that the total number of novels was 12. The next step was to divide these novels into smaller pieces of texts in order to have enough data instances (artificial documents) to train and test the probabilistic model. Researchers working on authorship attribution in literary texts have used different dividing strategies. For example, Hoover [10] decided to take just the first 10,000 words of each novel as a single text, while Argamon and Levitan [1] treated each chapter of each book as a separate text. In our experiment, we chose simply to chunk each novel into approximately equal parts of 2000 words, which is below the threshold proposed by Eder [5] specifying the smallest reasonable text size to achieve good attribution. This increases the degree of the difficulty of the task.

4.2 Verification Protocol

In our experiment, the corpus was POS tagged and function words were extracted. Each text is then represented by two vectors $R_n = \{r_1, r_2,...,r_n\}$, one for the normalized frequencies of occurrence of the top 30 function words in the corpus, and another for the normalized frequencies of occurrence of POS-tags. The normalization of the vectors of frequency representing a given text was done according to the size of the text. Then, for each author, we used 75 % of the data generated by texts written by this author to estimate the parameters of the model representing this author, and 20 % of the data from each author for testing it. The remaining 5 % data was merged with 5 % of

---

[1] http://www.gutenberg.org/
[2] http://gallica.bnf.fr/

the data (anomalous data) generated by each one of the other authors and was used to estimate the probability threshold α. To get a reasonable estimate of the expected generalization performance, we used a resampling with replacement method. The training and testing process was done 10 times. The overall authorship verification performance is taken as the average performance over these 10 runs. For evaluating the verification performance, we used the standard measures, calculating precision (P), recall (R), and F_β where:

$$F_\beta = \frac{(1+\beta^2)RP}{(\beta^2 R) + P} \tag{5}$$

We consider precision and recall to have the same value, so we set β equal to 1.

4.3 Baselines

To evaluate the effectiveness of the proposed method we used one-class SVM and binary SVMs classifier using RBF kernel (best performing). The one-class SVM was trained and tested on the same data used to train and test the multivariate Gaussian model respectively. The binary SVM classifier was trained on both the data used to train the probabilistic model and the data used to estimate the probability threshold, and it was tested on the same data as our probabilistic model. The overall baseline classification performance is taken as the average performance over the 10 runs.

4.4 Results

The results of measuring the verification performance for the two different style markers presented in our experimental validation are summarized in Table 2 for function words and in Table 3 for POS tags. These results show in general the superiority of the proposed method over the baselines in terms of F_1 score and recall. These results also show in general a better performance when using frequent function words than POS-tag for both the proposed method and the baselines.

Our study here indicates that the multivariate Gaussian model for anomaly detection combined with features based on frequent function words can achieve a high verification performance (e.g., F1 = 0.85). By contrast, the one-class SVM performs particularly poorly on this task. The binary SVM achieved relatively good results but doesn't outperform the probabilistic model; this shows that the authorship verification problem should not be handled as a binary class problem unless a sufficient amount of representative negative data is present to avoid the class imbalance problem.

Table 2. Results of the authorship verification using frequent function words

Method	P	R	F_1
One-class SVMs	0,34	0,50	0,40
Binary SVMs	**0,86**	0,75	0,80
Multivariate Gaussian model	0,82	**0,88**	**0,85**

Table 3. Results of the authorship verification using frequent POS-tags

Method	P	R	F_1
One-class SVMs	0,51	0,45	0,48
Binary SVMs	**0,81**	0,58	0,67
Multivariate Gaussian model	0,69	**0,89**	**0,77**

Finally, these results are in line with previous work that claimed that semi-supervised anomaly detection approaches, originating from a supervised classifier, are inappropriate and hardly detect new and unknown anomalies, and that semi-supervised anomaly detection needs to be grounded in the unsupervised learning paradigm [7].

5 A Classic French Literary Mystery: "Le Roman de Violette"

In this section, we apply our probabilistic method to settle one of the classic French literary mysteries. "Le Roman de Violette"[3] is a novel published in 1883. The authorship of this novel has still not been determined. Even though the novel was edited under the name of Alexandre Dumas, some literary critics state that a serious candidate for its authorship is "La Marquise de Mannoury d'Ectot". But this hypothesis cannot be definitely proved, partly because there is only one known book written by that author, which limits the quantity of text available to validate the computational authorship identification methods including our method.

We applied our proposed authorship verification method to handle this case. Since there is not enough available text written by "La Marquise de Mannoury d'Ectot" to verify whether she is the writer of "Le Roman de Violette" or not, we set Alexandre Dumas as the author candidate that we want to verify as the writer or not. We trained the probabilistic model based on frequent function words on texts written by Alexandre Dumas. The only known book written by "La Marquise de Mannoury d'Ectot" was used as the representative anomalous text to set the probability threshold. Finally, the verification test was performed on the "Roman de Violette". The authorship probability produced by the novel using our proposed method is under the threshold needed to validate the authorship. This result suggests that the novel "Le Roman de Violette" was not written by Alexandre Dumas.

6 Conclusion

In this paper, we have presented a study on using an anomaly detection method for the authorship verification task. We have considered a weakly supervised probabilistic model based on a multivariate Gaussian distribution. To evaluate the effectiveness of

---

[3] http://ero.corneille-moliere.com/?p=page52&m=ero&l=fra

the proposed method, we conducted experiments on a classic French literary corpus. Our preliminary results show that the probabilistic method can achieve a high verification performance that can reach an F1 score of 85 %.

Based on the current study, we have identified several future research directions. First, we will explore incorporating the non-verification option into our probabilistic model. In fact, in the field of authorship identification, the non-attribution option is better than a false attribution. Second, this study will be expanded to include more style markers. Third, we intend to experiment with other languages and text sizes using standard corpora employed in the field at large.

References

1. Argamon, S., Levitan, S.: Measuring the usefulness of function words for authorship attribution. In: Proceedings of the Joint Conference of the Association for Computers and the Humanities and the Association for Literary and Linguistic Computing (2005)
2. Baayen, H., van Halteren, H., Neijt, A., Tweedie, F.: An experiment in authorship attribution. In: 6th JADT, pp. 29–37 (2002)
3. Chandola, V., Banerjee, A., Kumar, V.: Anomaly detection: a survey. ACM Comput. Surv. (CSUR) **41**(3), 15 (2009)
4. Chung, C., Pennebaker, J.W.: The psychological functions of function words. In: Fielder, K. (ed.) Social Communication, pp. 343–359. Psychology Press, New York (2007)
5. Eder, M.: Does size matter? Authorship attribution, small samples, big problem. Lit. Linguist. Comput. fqt066 (2013)
6. Gamon, M.: Linguistic correlates of style: authorship classification with deep linguistic analysis features. In: Proceedings of the 20th International Conference on Computational Linguistics, p. 611 (2004)
7. Görnitz, N., Kloft, M.M., Rieck, K., Brefeld, U.: Toward supervised anomaly detection (2014) *arXiv Preprint arXiv:1401.6424*
8. Heller, K., Svore, K., Keromytis, A.D., Stolfo, S.: One class support vector machines for detecting anomalous windows registry accesses. In: Workshop on Data Mining for Computer Security (DMSEC), Melbourne, FL, 19 November 2003, pp. 2–9 (2003)
9. Holmes, D.I., Robertson, M., Paez, R.: Stephen Crane and the New-York tribune: a case study in traditional and non-traditional authorship attribution. Comput. Humanit. **35**(3), 315–331 (2001)
10. Hoover, D.L.: Frequent collocations and authorial style. Lit. Linguist. Comput. **18**(3), 261–286 (2003)
11. Kešelj, V., Peng, F., Cercone, N., Thomas, C.: N-gram-based author profiles for authorship attribution. In: Proceedings of the Conference Pacific Association for Computational Linguistics, PACLING, vol. 3, pp. 255–264 (2003)
12. Koppel, M., Schler, J.: Authorship verification as a one-class classification problem. In: Proceedings of the Twenty-First International Conference on Machine Learning, p. 62 (2004)
13. Koppel, M., Schler, J., Argamon, S.: Computational methods in authorship attribution. J. Am. Soc. Inf. Sci. Technol. **60**(1), 9–26 (2009)
14. Kukushkina, O.V., Polikarpov, A.A., Khmelev, V.: Using literal and grammatical statistics for authorship attribution. Probl. Inf. Transm. **37**(2), 172–184 (2001)

15. Markou, M., Singh, S.: Novelty detection: a review—part 1: statistical approaches. Sig. Process. **83**(12), 2481–2497 (2003)
16. Sebastiani, F.: Machine learning in automated text categorization. ACM Comput. Surv. (CSUR) **34**(1), 1–47 (2002)
17. Stamatatos, E.: A survey of modern authorship attribution methods. J. Am. Soc. Inform. Sci. Technol. **60**(3), 538–556 (2009)
18. Wressnegger, C., Schwenk, G., Arp, D., Rieck, K.: A close look on n-grams in intrusion detection: anomaly detection vs. classification. In: Proceedings of the 2013 ACM Workshop on Artificial Intelligence and Security, pp. 67–76 (2013)
19. Yule, G.U.: The Statistical Study of Literary Vocabulary. CUP Archive, Cambridge (1944)
20. Zhao, Y., Zobel, J.: Effective and scalable authorship attribution using function words. In: Lee, G.G., Yamada, A., Meng, H., Myaeng, S.H. (eds.) Information Retrieval Technology. LNCS, vol. 3689, pp. 174–189. Springer, Heidelberg (2005)

Measuring Global Similarity Between Texts

Uli Fahrenberg[1]([✉]), Fabrizio Biondi[1], Kevin Corre[1], Cyrille Jegourel[1],
Simon Kongshøj[2], and Axel Legay[1]

[1] Inria/IRISA Rennes, Rennes, France
ulrich.fahrenberg@irisa.fr
[2] University College of Northern Denmark, Aalborg, Denmark

Abstract. We propose a new similarity measure between texts which,
contrary to the current state-of-the-art approaches, takes a global view
of the texts to be compared. We have implemented a tool to compute our
textual distance and conducted experiments on several corpuses of texts.
The experiments show that our methods can reliably identify different
global types of texts.

1 Introduction

Statistical approaches for comparing texts are used for example in *machine translation* for assessing the quality of machine translation tools [19,20,23], or in *computational linguistics* in order to establish authorship [3,15,18,25,26,31] or to detect "fake", *i.e.*, automatically generated, scientific papers [17,22].

Generally speaking, these approaches consist in computing *distances*, or *similarity measures*, between texts and then using statistical methods such as, for instance, hierarchical clustering [11] to organize the distance data and draw conclusions.

The distances between texts which appear to be the most popular, *e.g.*, [15,23], are all based on measuring differences in *1-gram frequencies*: For each 1-gram (token, or word) w in the union of A and B, its absolute frequencies in both texts are calculated, *i.e.*, $F_A(w)$ and $F_B(w)$ are the numbers of occurrences of w in A and B, respectively, and then the distance between A and B is defined to be the sum, over all words w in the union of A and B, of the absolute differences $|F_A(w) - F_B(w)|$, divided by the combined length of A and B for normalization. When the texts A and B have different length, some adjustments are needed; also, some algorithms [19,23] take into account also 2-, 3- and 4-grams.

These distances are thus based on a *local* model of the texts: they measure differences of the multisets of n-grams for n between 1 and 4. Borrowing techniques from economics and theoretical computer science, we will propose below a new distance which instead builds on the *global* structure of the texts. It simultaneously measures differences in occurrences of n-grams for all n and uses a discounting parameter to balance the influence of long n-grams versus short n-grams.

Following the example of [17], we then use our distance to automatically identify "fake" scientific papers. These are "papers" which are automatically

L. Besacier et al. (Eds.): SLSP 2014, LNAI 8791, pp. 220–232, 2014.
DOI: 10.1007/978-3-319-11397-5_17

generated by some piece of software and are hence devoid of any meaning, but which, at first sight, have the *appearance* of a genuine scientific paper.

We can show that using our distance and hierarchical clustering, we are able to automatically identify such fake papers, also papers generated by other methods than the ones considered in [17], and that, importantly, some parts of the analysis become more reliable the higher the discounting factor. We conclude that measuring *global* differences between texts, as per our method, can be a more reliable way than the current state-of-the-art methods to automatically identify fake scientific papers. We believe that this also has applications in other areas such machine translation or computational linguistics.

Due to space constraints, some of the proofs and of the statistical analysis of results announced in this paper are available in a companion technical report [5].

2   Inter-textual Distances

For the purpose of this paper, a *text* A is a sequence $A = (a_1, a_2, \ldots, a_{N_A})$ of words. The number N_A is called the *length* of A. As a vehicle for showing idealized properties, we may sometimes also speak of *infinite* texts, but most commonly, texts are finite and their length is a natural number. Note that we pay no attention to punctuation, structure such as headings or footnotes, or non-textual parts such as images.

2.1   1-Gram Distance

Before introducing our global distance, we quickly recall the definition of standard 1-gram distance, which stands out as a rather popular distance in computational linguistics and other areas [3, 15–18, 25, 26, 31].

For a text $A = (a_1, a_2, \ldots, a_{N_A})$ and a word w, the natural number $F_A(w) = |\{i \mid a_i = w\}|$ is called the *absolute frequency* of w in A: the number of times (which may be 0) that w appears in A. We say that w is *contained* in A and write $w \in A$ if $F_A(w) \geq 1$.

For texts $A = (a_1, a_2, \ldots, a_{N_A})$, $B = (b_1, b_2, \ldots, b_{N_B})$, we write $A \circ B = (a_1, \ldots, a_{N_A}, b_1, \ldots, b_{N_B})$ for their *concatenation*. With this in place, the *1-gram distance* between texts A and B *of equal length* is defined to be

$$d_1(A, B) = \frac{\sum_{w \in A \circ B} |F_A(w) - F_B(w)|}{N_A + N_B},$$

where $|F_A(w) - F_B(w)|$ denotes the absolute difference between the absolute frequencies $F_A(w)$ and $F_B(w)$.

For texts A and B which are not of equal length, scaling is used: for $N_A < N_B$, one lets

$$d_1(A, B) = \frac{\sum_{w \in A \circ B} |F_A(w) - F_B(w) \frac{N_A}{N_B}|}{2N_A}.$$

By counting occurrences of n-grams instead of 1-grams, similar n-gram distances may be defined for all $n \geq 1$. The BLEU distance [23] for example, popular for evaluation of machine translation, computes n-gram distance for n between 1 and 4.

2.2   Global Distance

To compute our global inter-textual distance, we do not compare word frequencies as above, but *match n-grams* in the two texts *approximately*. Let $A = (a_1, a_2, \ldots, a_{N_A})$ and $B = (b_1, b_2, \ldots, b_{N_B})$ be two texts, where we make no assertion about whether $N_A < N_B$, $N_A = N_B$ or $N_A > N_B$. Define an indicator function $\delta_{i,j}$, for $i \in \{1, \ldots, N_A\}$, $j \in \{1, \ldots, N_B\}$, by

$$\delta_{i,j} = \begin{cases} 0 & \text{if } a_i = b_j, \\ 1 & \text{otherwise} \end{cases} \tag{1}$$

(this is the *Kronecker delta* for the two sequences A and B). The symbol $\delta_{i,j}$ indicates whether the i-th word a_i in A is the same as the j-th word b_j in B. For ease of notation, we extend $\delta_{i,j}$ to indices above i, j, by declaring $\delta_{i,j} = 1$ if $i > N_A$ or $j > N_B$.

Let $\lambda \in \mathbb{R}$, with $0 \leq \lambda < 1$, be a *discounting factor*. Intuitively, λ indicates how much weight we give to the length of n-grams when matching texts: for $\lambda = 0$, we match 1-grams only (see also Theorem 1 below), and the higher λ, the longer the n-grams we wish to match. Discounting is a technique commonly applied for example in economics, when gauging the long-term effects of economic decisions. Here we remove it from its time-based context and apply it to *n-gram length* instead: We define the *position match* from any position index pair (i, j) in the texts by

$$d_{\mathrm{pm}}(i, j, \lambda) = \delta_{i,j} + \lambda \delta_{i+1,j+1} + \lambda^2 \delta_{i+2,j+2} + \cdots$$
$$= \sum_{k=0}^{\infty} \lambda^k \delta_{i+k,j+k}. \tag{2}$$

This measures how much the texts A and B "look alike" when starting with the tokens a_i in A and b_j in B. Note that it takes values between 0 (if a_i and b_j are the starting points for two equal infinite sequences of tokens) and $\frac{1}{1-\lambda}$. Intuitively, the more two token sequences are alike, and the later they become different, the smaller their distance. Table 1 shows a few examples of position match calculations.

This gives us an N_A-by-N_B matrix $D_{\mathrm{pm}}(\lambda)$ of position matches; see Table 2 for an example. We now need to consolidate this matrix into *one* global distance value between A and B. Intuitively, we do this by averaging over position matches: for each position a_i in A, we find the position b_j in B which best matches a_i, i.e., for which $d_{\mathrm{pm}}(i, j, \lambda)$ is minimal, and then we average over these matchings.

Table 1. Position matches, starting from index pair $(1,1)$ and scaled by $1 - \lambda$, of different example texts, for general discounting factor λ and for $\lambda = .8$. Note that the last two example texts are infinite.

Text A	Text B	$(1 - \lambda)d_{\mathrm{pm}}(1,1,\lambda)$	$\lambda = .8$
"man"	"dog"	1	1
"dog"	"dog"	λ	.8
"man bites dog"	"man bites dog"	λ^3	.51
"man bites dog"	"dog bites man"	$1 - \lambda + \lambda^2$	.84
"the quick brown fox jumps over the lazy dog"	"the quick white fox crawls under the high dog"	$\lambda^2 - \lambda^3 + \lambda^4 - \lambda^6 + \lambda^7 - \lambda^8 + \lambda^9$	.45
"me me me me..."	"me me me me..."	0	0

Table 2. Position match matrix example, with discounting factor $\lambda = .8$.

	the	quick	fox	jumps	over	the	lazy	dog
the	0.67	1.00	1.00	1.00	1.00	0.64	1.00	1.00
lazy	1.00	0.84	1.00	1.00	1.00	1.00	0.80	1.00
fox	1.00	1.00	0.80	1.00	1.00	1.00	1.00	1.00

Formally, this can be stated as an *assignment problem*: Assuming for now that $N_A = N_B$, we want to find a matching of indices i to indices j which minimizes the sum of the involved $d_{\mathrm{pm}}(i,j)$. Denoting by S_{N_A} the set of all permutations of indices $\{1,\ldots,N_A\}$ (the *symmetric group* on N_A elements), we hence define

$$d_2(A, B, \lambda) = (1 - \lambda)\frac{1}{N_A} \min_{\phi \in S_{N_A}} \sum_{i=1}^{N_A} d_{\mathrm{pm}}(i, \phi(i), \lambda).$$

This is a conservative extension of 1-gram distance, in the sense that for discounting factor $\lambda = 0$ we end up computing d_1:

Theorem 1. *For all texts A, B with equal lengths, $d_2(A, B, 0) = d_1(A, B)$.*

Proof. For $\lambda = 0$, the entries in the phrase distance matrix are $d_{\mathrm{pm}}(i, j, 0) = \delta_{i,j}$. Hence a *perfect match* in D_{pm}, with $\sum_{i=1}^{N_A} d_{\mathrm{pm}}(i, \phi(i), 0) = 0$, matches each word in A with an equal word in B and vice versa. This is possible if, and only if, $F_A(w) = F_B(w)$ for each word w. Hence $d_2(A, B, 0) = 0$ iff $d_1(A, B) = 0$. The proof of the general case is in [5]. $\square$

There are, however, some problems with the way we have defined d_2. For the first, the assignment problem is computationally rather expensive: the best know algorithm (the *Hungarian algorithm* [13]) runs in time cubic in the size of the matrix, which when comparing large texts may result in prohibitively long running times. Secondly, and more important, it is unclear how to extend this definition to texts which are not of equal length, *i.e.*, for which $N_A \neq N_B$. (The scaling approach does not work here.)

Hence we propose a different definition which has shown to work well in practice, where we abandon the idea that we want to match phrases *uniquely*. In the definition below, we simply match every phrase in A with its best equivalent in B, and we do not take care whether we match two different phrases in A with the same phrase in B. Hence,

$$d_3(A, B, \lambda) = \frac{1}{N_A} \sum_{i=1}^{N_A} \min_{j=1,\dots,N_B} d_{\mathrm{pm}}(i, j, \lambda).$$

Note that $d_3(A, B, \lambda) \leq d_2(A, B, \lambda)$, and that contrary to d_2, d_3 is *not symmetric*. We can fix this by taking as our final distance measure the symmetrization of d_3:

$$d_4(A, B, \lambda) = \max(d_3(A, B, \lambda), d_3(B, A, \lambda)).$$

3   Implementation

We have written a C program and some `bash` helper scripts which implement the computations above. All our software is available at http://textdist.gforge. inria.fr/.

The C program, `textdist.c`, takes as input a list of `txt`-files $A_1, A_2, \dots, A_k$ and a discounting factor λ and outputs $d_4(A_i, A_j, \lambda)$ for all pairs $i, j = 1, \dots, k$. With the current implementation, the `txt`-files can be up to 15,000 words long, which is more than enough for all texts we have encountered. On a standard 3-year-old business laptop (Intel® Core™ i5 at 2.53GHz×4), computation of d_4 for takes less than one second for each pair of texts.

We preprocess texts to convert them to `txt`-format and remove non-word tokens. The `bash`-script `preprocess-pdf.sh` takes as input a `pdf`-file and converts it to a text file, using the `poppler` library's `pdftotext` tool. Afterwards, `sed` and `grep` are used to convert whitespace to newlines and remove excessive whitespace; we also remove all "words" which contain non-letters and only keep words of at least two letters.

The `bash`-script `compareall.sh` is used to compute mutual distances for a corpus of texts. Using `textdist.c` and taking λ as input, it computes $d_4(A, B, \lambda)$ for all texts (`txt`-files) A, B in a given directory and outputs these as a matrix. We then use R and `gnuplot` for statistical analysis and visualization.

We would like to remark that all of the above-mentioned tools are free or open-source software and available without charge. One often forgets how much science has come to rely on this free-software infrastructure.

4   Experiments

We have conducted two experiments using our software. The data sets on which we have based these experiments are available on request.

4.1   Types of Texts Used

We have run our experiments on papers in computer science, both genuine papers and automatically generated "fake" papers. As to the genuine papers, for the first experiment, we have used 42 such papers from within theoretical computer science, 22 from the proceedings of the FORMATS 2011 conference [9] and 20 others which we happened to have around. For the second experiment, we collected 100 papers from arxiv.org, by searching their Computer Science repository for authors named "Smith" (`arxiv.org` strives to prevent bulk paper collection), of which we had to remove three due to excessive length (one "status report" of more than 40,000 words, one PhD thesis of more than 30,000 words, and one "road map" of more than 20,000 words).

We have employed three methods to collect automatically generated "papers". For the first experiment, we downloaded four fake publications by "Ike Antkare". These are out of a set of 100 papers by the same "author" which have been generated, using the SCIgen paper generator, for another experiment [14]. For the purpose of this other experiment, these papers all have the same bibliography, each of which references the other 99 papers; hence not to skew our results (and like was done in [17]), we have stripped their bibliography.

SCIgen[1] is an automatic generator of computer science papers developed in 2005 for the purpose of exposing "fake" conferences and journals (by submitting generated papers to such venues and getting them accepted). It uses an elaborate grammar to generate random text which is devoid of any meaning, but which to the untrained (or inattentive) eye looks entirely legitimate, complete with abstract, introduction, figures and bibliography. For the first experiment, we have supplemented our corpus with four SCIgen papers which we generated on their website. For the second experiment, we modified SCIgen so that we could control the length of generated papers and then generated 50 papers.

For the second experiment, we have also employed another paper generator which works using a simple Markov chain model. This program, `automogensen`[2], was originally written to expose the lack of meaning of many of a certain Danish political commentator's writings, the challenge being to distinguish genuine *Mogensen* texts from "fake" `automogensen` texts. For our purposes, we have modified `automogensen` to be able to control the length of its output and fed it with a 248,000-word corpus of structured computer science text (created by concatenating all 42 genuine papers from the first experiment), but otherwise, its functionality is rather simple: It randomly selects a 3-word starting phrase from the corpus and then, recursively, selects a new word from the corpus based on the last three words in its output and the distribution of successor words of this three-word phrase in the corpus.

---

[1] http://pdos.csail.mit.edu/scigen/
[2] http://www.kongshoj.net/automogensen/

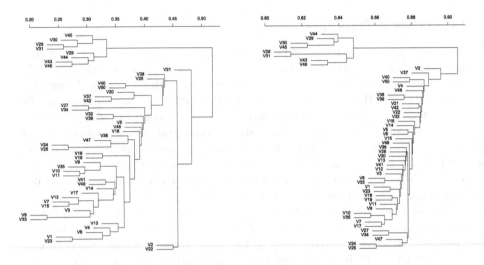

Fig. 1. Dendrograms for Experiment 1, using average clustering, for discounting factors 0 (left) and .95 (right), respectively. Fake papers are numbered 28-31 (Antkare) and 43-46 (SCIgen), the others are genuine.

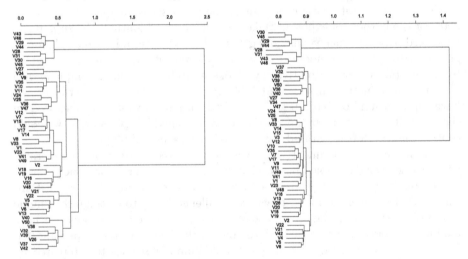

Fig. 2. Dendrograms for Experiment 1, using Ward clustering, for discounting factors 0 (left) and .95 (right), respectively. Fake papers are numbered 28-31 (Antkare) and 43-46 (SCIgen), the others are genuine.

4.2   First Experiment

The first experiment was conducted on 42 genuine papers of lengths between 3,000 and 11,000 words and 8 fake papers of lengths between 1500 and 2200 words. Figure 1 shows two dendrograms with average clustering created from the collected distances; more dendrograms are available in [5]. The left dendrogram

Table 3. Minimal and maximal distances between different types of papers depending on the discounting factor.

Type	Discounting	0	.1	.2	.3	.4	.5	.6	.7	.8	.9	.95
genuine/genuine	min	.23	.26	.30	.35	.40	.45	.52	.59	.68	.79	.86
	max	.55	.56	.57	.59	.61	.64	.67	.72	.78	.85	.90
fake/fake	min	.26	.28	.31	.35	.39	.43	.49	.55	.63	.73	.81
	max	.38	.40	.43	.46	.49	.53	.58	.64	.71	.80	.86
fake/genuine	min	.44	.46	.49	.52	.55	.59	.64	.70	.76	.84	.89
	max	.58	.60	.62	.64	.66	.68	.72	.76	.80	.87	.92

was computed for discounting factor $\lambda = 0$, *i.e.*, word matching only. One clearly sees the fake papers grouped together in the top cluster and the genuine papers in cluster below. In the right dendrogram, with very high discounting ($\lambda = .95$), this distinction is much more clear; here, the fake cluster is created (at height .85) while all the genuine papers are still separate. The dendrograms in Fig. 2, created using Ward clustering, clearly show that one should distinguish the data into *two* clusters, one which turns out to be composed only of fake papers, the other only of genuine papers.

We want to call attention to two other interesting observations which can be made from the dendrograms in Fig. 1. First, papers 2, 21 and 22 seem to stick out from the other genuine papers. While all other genuine papers are technical papers from within theoretical computer science, these three are not. Paper 2 [10] is a non-technical position paper, and papers 21 [24] and 22 [12] are about applications in medicine and communication. Note that the $\lambda = .95$ dendrogram more clearly distinguishes the position paper [10] from the others.

Another interesting observation concerns papers 8 [2] and 33 [1]. These papers share an author (E. Asarin) and are within the same specialized area (topological properties of timed automata), but published two years apart. When measuring only word distance, *i.e.*, with $\lambda = 0$, these papers have the absolutely lowest distance, .23, even below any of the fake papers' mutual distances, but increasing the discounting factor increases their distance much faster than any of the fake papers' mutual distances. At $\lambda = .95$, their distance is .87, above any of the fake papers' mutual distances. A conclusion can be that these two papers may have *word* similarity, but they are distinct in their *phrasing*.

Finally, we show in Table 3 (see also Fig. 10 in [5] for a visualization) how the mutual distances between the 50 papers evolve depending on the discounting factor. One can see that at $\lambda = 0$, the three types of mutual distances are overlapping, whereas at $\lambda = .95$, they are almost separated into three bands: .81-.86 for fake papers, .86-.90 for genuine papers, and .89-.92 for comparing genuine with fake papers.

Altogether, we conclude from the first experiment that our inter-textual distance can achieve a safe separation between genuine and fake papers in our corpus, and that the separation is stronger for higher discounting factors.

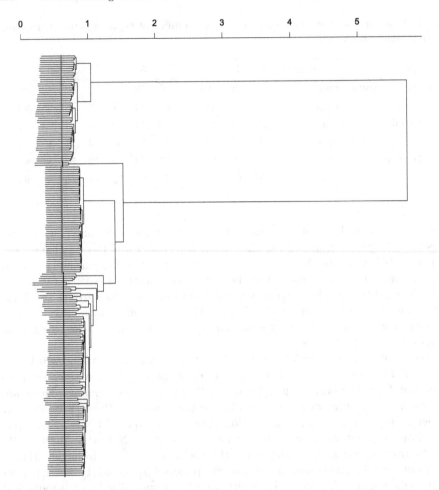

Fig. 3. Dendrogram for Experiment 2, using Ward clustering, for discounting factor .95. Black dots mark `arxiv` papers, green marks SCIgen papers, and `automogensen` papers are marked red (Color figure online).

4.3 Second Experiment

The second experiment was conducted on 97 papers from `arxiv.org`, 50 fake papers generated by a modified SCIgen program, and 50 fake papers generated by `automogensen`. The `arxiv` papers were between 1400 and 15,000 words long, the SCIgen papers between 2700 and 12,000 words, and the `automogensen` papers between 4,000 and 10,000 words. The distances were computed for discounting factors 0, .4, .8 and .95; with our software, computations took about four hours for each discounting factor.

We show the dendrograms using average clustering in Figs. 11 to 14 in [5]; they appear somewhat inconclusive. One clearly notices the SCIgen and `automogensen` parts of the corpus, but the `arxiv` papers have wildly varying distances and disturb the dendrogram. One interesting observation is that with discounting factor 0, the automogensen papers have small mutual distances compared to the arxiv corpus, comparable to the SCIgen papers' mutual distances, whereas with high discounting (.95), the automogensen papers' mutual distances look more like the arxiv papers'. Note that the difficulties in clustering appear also with discounting factor 0, hence also when only matching words.

The dendrograms using Ward clustering, however, do show a clear distinction between the three types of papers. We can only show one of them here, for $\lambda = .95$ in Fig. 3; the rest are available in [5]. One clearly sees the SCIgen cluster (top) separated from all other papers, and then the automogensen cluster (middle) separated from the arxiv cluster.

There is, though, one anomaly: two arxiv papers have been "wrongly" grouped into their own cluster (between the SCIgen and the automogensen clusters). Looking at these papers, we noticed that here our pdf-to-text conversion had gone wrong: the papers' text was all garbled, consisting only of "AOUOO OO AOO EU OO OU AO" etc. The dendrograms rightly identify these two papers in their own cluster; in the dendrograms using average clustering, this garbled cluster consistently has distance 1 to the other clusters.

We also notice in the dendrogram with average clustering and discounting factor .95 (Fig. 14 in [5]) that some of the arxiv papers with small mutual distances have the same authors and are within the same subject. This applies to [27] vs. [28] and to [32] vs. [33]. These similarities appear much more clearly in the $\lambda = .95$ dendrogram than in the ones with lower discounting factor.

As a conclusion from this experiment, we can say that whereas average clustering had some difficulties in distinguishing between fake and arxiv papers, Ward clustering did not have any problems. The only effect of the discounting factor we could see was in identifying similar arxiv papers. We believe that one reason for the inconclusiveness of the dendrograms with average clustering is the huge variety of the arxiv corpus. Whereas the genuine corpus of the first experiment included only papers from the verification sub-field of theoretical computer science, the arxiv corpus is comprised of papers from a diverse selection of research areas within computer science, including robotics, network detection, computational geometry, constraint programming, numerical simulation and many others. Hence, the intra-corpus variation in the arxiv corpus hides the inter-corpus variations.

5   Conclusion and Further Work

We believe we have collected enough evidence that our global inter-textual distance provides an interesting alternative, or supplement, to the standard 1-gram distance. In our experiments, we have seen that measuring inter-textual distance with high discounting factor enables us to better differentiate between similar and dissimilar texts. More experiments will be needed to identify areas where our global matching provides advantages over pure 1-gram matching.

With regard to identifying fake scientific papers, we remark that, according to [17], "[u]sing [the 1-gram distance] to detect SCIgen papers relies on the fact that [...] the SCIgen vocabulary remains quite poor". Springer has recently announced [30] that they will integrate "[a]n automatic SCIgen detection system [...] in [their] submission check system", but they also notice that the "intention [of fake papers' authors] seems to have been to increase their publication numbers and [...] their standing in their respective disciplines and at their institutions"; of course, auto-detecting SCIgen papers does not change these motivations. It is thus reasonable to expect that generators of fake papers will get better, so that also better tools will be needed to detect them. We propose that our phrase-based distance may be such a tool.

There is room for much improvement in our distance definition. For once, we perform no tagging of words which could identify different spellings or inflections of the same word. This could easily be achieved by, using for example the Wordnet database[3], replacing our binary distance between words in Eq. (1) with a quantitative measure of *word similarity*. For the second, we take no consideration of *omitted* words in a phrase; our position match calculation in Eq. (2) cannot see when two phrases become one-off like in "the quick brown fox jumps..." vs. "the brown fox jumps...".

Our inter-textual distance is inspired by our work in [6–8] and other papers, where we define distances between arbitrary *transition systems*. Now a text is a very simple transition system, but so is a text with "one-off jumps" like the one above. Similarly, we can incorporate *swapping* of words into our distance, so that we would be computing a kind of *discounted Damerau-Levenshtein distance* [4] (related approaches, generally without discounting, are used for *sequence alignment* in bioinformatics [21,29]). We have integrated this approach in an experimental version of our `textdist` tool.

References

1. Asarin, E., Degorre, A.: Volume and entropy of regular timed languages. hal (2009). http://hal.archives-ouvertes.fr/hal-00369812
2. Basset, N., Asarin, E.: Thin and thick timed regular languages. In: Fahrenberg and Tripakis [9], pp. 113–128
3. Cortelazzo, M.A., Nadalutti, P., Tuzzi, A.: Improving Labbé's intertextual distance: testing a revised version on a large corpus of italian literature. J. Quant. Linguist. **20**(2), 125–152 (2013)
4. Damerau, F.: A technique for computer detection and correction of spelling errors. Commun. ACM **7**(3), 171–176 (1964)
5. Fahrenberg, U., Biondi, F., Corre, K., Jegourel, C., Kongshøj, S., Legay, A.: Measuring global similarity between texts. Technical report, arxiv (2014). http://arxiv.org/abs/1403.4024
6. Fahrenberg, U., Legay, A.: Generalized quantitative analysis of metric transition systems. In: Shan, C. (ed.) APLAS 2013. LNCS, vol. 8301, pp. 192–208. Springer, Heidelberg (2013)
7. Fahrenberg, U., Legay, A.: The quantitative linear-time-branching-time spectrum. Theor. Comput. Sci. (2013). http://dx.doi.org/10.1016/j.tcs.2013.07.030

---

[3] http://wordnet.princeton.edu/

8. Fahrenberg, U., Legay, A., Thrane, C.R.: The quantitative linear-time-branching-time spectrum. In: Chakraborty, S., Kumar, A. (eds.) FSTTCS. vol. 13 of LIPIcs, pp. 103–114 (2011)
9. Fahrenberg, U., Tripakis, S. (eds.): FORMATS 2011. LNCS, vol. 6919. Springer, Heidelberg (2011)
10. Haverkort, B.R.: Formal modeling and analysis of timed systems: Technology push or market pull? In: Fahrenberg and Tripakis [9], pp. 18–24
11. Kaufman, L., Rousseeuw, P.J.: Finding Groups in Data: An Introduction to Cluster Analysis. Wiley, New York (1990)
12. Kharmeh, S.A., Eder, K., May, D.: A design-for-verification framework for a configurable performance-critical communication interface. In: Fahrenberg and Tripakis [9], pp. 335–351
13. Kuhn, H.W.: The Hungarian method for the assignment problem. Nav. Res. Logist. Q. **2**(1–2), 83–97 (1955)
14. Labbé, C.: Ike Antkare, one of the great stars in the scientific firmament. ISSI Newsl. **6**(2), 48–52 (2010). http://hal.archives-ouvertes.fr/hal-00713564
15. Labbé, C., Labbé, D.: Inter-textual distance and authorship attribution Corneille and Molière. J. Quant. Linguist. **8**(3), 213–231 (2001)
16. Labbé, C., Labbé, D.: A tool for literary studies: intertextual distance and tree classification. Literary Linguist. Comp. **21**(3), 311–326 (2006)
17. Labbé, C., Labbé, D.: Duplicate and fake publications in the scientific literature: how many SCIgen papers in computer science? Scientometrics **94**(1), 379–396 (2013)
18. Labbé, D.: Experiments on authorship attribution by intertextual distance in English. J. Quant. Linguist. **14**(1), 33–80 (2007)
19. Lin, C.Y., Hovy, E.H.: Automatic evaluation of summaries using n-gram co-occurrence statistics. In: HLT-NAACL (2003)
20. Lin, C.Y., Och, F.J.: Automatic evaluation of machine translation quality using longest common subsequence and skip-bigram statistics. In: Scott, D., Daelemans, W., Walker, M.A. (eds.) ACL. pp. 605–612. ACL (2004)
21. Needleman, S.B., Wunsch, C.D.: A general method applicable to the search for similarities in the amino acid sequence of two proteins. J. Mol. Biol. **48**(3), 443–453 (1970)
22. Noorden, R.V.: Publishers withdraw more than 120 gibberish papers. Nature News & Comment, February 2014. http://dx.doi.org/10.1038/nature.2014.14763
23. Papineni, K., Roukos, S., Ward, T., Zhu, W.J.: BLEU: a method for automatic evaluation of machine translation. In: ACL. pp. 311–318. ACL (2002)
24. Sankaranarayanan, S., Homaei, H., Lewis, C.: Model-based dependability analysis of programmable drug infusion pumps. In: Fahrenberg and Tripakis [9], pp. 317–334
25. Savoy, J.: Authorship attribution: a comparative study of three text corpora and three languages. J. Quant. Linguist. **19**(2), 132–161 (2012)
26. Savoy, J.: Authorship attribution based on specific vocabulary. ACM Trans. Inf. Syst. **30**(2), 12 (2012)
27. Smith, S.T., Kao, E.K., Senne, K.D., Bernstein, G., Philips, S.: Bayesian discovery of threat networks. CoRR abs/1311.5552v1 (2013)
28. Smith, S.T., Senne, K.D., Philips, S., Kao, E.K., Bernstein, G.: Network detection theory and performance. CoRR abs/1303.5613v1 (2013)
29. Smith, T., Waterman, M.: Identification of common molecular subsequences. J. Mol. Biol. **147**(1), 195–197 (1981)

30. Springer second update on SCIgen-generated papers in conference proceedings. Springer Statement, April 2014. http://www.springer.com/about+springer/media/statements?SGWID=0-1760813-6-1460747-0

31. Tomasi, F., Bartolini, I., Condello, F., Degli Esposti, M., Garulli, V., Viale, M.: Towards a taxonomy of suspected forgery in authorship attribution field. A case: Montale's Diario Postumo. In: DH-CASE. pp. 10:1–10:8. ACM (2013)

32. Ulusoy, A., Smith, S.L., Ding, X.C., Belta, C.: Robust multi-robot optimal path planning with temporal logic constraints. CoRR abs/1202.1307v2 (2012)

33. Ulusoy, A., Smith, S.L., Ding, X.C., Belta, C., Rus, D.: Optimal multi-robot path planning with temporal logic constraints. CoRR abs/1107.0062v1 (2011)

Identifying and Clustering Relevant Terms in Clinical Records Using Unsupervised Methods

Borbála Siklósi[2]([⊠]) and Attila Novák[1,2]

[1] MTA-PPKE Hungarian Language Technology Research Group,
Pázmány Péter Catholic University, 50/a Práter Street, Budapest 1083, Hungary
[2] Faculty of Information Technology and Bionics,
Pázmány Péter Catholic University, 50/a Práter Street, Budapest 1083, Hungary
{siklosi.borbala,novak.attila}@itk.ppke.hu

Abstract. The automatic processing of clinical documents created at clinical settings has become a focus of research in natural language processing. However, standard tools developed for general texts are not applicable or perform poorly on this type of documents. Moreover, several crucial tasks require lexical resources and relational thesauri or ontologies to identify relevant concepts and their connections. In the case of less-resourced languages, such as Hungarian, there are no such lexicons available. The construction of annotated data and their organization requires human expert work. In this paper we show how applying statistical methods can result in a preprocessed, semi-structured transformation of the raw documents that can be used to aid human work. The modules detect and resolve abbreviations, identify multiword terms and derive their similarity, all based on the corpus itself.

Keywords: Clinical text processing · Abbreviations · Multiword terms · Distributional similarity · Less-resourced languages · Unsupervised methods

1 Introduction

Clinical records are documents created at clinical settings with the purpose of documenting every-day clinical cases or treatments. The quality of this type of text stays far behind that of biomedical texts, which are also the object of several studies. Biomedical texts, mainly written in English, are the ones that are published in scientific journals, books, proceedings, etc. These are written in the standard language, in accordance with orthographic rules [11,16]. On the contrary, clinical records are created as unstructured texts without using any proofing tools, resulting in texts full of spelling errors and nonstandard use of word forms in a language that is usually a mixture of the local language (Hungarian in our case) and Latin [18,19]. These texts are also characterized by a high ratio of abbreviated forms, most of them used in an arbitrary manner. Moreover, in many cases, full statements are written in a special notational

© Springer International Publishing Switzerland 2014
L. Besacier et al. (Eds.): SLSP 2014, LNAI 8791, pp. 233–243, 2014.
DOI: 10.1007/978-3-319-11397-5_18

language [1] that is often used in clinical settings, consisting only, or mostly of abbreviated forms.

Another characteristics of clinical records is that the target readers are usually the doctors themselves, thus using their own unique language and notational habits is not perceived to cause any loss in the information to be stored and retrieved. However, beyond the primary aim of recording patient history, these documents contain much more information which, if extracted, could be useful for other fields of medicine as well. In order to access this implicit knowledge, an efficient representation of the facts and statements recorded in the texts should be created.

Several attempts have been made to apply general text processing tools to clinical notes, but their performance is much worse on these special texts, than on general, well-formed documents (e.g. see [15]). Moreover, applications used for processing domain-specific texts are usually supported by some hand-made lexical resources, such as ontologies or vocabularies. In the case of less-resourced languages, there are very few such datasets and their construction needs quite an amount of human work. Furthermore, as facts and statements in clinical records are from a narrow domain, applications of the sublanguage theory [8] have been used in similar approaches, which also requires a domain-specific categorization of words of the specific sublanguage [7,9,16].

In order to be able to support the adaptation of existing tools, and the building of structured resources, we examined a corpus of Hungarian ophthalmology notes. In this study, statistical methods are applied to the corpus in order to capture as much information as possible based on the raw data. Even though the results of each module are not robust representations of the underlying information, these groups of semi-structured data can be used in the real construction process.

In this paper, these preprocessing methods are applied to a Hungarian corpus of ophthalmology notes. The core of each module is based on statistical observations from the corpus itself, augmented by some linguistic rules or resources at just a very few points. First, we describe the clinical corpus and compare it to a general Hungarian corpus along several aspects in Sect. 2. Then, in Sect. 3 the applied methods are described. Finally, the results of the transformation are presented, which is followed by some concluding thoughts.

2   The Corpus

In this research, anonymized clinical documents from the ophthalmology department of a Hungarian clinic were used. This corpus contains 334 546 tokens (34 432 sentences). The models were built using this set, and tested on another set of documents, which contained 5 599 tokens (693 sentences). The state of the corpus before processing was a structured high-level xml as described in [19]. It was also segmented into sentences and tokens and pos-tagged applying the methods described in [14,15]. Though such preprocessing tasks are considered to be solved for most languages in the case of general texts, they perform significantly worse in the case of clinical documents as discussed in the aforementioned publications. Still, this level of preprocessing was unavoidable.

When our clinical notes are compared to a general Hungarian corpus, we find reasonable differences between the two domains. This explains some of the difficulties that prevent tools developed for general texts working in the clinical domain. These differences are not only present in the semantics of the content, but in the syntax and even in the surface form of the texts and fall into three main categories discussed in the following subsections. The corpus used in the comparison as general text was the Szeged Corpus [4], containing 1 194 348 tokens (70 990 sentences) and the statistics related to this corpus was taken from [20].

2.1 Syntactic Behaviour

The length of the sentences used in a language reflects the complexity of the syntactic behaviour of utterances. In the general corpus, the average length of sentences is 16.82 tokens, while in the clinical corpus it is 9.7. Doctors tend to use shorter and rather incomplete and compact statements. This habit makes the creation of the notes faster, but the high frequency of ellipsis of crucial grammatical constituents makes most parsers fail when trying to process them.

Regarding the distribution of part-of-speech (pos) in the two domains, there are also significant differences. While in the general corpus, the three most frequent types are nouns, verbs and adjectives, in the clinical domain nouns are followed by adjectives and numbers in the frequency ranking, while the number of verbs in this corpus is just one third of the number of the latter two. Another significant difference is that in the clinical domain, determiners, conjunctions, and pronouns are also ranked lower in the frequency list. These occurrence ratios are not surprising, since a significant portion of clinical documents record a statement (*something has a property*, which is expressed in Hungarian with a phrase containing only a noun and an adjective), or the result of an examination (*the value of something is some amount*, i.e. a noun and a number). Furthermore, most of the numbers in the clinical corpus are numerical data. Table 1 shows the detailed statistics and ranking of pos tags in the two corpora.

Table 1. The distribution and ranking of part-of-speech in the clinical corpus (CLIN) and the general Szeged Corpus (SZEG)

	NOUN	ADJ	NUM	VERB	ADV	PRN	DET	POSTP	CONJ
CLIN	43,02 %	13,87 %	12,33 %	3,88 %	2,47 %	2,21 %	2,12 %	1,03 %	0,87 %
SZEG	21,96 %	9,48 %	2,46 %	9,55 %	7,60 %	3,85 %	9,39 %	1,24 %	5,58 %
	NOUN	ADJ	NUM	VERB	ADV	PRN	DET	POSTP	CONJ
CLIN	1	2	3	4	5	6	7	8	9
SZEG	1	3	8	2	5	7	4	9	6

2.2   Spelling Errors

Clinical documents are usually created in a rush without proofreading. The medical records creation and archival tools used at most Hungarian hospitals provide no proofing or structuring tools. Thus, the number of spelling errors is very high, and a wide variety of error types occur [18]. These errors are not only due to the complexity of the Hungarian language and orthography, but also to characteristics typical of the medical domain and the situation in which the documents are created. The most frequent types of errors are the following:

- mistyping, accidentally swapping letters, inserting extra letters or just missing some,
- lack or improper use of punctuation (e.g. no sign of sentence boundaries, missing commas, no space between punctuation and the neighboring words),
- grammatical errors,
- sentence fragments,
- domain-specific and often ad hoc abbreviations, which usually do not correspond to any standard
- Latin medical terminology not conforming to orthographical standards.

A common characteristic of these phenomena is that the prevailing errors vary with the doctor or assistant typing the text. Thus it can occur that a certain word is mistyped and should be corrected in one document while the same word is a specific abbreviation in another one, which does not correspond to the same concept as the corrected one. Latin medical terms usually have two standard forms, one based on Latin and another based on Hungarian orthography, however what we find in the documents is often an inconsistent mixture of the two (e.g. tensio/tenzio/tensió/tenzió). Even though the spelling of these forms is standardized, doctors tend to develop their own customs which they use inconsistently.

The ratio of misspelled words is 0.27 % in the general Hungarian texts in the Szeged Corpus, while it is 8.44 % in the clinical notes. Moreover, the general corpus has several subcorpora, including one of primary school essays, which still has only an 0.87 % error rate.

2.3   Abbreviations and Word Forms

The use of a kind of notational text is very common in clinical documents. This dense form of documentation contains a high ratio of standard or arbitrary abbreviations and symbols, some of which may be specific to a special domain or even to a doctor or administrator. These short forms might refer to clinically relevant concepts or to some common phrases that are very frequent in the specific domain. For the clinicians, the meaning of these common phrases is as trivial as the standard shortened forms of clinical concepts due to their expertise and familiarity with the context. The difference in the ratio of abbreviations in the general and clinical corpora is also significant, being 0.08 % in the Szeged

Corpus, while 7.15 % in the clinical corpus, which means that the frequency of abbreviations is two orders of magnitude larger in clinical documents than in general language.

3   Applied Methods

In this section, three methods are described which are later combined in order to provide a semi-structured overview of a clinical record. The first module is responsible for resolving abbreviations and acronyms in the texts. The second one extracts multiword term candidates from the document and ranks them according to the proposed metric. The last one assigns similarity measures to pairs of words, which is in our case applied to the multiword terms. In all of the three modules, the emphasis is put on statistical characteristics of the corpus.

3.1   Resolving Abbreviations

The task of abbreviation resolution is often treated as word sense disambiguation (WSD) [13]. The best-performing approaches of WSD use supervised machine learning techniques. In the case of less-resourced languages, however, neither manually annotated data, nor an inventory of possible senses of abbreviations is available, which are prerequisites of supervised algorithms [12]. On the other hand, unsupervised WSD methods are composed of two phases: word sense induction (WSI) must precede the disambiguation process. Possible senses for words or abbreviations can be induced from a corpus based on contextual features. However, such methods require large corpora to work properly, especially if the ratio of ambiguous terms and abbreviations is as high as in the case of clinical texts. Due to confidentiality issues and quality problems, this approach is not promising either.

Thus, in this research, a corpus-based approach was applied for the resolution of abbreviations with using the very few lexical resources available in Hungarian. Even though the first approach was based on the corpus itself, it did not provide acceptable results, thus the construction of a domain-specific lexicon was unavoidable. But, instead of trying to create huge resources covering the whole field of medical expressions, it was shown in [2] that a small domain-specific lexicon is satisfactory, and the abbreviations to be included can be derived from the corpus itself.

Having this lexicon and the abbreviated tokens detected, the resolution was based on series of abbreviations. Even though standalone abbreviated tokens are highly ambiguous, they more frequently occur as members of multiword abbreviated phrases, in which they are usually easier to interpret unambiguously. For example *o.* could stand for any word either in Hungarian or in Latin, starting with the letter *o*, even if limited to the medical domain. However, in the ophthalmology reports, *o.* is barely used by itself, but together with a laterality indicator, i.e. in forms such as *o. s.*, *o. d.*, or *o. u.* meaning *oculus sinister* 'left eye', *oculus dexter* 'right eye', or *oculi utriusque* 'both eyes', respectively. In such contexts, the meaning of the abbreviated *o.* is unambiguous. It should be noted, that these are not the only representations for these abbreviated phrases, for example *oculus sinister* is also abbreviated as *o. sin., os, OS,* etc. Thus, when performing the resolution of abbreviations, we considered series of such shortened forms instead of single tokens. A series is defined as a continuous sequence of shortened forms without any unabbreviated word breaking the sequence.

Moreover, in order to save mixed phrases (when only some parts of a multi-word phrase is abbreviated) and to keep the information relevant for the resolution of multiword abbreviations, the context of a certain length was attached to the detected series. Beside completing such mixed phrases, the context also plays a role in the process of disambiguation. The meaning (i.e. the resolution) of abbreviations of the same surface form might vary in different contexts.

These abbreviation series are then matched against the corpus, looking for resolution candidates, and only unresolved fragments are completed based on searching in the lexicon. The details of the algorithm and the results are published in [2,17]. It is shown there that having the corpus as the primary source is though insufficient, but provides more adequate resolutions in the actual domain, resulting in a performance of 96.5 % f-measure in the case of abbreviation detection and 80.88 % f-measure when resolving abbreviations of any length, while 88.05 % for abbreviation series of more than one token.

3.2   Extracting Multiword Terms

In the clinical language (or in any other domain-specific or technical language), there are certain multiword terms that express a single concept. These are important to be recognized, because a disease, a treatment, a part of the body, or other relevant information can be in such a form. Moreover, these terms in the clinical reports could not be covered by a standard lexicon. For example, the word *eye* is a part of the body, but by itself it does not say too much about the actual case. Thus, in this domain the terms *left eye, right eye* or *both eyes* are single terms, referring to the exact target of the event the note is about. Moreover, the word *eye* seldom occurs in the corpus without a modifier. This would indicate the need to use some common method for collocation identification, e.g. one based on mutual information or another association measure. After a review of such methods, in [6] a c-value approach is described for multiword term extraction, emphasising the recognition of nested terms.

We used a modified version of this c-value algorithm. First, a linguistic filter is applied in order to ensure that the resulting list of terms contains only well-formed phrases. Phrases of the following forms were allowed:

$$(Noun|Adjective|PresentParticiple|Past(passive)Participle)^{+}Noun$$

This pattern ensures that only noun phrases are extracted and excludes fragments of frequent cooccurrences. It should be noted that other types of phrases, such as verb phrases, might be relevant as well, however, as described in Sect. 2.1, the ratio of verbs is much lower in the clinical corpus, than in a general one. Thus, having only a relatively small corpus of this domain, statistical methods would be inefficient to build accurate models.

After collecting all n-grams matching this pattern, the corresponding c-value is calculated for each of them, which is an indicator of the termhood of a phrase. The c-value is based on four components: the frequency of the candidate phrase; the frequency of the candidate phrase as a subphrase of a longer one; the number of these longer phrases; and the length of the candidate phrase. These statistics are derived from the whole corpus of clinical notes. The details of the algorithm are found in [6].

3.3 Distributional Semantic Models

Creating groups of relevant terms in the corpus requires a similarity metric measuring the closeness of two terms. Instead of using an ontology for retrieving similarity relations between words, the unsupervised method of distributional semantics was applied. Thus, the similarity of terms is based on the way they are used in the specific corpus.

The theory behind distributional semantics is that semantically similar words tend to occur in similar contexts [5] i.e. the similarity of two concepts is determined by their shared contexts. The context of a word is represented as a set of features, each feature consisting of a relation (r) and the related word (w'). In other studies these relations are usually grammatical relations, however in the case of clinical texts, the grammatical analysis performs poorly, resulting in a rather noisy model. In [3], Carroll et al. suggest using only the occurrences of surface word forms within a small window around the target word as features. In this research, a mixture of these ideas was used by applying the following relations to determine the features for a certain word:

- prev_1: the previous word
- prev_w: words preceding the target word within a distance of 2 to 4
- next_1: the following word
- next_w: words following the target word within a distance of 2 to 4
- pos: the part-of-speech tag of the actual word
- prev_pos: the part-of-speech tag of the preceding word
- next_pos: the part-of-speech tag of the following word

Each feature is associated with a frequency determined from the corpus. From these frequencies the amount of information contained in a tuple of (w, r, w') can be computed by using maximum likelihood estimation. This is equal to the mutual information between w and w'. Then, to determine the similarity between two words (w_1 and w_2) the similarity measure described in [10] was used, i.e.:

$$\frac{\sum_{(r,w)\in T(w_1)\cap T(w_2)}(I(w_1,r,w) + I(w_2,r,w))}{\sum_{(r,w)\in T(w_1)} I(w_1,r,w) + \sum_{(r,w)\in T(w_2)} I(w_2,r,w)}$$

where $T(w)$ is the set of pairs (r, w') such that $I(w, r, w')$ is positive.

Having this metric, the pairwise distributional similarity of any two terms can be counted. Though in theory it could be applied to any two words, it does not make sense to compare a verb to a noun or an adjective, or vice versa. This makes the comparison of multiword terms a more complex task. However, as in the present state of this research, these terms are defined as nouns with some modifiers, they fall into the category of nouns. Thus, the similarity of these multiword terms corresponds to the similarity of the last noun in the phrase.

4   Results

The aim of this research was to create a transformation of clinical documents into a semi-structured form to aid the construction of hand-made resources and the annotation of clinical texts. Thus, the results of each module were to be presented in various formats for human judgement. For a set of randomly selected documents taken one by one, the modules described above were applied. First, abbreviations were detected, collected and resolved. The resolutions were expanded with Latin and Hungarian variants as well. Then, multiword terms were identified and ranked for each document by the corresponding c-value. Finally, the pairwise similarity values for these terms were displayed in a heatmap and by listing the groups of the most similar terms.

An example for a processed document is shown in Fig. 1. The similarity of terms reveals that *tiszta törőközeg*, 'clean refractive media' and *békés elülső szegmentum*, 'calm anterior segment' behave very much alike, while they are different from *bal szem*, 'left eye'. This differentiation could indicate two types of annotation or two clusters of these three terms in this small example. These clusters can then be populated from more related terms extracted from other documents as well. Moreover, standalone words (nouns in our case) can be added to these clusters, when they are not part of any longer terms.

Other reasonable clusters that arise during the processing are ones that collect measurement units, such as *d sph* and *d cyl*. These terms appear as abbreviations as well in such documents, thus their resolution can also be linked to the cluster. Names of diseases are also grouped automatically, such as *asteroid hyalosis* and *cat. incip.* which are found in the same document. Note that the term *cat. incip.* is also an abbreviated form and is correctly recognized by both the multiword term extractor as a single term, and the abbreviation resolver, which generated the correct resolution as *cataracta inicipiens*. Another phenomenon that can be observed is the collection of word form or even phrase form variants, such as misspelled forms.

Being presented with such preprocessed, semi-structured transformation of the raw documents, human experts are more willing to complete the annotation. For example, if a group of semantically related terms are collected (such as diseases, treatments, etc.), the human annotator can just name the group and the label will be assigned to all items. Moreover, similarity values of looser relations can also be used as an initial setting when building a relational thesaurus.

Term	English translation	c-value
bal szem	'left eye'	2431.708
ép papilla	'intact papilla'	1172.0
tiszta törőközeg	'clean refractive media'	373.0
békés elülső szegmentum	'calm anterior segment'	160.08
hátsó polus	'posterior pole'	47.5
tompa sérülés	'faint damage'	12.0

Abbreviation	Resolutions	English translation
mydr	mydrum	mydrum
mksz	mindkét szem; oculi utriusque	both eyes
V	visus	visus
D	dioptria	dioptre
mou	méterről olvas ujjat	reads fingers from meters of
ünj	üveg nem javít	glasses do not help
o. u	oculi utriusque; mindkét szem	both eyes
F	szemfenék; fundus oculi; fundus	fundus
j.o.	jobb oldal	right side

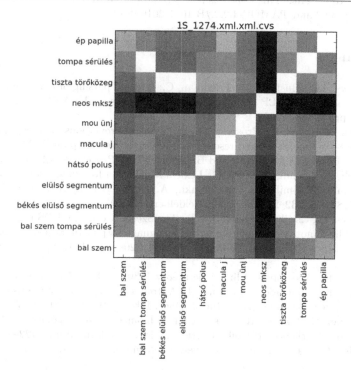

Fig. 1. A fraction of a processed document with some examples of multiword terms, resolved abbreviations and the heatmap of the similarities between terms in the actual document. The lighter a square is, the more similar the two terms are.

5   Conclusion

Clinical documents represent a sublanguage regarding both the content and the language used to record them. However, one of the main characteristics of these texts is the high ratio of noise due to misspellings, abbreviations and incomplete syntactic structures. It has been shown that for a less-resourced language, such as Hungarian, there is a lack of lexical resources, which are used in similar studies to identify relevant concepts and relations for other languages. Thus, such lexicons should be built manually by human experts. However, an initial preprocessed transformation of the raw documents makes the task more efficient. Due to the availability of efficient implementations, statistical methods can be applied to a wide variety of text processing tasks. That is why, in this paper, we have shown that corpus-based approaches (augmented with some linguistic restrictions) perform well on abbreviation resolution, multiword term extraction and distributional similarity measures. Applying such methods can result in a semi-structured representation of clinical documents, appropriate for further human analyses.

Acknowledgement. This work was partially supported by TÁMOP – 4.2.1.B – 11/2/ KMR-2011-0002 and TÁMOP-4.2.2./B-10/1-2010-0014.

References

1. Barrows, J.R., Busuioc, M., Friedman, C.: Limited parsing of notational text visit notes: ad-hoc vs. NLP approaches. In: Proceedings of the AMIA Annual Symposium, pp. 51–55 (2000)
2. Siklósi, B., Novák, A., Prószéky, G.: Resolving abbreviations in clinical texts without pre-existing structured resources. In: Fourth Workshop on Building and Evaluating Resources for Health and Biomedical Text Processing, LREC 2014 (2014)
3. Carroll, J., Koeling, R., Puri, S.: Lexical acquisition for clinical text mining using distributional similarity. In: Gelbukh, A. (ed.) CICLing 2012, Part II. LNCS, vol. 7182, pp. 232–246. Springer, Heidelberg (2012)
4. Csendes, D., Csirik, J., Gyimóthy, T.: The Szeged Corpus: a POS tagged and syntactically annotated Hungarian natural language corpus. In: Sojka, P., Kopeček, I., Pala, K. (eds.) TSD 2004. LNCS (LNAI), vol. 3206, pp. 41–47. Springer, Heidelberg (2004)
5. Firth, J.R.: A synopsis of linguistic theory 1930–55, 1952–59, pp. 1–32 (1957)
6. Frantzi, K., Ananiadou, S., Mima, H.: Automatic recognition of multi-word terms: the c-value/nc-value method. Int. J. Digit. Libr. **3**(2), 115–130 (2000)
7. Friedman, C., Kra, P., Rzhetsky, A.: Two biomedical sublanguages: a description based on the theories of Zellig Harris. J. Biomed. Inform. **35**(4), 222–235 (2002)
8. Harris, Z.S.: The structure of science information. J. Biomed. Inform. **35**(4), 215–221 (2002)
9. Kate, R.J.: Unsupervised grammar induction of clinical report sublanguage. J. Biomed. Semant. **3**(S-3), S4 (2012)

10. Lin, D.: Automatic retrieval and clustering of similar words. In: Proceedings of the 17th International Conference on Computational Linguistics, COLING '98, vol. 2, pp. 768–774. Association for Computational Linguistics, Stroudsburg, PA, USA (1998)
11. Meystre, S., Savova, G., Kipper-Schuler, K., Hurdle, J.: Extracting information from textual documents in the electronic health record: a review of recent research. Yearb. Med. Inform. **35**, 128–144 (2008)
12. Nasiruddin, M.: A state of the art of word sense induction: a way towards word sense disambiguation for under-resourced languages. In: CoRR abs/1310.1425 (2013)
13. Navigli, R.: A quick tour of word sense disambiguation, induction and related approaches. In: Bieliková, M., Friedrich, G., Gottlob, G., Katzenbeisser, S., Turán, G. (eds.) SOFSEM 2012. LNCS, vol. 7147, pp. 115–129. Springer, Heidelberg (2012)
14. Orosz, Gy., Novák, A., Prószéky, G.: Hybrid text segmentation for Hungarian clinical records. In: Castro, F., Gelbukh, A., González, M. (eds.) MICAI 2013, Part I. LNCS (LNAI), vol. 8265, pp. 306–317. Springer, Heidelberg (2013)
15. Orosz, Gy., Novák, A., Prószéky, G.: Lessons learned from tagging clinical Hungarian. Int. J. Comput. Linguist. Appl. **5**(1), 159–176 (2014)
16. Sager, N., Lyman, M., Bucknall, C., Nhan, N., Tick, L.J.: Natural language processing and the representation of clinical data. J. Am. Med. Inform. Assoc. **1**(2), 142–160 (1994)
17. Siklósi, B., Novák, A.: Detection and expansion of abbreviations in Hungarian clinical notes. In: Castro, F., Gelbukh, A., González, M. (eds.) MICAI 2013, Part I. LNCS (LNAI), vol. 8265, pp. 318–328. Springer, Heidelberg (2013)
18. Siklósi, B., Novák, A., Prószéky, G.: Context-aware correction of spelling errors in Hungarian medical documents. In: Dediu, A.-H., Martín-Vide, C., Mitkov, R., Truthe, B. (eds.) SLSP 2013. LNCS (LNAI), vol. 7978, pp. 248–259. Springer, Heidelberg (2013)
19. Siklósi, B., Orosz, Gy., Novák, A., Prószéky, G.: Automatic structuring and correction suggestion system for Hungarian clinical records. In: De Pauw, G., De Schryver, G.-M., Forcada, M.L., Sarasola, K., Tyers, F.M., Wagacha, P.W. (eds.) 8th SaLTMiL Workshop on Creation and use of Basic Lexical Resources for Less-Resourced Languages, pp. 29–34 (2012)
20. Vincze, V.: Dom\u00e9nek közti hasonlóságok és különbségek a szófajok és szintaktikai viszonyok eloszlásában. In: IX. Magyar Számítógépes Nyelvészeti Konferencia, pp. 182–192 (2013)

Mining Text

Predicting Medical Roles in Online Health Fora

Amine Abdaoui[1(✉)], Jérôme Azé[1], Sandra Bringay[1], Natalia Grabar[2],
and Pascal Poncelet[1]

[1] LIRMM UM2 CNRS, UMR 5506, 161 Rue Ada, 34095 Montpellier, France
abdaoui@lirmm.fr
[2] STL UMR 8163 CNRS, Université Lille 3, Lille 1, France

Abstract. Online health fora are increasingly visited by patients to get help and information related to their health. However, these fora are not limited to patients: a significant number of health professionals actively participate in many discussions. As experts their posted information are very important since, they are able to well explain the problems, the symptoms, correct false affirmations and give useful advices, etc. For someone interested in trusty medical information, obtaining only these kinds of posts can be very useful and informative. Unfortunately, extracting such knowledge needs to navigate over the fora in order to evaluate the information. Navigation and selection are time consuming, tedious, difficult and error-prone activities when done manually. It is thus important to propose a new method for automatically categorize information proposed both by non-experts as well as by professionals in online health fora. In this paper, we propose to use a supervised approach to evaluate what are the most representative components of a post considering vocabularies, uncertainty markers, emotions, misspellings and interrogative forms to perform efficiently this categorization. Experiments have been conducted on two real fora and shown that our approach is efficient for extracting posts done by professionals.

Keywords: Text categorization · Text mining · Online health fora

1 Introduction

The Text Mining and Natural Language Processing communities have extensively investigated the huge amount of data on online health fora for different purposes, such as: classifying lay requests to an internal medical expert [1], assisting moderators on online health fora [2], identifying sentiments and emotions [3], identifying the targets of the emotions [4], etc. Indeed, online health fora are increasingly visited by both sick and healthy users to get help and information related to their health [2]. However, these fora are not limited to non-health professional users. More and more frequently, significant number of medical experts is involved in online discussions. For example, many websites, also called "Ask the doctor" services, allow non-health expert users to interact with medical experts [1].

For users searching medical information in online health fora, it may be interesting to automatically distinguish between posts made by health professionals and those

© Springer International Publishing Switzerland 2014
L. Besacier et al. (Eds.): SLSP 2014, LNAI 8791, pp. 247–258, 2014.
DOI: 10.1007/978-3-319-11397-5_19

made by lay men. For instance, users may be more interested by posts made by health professionals, who should give more precise and trustier answers. Some medical websites hire health experts and indicate explicitly their health role. The main purpose of this study is to use such websites to build classification models that can be used to predict roles of users (medical experts vs patients) on websites while this information is not explicitly indicated. Indeed, according to personal indications we obtained, medical experts confirmed that they usually post messages to help non health professional users online, although their medical expert role may not be indicated.

Health professional and non-health professional posts present some differences that are related, for instance, to the used vocabulary, to the practice of subjectivity markers (emotion and uncertainty) and to the nature and the quality of the produced text (question forms and misspellings). We assume that health professionals may use a different vocabulary by comparison with non-health professionals. Then, lay men may show their emotions more easily than health experts, for example to express their sadness due to their illness: the pain was so bad, etc., while health professionals may use more uncertainty words, for example to make an uncertain diagnosis: you may have an arteritis, etc. Finally, non-health professionals may ask more questions and make more misspellings. In our work, we propose to consider these differences to evaluate what are the most representative components of a forum post to perform efficiently medical role categorization in online health fora.

Several studies have been proposed for user profiling [5] as well as the studies proposed for the identification of user roles on social media [6, 7], etc. but less works are concerned with the identification of medical roles. Among the works done to automatically categorize the discourse of doctors and the discourse of patients, Chauveau-Thoumelin and Grabar [8] have proposed to use subjectivity markers (emotions and uncertainty markers) in a supervised approach. The discourse of doctors was obtained from scientific papers and clinical reports, while the discourse of patients was obtained from fora posts. The results obtained by the Random Forests algorithm [9] showed high F-scores (from 0.91 to 0.95 for bi-class classification and from 0.88 to 0.90 for tri-class classification). A medical consultations transcriptions corpus has been used by Tanguy et al. [10]. Using linguistic and statistical techniques, the authors have highlighted some characteristics (for example the length of the discourse, the used vocabulary, the gender, the proportion of questions, etc.) that can be interesting for improving our categorization task.

The rest of this paper is organized as follows: Sect. 2 presents two fora that have been used for evaluating the most significant components of a post. Section 3 introduces our categorization method. Section 4 presents the results obtained and a discussion is proposed in Sect. 5. Finally, Sect. 6 concludes and proposes future work.

2   Studied Corpora

Two French corpora from two different fora have been collected and cleaned as described below.

2.1   Data Collection

Posts from two French websites have been collected.

AlloDocteurs. *AlloDocteur* is a French health forum with more than 16,000 posts[1] covering a large number of topics related to health issues like potentially dangerous medicines, alcoholism, diseases, pregnancy, and sexuality. The forum contains two categories of users: health professional users and non-health professional users. The health professional category may include professional physicians or medical students. Even if their number is limited (16 health professional users are indicated to participate in the forum discussions), their participation in the forum exchanges is important. Indeed, they posted more than 3,000 posts among the 16,000 collected.

MaSanteNet. *MaSanteNet* is an online 'ask the doctor service' subject to charges, that allows users to ask one or more questions to two doctors. The range of topics covered is also large. Users can ask questions on more than 20 different topics such as nutrition, dermatology, and pregnancy. All the questions published on the website have answers. More than 12,000 posts[2] have been collected from this website equitably divided between patient questions and doctor answers.

2.2   Data Cleaning

Once the two corpora collected, a cleaning step has been applied in order to improve their quality. First, all posts containing quotes have been filtered out. Indeed, some health professionals repeat the questions before answering them, which may introduce patient statements into health professional posts. Furthermore, all pieces of texts such as author signatures and date of the last modification have been deleted. Finally, posts with less than 10 words have been considered as irrelevant and therefore removed.

2.3   Data Preparation

After this cleaning step, we obtained two datasets with more or less balanced data from health professional posts and non-health professional posts.

Table 1 shows that the first corpus has fewer posts than the second (about 4,400 posts from AlloDocteurs and about 12,000 posts from MaSanteNet) but more words per post (on average 85 words per post of AlloDocteurs and 57 words per post of MaSanteNet). It also shows that in both datasets non-health professional posts are longer than health professional posts.

3   Methods

The proposed and implemented method consists in three main steps: annotation, pre-processing and classification.

---

[1] www.allodocteurs.fr/forum-rubrique.asp [collected on: 19-11-2013].

[2] www.masantenet.com/questions.php [collected on: 18-02-2014].

Table 1. The number of words and the number of posts in the two datasets

	AlloDocteurs		MaSanteNet	
	Health professionals	Non health professionals	Health professionals	Non health professionals
Number of words	147,419	222,463	233,565	452,453
Number of posts	2,193	2,179	5,876	6,136
Mean words/posts	67	102	40	74

3.1 Annotation

The Ogmios platform [11] was used to perform the following annotations:

Medical Concepts. Terms belonging to three semantic types (diseases, treatments and procedures) have been detected as medical concepts using the following medical terminologies and classifications:

- The Systematized Nomenclature of Human and Veterinary Medicine[3]
- The Thériaque database[4]
- The Unified Medical Language System[5]
- The list of authorized medication that can be marketed in France.

Two lists of all medical terms detected in each corpus have been extracted for a later use.

Emotions. A French emotion lexicon [12], containing about 1,200 words, was used to annotate adjectives, verbs and nouns conveying emotions (joy, sadness, anger, fear, surprise, etc.). In addition to this lexicon, some non-lexical expressions of emotions, such as repeated letters, repeated punctuation signs, smileys, slang and capital letters, have been detected and annotated with specifically designed regular expressions.

Uncertainty. A set of 101 uncertainty words, built in previous study [7], has been used to annotate verbs, nouns, adjectives and even adverbs conveying uncertainty meaning in our corpus.

3.2 Pre-processing

As observed by Balahur [13], fora posts have several linguistic peculiarities that may influence the classification performance. For this reason the following pre-processing steps have been applied:

---

[3] www.ihtsdo.org/snomed-ct [last access: 06-05-2014].

[4] www.theriaque.org [last access: 06-05-2014].

[5] www.nlm.nih.gov/research/umls [last access: 06-05-2014].

Slang Replacement. Some expressions are frequently used in Social Media (*"lol"*). They have been replaced by the corresponding standard text (*"lot of laugh"*).

Replacement of User Tags. All user tags have been identified in our corpora and replaced by the word "Tag" (for example *"@Laurie..."* becomes *"Tag Laurie ..."*).

Hyperlinks and Email Addresses. All the hypertext links have been replaced by the word *"link"* and all the email addresses have been replaced by the word *"mail"*.

Health Pseudonyms. The health professional pseudonyms, previously extracted from each website, are used to replace these pseudonyms in posts by the word *"fdoctor"*. Similarly, pseudonyms of non-health professionals have been extracted and used for their replacement by the word *"fpatient"*.

Lowercasing and Spelling Correction. All words have been lowercased and processed with the spell checker Aspell.[6] The default Aspell French dictionary was expanded with medical words extracted from our corpora during the annotation step. The number of misspellings has been computed for each post and used as attribute for the classification.

3.3 Classification

Supervised classifications to categorize health professional and non-health professional posts have been done as follows.

Descriptors Used. In order to detect the most discriminative features for our classification task, the number of occurrences of medical concepts, emotions, uncertainty markers, misspellings and question marks have been calculated in both health and non-health professional posts for the two websites processed.

From Table 2, we can note that medical words are used massively by both health and non-health professionals and that there is no significant difference between the two

Table 2. The number of occurrences of each feature group in both health and non-health professional posts for the two websites

	AlloDocteurs		MaSanteNet	
	Health professionals	Non-health professionals	Health professionals	Non-health professionals
Medical concepts	8,924	8,888	21,690	22,921
Emotions (EM)	554	2,137	865	2,962
Uncertainty markers (UM)	5,561	3,871	8,449	7,356
Misspellings (MI)	3,828	12,921	11,529	22,137
Question marks (QM)	560	2,594	509	16,991

---

[6] www.aspell.net [last access: 06-05-2014].

categories of users. Nevertheless, the other descriptors indicate that there is difference between these two kinds of users. Non-health professionals express their emotions more frequently than health professionals. Uncertainty markers are slightly more frequent in health professional posts. And as expected, there are also more misspellings and question marks in non-health professional posts.

According to these observations, emotions, uncertainty markers, misspellings and question marks have been chosen as descriptors in our classification task. For each feature, we compute the number of occurrences normalized by the corresponding post length. The length of each post corresponds to the number of words it contains.

In addition to the four features presented before, word ngrams have been considered. The following process has been applied to each corpus: First, all unigrams (words) and bigrams (two words sequences) that appear at least two times are extracted. Then, the number of occurrences of each considered ngram is computed for every post. This number is also normalized by the corresponding post length (number of words) and weighted by its tf-idf score (term frequency * inverse document frequency) [14]. Finally, ngrams obtained from the first corpus have been also used on the second corpus and those obtained from the second have been used on the first, which allowed us to test models learned on posts provided by one corpus with the posts from the other corpus. All these treatments were performed with the "StringToWordVector" filter from the Weka platform [15].

Feature Selection. A feature selection step has been applied to select the most discriminant features: those that frequently appear in one category of posts but not in the other one. Therefore, the selected features should characterize one category of users as compared to the other category. Another filter algorithm from the *Weka* platform, named *"InfoGainAttributeEval"*, has been used to perform the selection. The gain of each attribute on the classification task has been computed and features that have negative gain (i.e. those that don't improve the classification) have been removed. Table 3 indicates most discriminant ngrams (those that had the best gain scores) for each category in the two datasets.[7]

Table 3. Most discriminant unigrams and bigrams for the two categories of users in AlloDocteurs and in MaSanteNet

	AlloDocteurs		MaSanteNet	
	Non-health professionals	Health professionals	Non-health professionals	Health professionals
Unigrams (U)	*I, am, thanks, me, have, my*	*Cordially, hello, you, can*	*Am, I, thanks, hello, my*	*Fpatient, must, good, cordially, have*
Bigrams (B)	*I am, I have, thanks, my, that I*	*Your doctor, cordially, fdoctor, can you*	*I am, I have, thanks, that I, is it*	*Cordially, you must, you have, fpatient you*

---

[7] For readability reasons, ngrams have been translated from French to English.

Table 3 shows that each group uses a specific vocabulary, which is almost the same in both websites. Non-health professionals use the first person singular pronouns (*I, my*) while health professionals use the second person pronouns (*you, your*), which makes sense because the subject of the talk is often the patient and his illness. Besides, non-health professionals show their acknowledgment (*thanks*) while health professionals prefer using a more formal discourse (*cordially*).

Evaluation. Four classification algorithms implemented in *Weka* have been used to test our approach: SVM SMO [16], Naive Bayes [17], Random Forest [9], JRip [18]. For each algorithm, Weighted F-scores are computed with different combinations of features. F-score measures the accuracy of a class; it combines both precision and recall. Usually, it is computed as the harmonic mean of the precision and the recall of the class. Weighted F-score is the mean of all class F-scores weighted by the proportion of elements in each class.

4 Results

Four experiments have been tested: (1) 10-fold cross validation [19] on AlloDocteurs, (2) 10-fold cross validation on MaSanteNet, (3) AlloDocteurs as train set and MaSanteNet as test set and finally (4) MaSanteNet as train set and AlloDocteurs as test set.

4.1 10-Fold Cross Validation on AlloDocteurs

See Table 4.

Table 4. Weighted F-scores obtained with 10-fold cross validation on AlloDocteurs.

Feature group	Number of features	SVM SMO	Naive Bayes	Random Forest	JRip
U	1,120	0.938	0.869	0.901	0.892
U + B	2,160	0.921	0.865	0.902	0.889
EM	1	0.565	0.529	0.564	0.609
UM	1	0.682	0.660	0.657	0.689
MI	1	0.636	0.601	0.641	0.653
QM	1	0.560	0.516	0.613	0.653
EM + UM + MI + QM	4	0.751	0.66	0.725	0.751
U + EM + UM + MI + QM	1,124	**0.940**	0.872	0.901	0.900
U + B+EM + UM + MI + QM	2,164	0.927	0.866	0.906	0.897

4.2   10-Fold Cross Validation on MaSanteNet

See Table 5.

Table 5. Weighted F-scores obtained with 10-fold cross validation on MaSanteNet.

Feature group	Number of features	SVM SMO	Naive Bayes	Random Forest	JRip
U	3,096	**1.000**	0.935	0.999	**1.000**
U + B	4,567	**1.000**	0.949	**1.000**	**1.000**
EM	1	0.503	0.495	0.542	0.558
UM	1	0.678	0.653	0.690	0.680
MI	1	0.438	0.648	0.739	0.686
QM	1	0.748	0.715	0.773	0.773
EM + UM + MI + QM	4	0.761	0.741	0.858	0.851
U + EM + UM + MI + QM	3,100	**1.000**	0.942	0.999	**1.000**
U + B+EM + UM + MI + QM	4,571	**1.000**	0.953	**1.000**	0.999

4.3   AlloDocteurs as Train Set and MaSanteNet as Test Set

See Table 6.

Table 6. Weighted F-scores obtained by considering **AlloDocteurs as train set and MaSanteNet as test set**

Feature group	Number of features	SVM SMO	Naive Bayes	Random Forest	JRip
U	1,120	0.948	0.862	0.938	0.960
U + B	2,160	0.940	0.914	0.938	**0.970**
EM	1	0.558	0.460	0.504	0.558
UM	1	0.679	0.665	0.654	0.681
MI	1	0.436	0.453	0.44	0.371
QM	1	0.773	0.608	0.720	0.773
EM + UM + MI + QM	4	0.677	0.605	0.679	0.705
U + EM + UM+ MI + QM	1,124	0.930	0.866	0.946	0.975
U + B+EM + UM + MI + QM	2,164	0.954	0.915	**0.970**	0.961

4.4   MaSanteNet as Train Set and AlloDocteurs as Test Set

See Table 7.

Table 7. Weighted F-scores obtained by considering **MaSanteNet as train set and AlloDocteurs as test set**

Feature group	Number of features	SVM SMO	Naive Bayes	Random Forest	JRip
U	3,096	0.559	0.816	0.615	0.334
U + B	4,567	0.421	**0.841**	0.599	0.335
EM	1	0.610	0.555	0.579	0.610
UM	1	0.681	0.656	0.669	0.681
MI	1	0.313	0.438	0.490	0.440
QM	1	0.653	0.584	0.650	0.653
EM + UM + MI + QM	4	0.685	0.645	0.641	0.595
U + EM + UM + MI + QM	3,100	0.582	0.821	0.555	0.334
U + B+EM + UM + MI + QM	4,571	0.434	**0.841**	0.560	0.335

5 Discussion

Globally, the cross validations on both websites processed shows good results. First, the use of ngrams shows high F-scores (between 0.865 and 0.938 obtained on *Allo-Docteurs* and between 0.935 and 1 obtained on *MaSanteNet*) comparing to the use of emotions, uncertainty markers, misspellings and question marks which shows low and medium F-scores (between 0.516 and 0.751 obtained on *AlloDocteurs* and between 0.438 and 0.858 obtained on *MaSanteNet*). The combination of ngrams with the rest of the features increases slightly the classification performances (between 0.866 and 0.94 obtained on *AlloDocteurs* and between 0.942 and 1 obtained on *MaSanteNet*). This increase is so small (between 0.001 and 0.008) that it tends to be statistically insignificant.

The models learned on AlloDocteurs and tested on MaSanteNet shows similar results. Ngrams show high F-scores (between 0.862 and 0.97) while emotions, uncertainty markers, misspellings and question marks show low F-scores (between 0.371 and 0.773). The combination of all features doesn't improve the classification performances or improves them very little (F-scores obtained by considering all the features are between 0.866 and 0.97). These results tend to confirm the hypothesis according to which the models learned on one website can be efficiently used on other websites.

The models learned on MaSanteNet and tested on AlloDocteurs gives the worst results. Ngrams show low and medium F-scores if we do not consider Naive Bayes (between 0.334 and 0.615), but high F-scores using Naïve Bayes (between 0.816 and 0.841). Similarly, the results obtained with emotions, uncertainty markers, misspellings and question marks show low F-scores if we do not consider Naive Bayes (between

0.371 and 0.685), low and medium using it (between 0.438 and 0.821). The combination of all features doesn't improve the classification performances neither: the F-scores obtained by considering all the features are between 0.334 and 0.841.

The difference between the two last experiments can be explained by the fact that the first website is a forum, where 16 health professionals post messages in many threads. This makes the discourse of medical users more extensive and diversified, so that models learned on this website may cover the topics and medical discourse observed on the other website: these models have more chances to identify medical professional posts on other websites. On the other hand, the second website is an "Ask the doctor" service where only two medical experts answers the questions. Moreover, their answers are constrained and normalized, as they always answer in the same way. This makes the discourse of medical experts extremely specific to this website: for this reason it appears to be less adapted to learn language models that can be used on other data.

6 Conclusion and Perspectives

In this paper, we presented a supervised method that allows categorizing posts made by health professionals and those made by non-health professionals. Several features have been tested to perform the categorization: ngrams, emotions, uncertainty markers, misspellings and question marks. The experiments indicate that ngrams are the most efficient. The results indicate that models leaned on appropriate websites may be used efficiently on other websites. Moreover, models learned on more general and varied websites (like fora) where many health professionals are involved provide better data for the learning step.

The results obtained are very encouraging but they can be improved. First, the filter used in the feature selection step computes the gain of each feature independently from the other features and doesn't treat the case of redundancy between the features, which may influence the results of some classification algorithms (such as: Naive Bayes) which assume that the features are independent. Furthermore, we used a small French emotion lexicon (containing about 1,200 words). A more comprehensive emotion lexicon [20] is now under construction; we are translating and expanding to synonyms the English emotion lexicon NRC [21] with the help of a professional translator. Up to now, the new emotion lexicon contains more than 20,000 emotion words and we expect it will become even more extensive. The spell checking can also be improved either by considering grammar rules or by a more stringent human supervision of the correction process which also implies that we may obtain a more correct number of misspellings.

The question of detecting trustier and more precise posts in online health fora may be addressed with different methods. Indeed, trust models tested on other social media may be applied either by looking at the structure of the threads (computing scores based on the number of quotes, the number of likes, the number of posts between each post and its replies, etc.) [22, 23] or by inferring these information from the text [24]. In addition to these models, we plan to include the emotional reaction of users to a specific post while computing the trust scores (for example posts arousing the anger of the users).

Finally, we are interested in other applications of Natural Language Processing and Text Mining on online health fora. Currently, we are working on a recommendation system that suggests appropriate topics where the user should post his message. We exploit the content of the posts (title and body), the gender and the age of users, etc. An additional descriptor may be related to the topics where the user has already posted the messages: we assume it may improve the automatic system because the previous preferences of the users may be indicative of his current interests.

Acknowledgement. This paper is based on studies supported by the "Maison des Sciences de l'Homme de Montpellier" (MSH-M) within the framework of the French project "Patient's mind".[8]

References

1. Himmel, W., Reincke, U., Michelmann, H.W.: Text mining and natural language processing approaches for automatic categorization of lay requests to web-based expert forums. J. Med. Internet Res. **11**(3), 1 (2009)
2. Huh, J., Yetisgen-Yildiz, M., Pratt, W.: Text classification for assisting moderators in online health communities. J. Biomed. Inform. **46**(6), 998–1005 (2013)
3. Melzi, S., Abdaoui, A., Azé, J., Bringay, S., Poncelet, P., Galtier, F.: Patient's rationale: patient knowledge retrieval from health forums. In: ETELEMED 2014, The Sixth International Conference on eHealth, Telemedicine, and Social Medicine, 2014, pp. 140–145 (2014)
4. Bringay, S., Kergosien, E., Pompidor, P., Poncelet, P.: Identifying the targets of the emotions expressed in health forums. In: Gelbukh, A. (ed.) CICLing 2014, Part II. LNCS, vol. 8404, pp. 85–97. Springer, Heidelberg (2014)
5. Rangel, F., Rosso, P., Koppel, M., Stamatatos, E., Inches, G.: Overview of the author profiling task at PAN 2013. Notebook Papers of CLEF, pp. 23–26 (2013)
6. Bouguessa, M., Dumoulin, B., Wang, S.: Identifying authoritative actors in question-answering forums: the case of Yahoo! answers. In: Proceedings of the 14th ACM SIGKDD International Conference on Knowledge Discovery and Data Mining, New York, NY, USA, pp. 866–874 (2008)
7. Fisher, D., Smith, M., Welser, H.T.: You are who you talk to: detecting roles in usenet newsgroups. In: Proceedings of the 39th Annual Hawaii International Conference on System Sciences, 2006, HICSS '06, vol. 3, p. 59b (2006)
8. Thoumelin, P.C., Grabar, N.: La subjectivité dans le discours médical: sur les traces de l'incertitude et des émotions. Rev. Nouv. Technol. Inf., Extraction et Gestion des Connaissances, RNTI-E-26, pp. 455–466 (2014)
9. Breiman, L.: Random forests. Mach. Learn. **45**(1), 5–32 (2001)
10. Tanguy, L., Fabre, C., Ho-Dac, L.-M., Rebeyrolle, J.: Caractérisation des échanges entre patients et médecins : approche outillée d'un corpus de consultations médicales. Corpus **10**, 137–154 (2012)
11. Hamon, T., Nazarenko, A.: Le développement d'une plate-forme pour l'annotation spécialisée de documents Web: retour d'expérience. Trait. Autom. Lang. **49**(2), 127–154 (2008)

---

[8] https://www.lirmm.fr/patient-mind/pmwiki/pmwiki.php?n=Site.Accueil

12. Augustyn, M., Hamou, S.B., Bloquet, G., Goossens, V., Loiseau, M., Rinck, F.: Lexique des affects: constitution de ressources pédagogiques numériques.. In: Autour du langage et des langues: perspective pluridisciplinaire, Sélection d'articles du Colloque International des étudiants-chercheurs en didactique des langues et linguistique. (2008)
13. Balahur, A.: Sentiment analysis in social media texts. In: 4th Workshop on Computational Approaches to Subjectivity, Sentiment and Social Media Analysis, Atlanta, Georgia, pp. 120–128 (2013)
14. Salton, G.: Developments in automatic text retrieval. Science 253(5023), 974–980 (1991)
15. Hall, M., Frank, E., Holmes, G., Pfahringer, B., Reutemann, P., Witten, I.H.: The WEKA data mining software: an update. SIGKDD Explor. Newsl. 11(1), 10–18 (2009)
16. Platt, J.C.: Fast training of SVMs using sequential minimal optimization. In: Schölkopf, B., Burges, C.J.C., Smola, A.J. (eds.) Advances in Kernel Methods, pp. 185–208. MIT Press, Cambridge (1999)
17. John, G.H. Langley, P.: Estimating continuous distributions in Bayesian classifiers. In: Eleventh Conference on Uncertainty in Artificial Intelligence, San Mateo, pp. 338–345 (1995)
18. Cohen, W.W.: Fast Effective Rule Induction. In: Twelfth International Conference on Machine Learning, pp. 115–123 (1995)
19. Cross-validation and selection of priors. Statistical Modeling, Causal Inference, and Social Science [Online]. http://andrewgelman.com/2006/03/24/crossvalidation_2/. Accessed 7 May 2014
20. Lexique des sentiments et des émotions français
21. Mohammad, S.M., Turney, P.D.: Emotions evoked by common words and phrases: using mechanical turk to create an emotion Lexicon. In Workshop on Computational Approaches to Analysis and Generation of Emotion in Text, Stroudsburg, PA, USA, pp. 26–34 (2010)
22. Skopik, F., Truong, H.-L., Dustdar, S.: Trust and reputation mining in professional virtual communities. In: Gaedke, M., Grossniklaus, M., Díaz, O. (eds.) ICWE 2009. LNCS, vol. 5648, pp. 76–90. Springer, Heidelberg (2009)
23. Wanas, N., El-Saban, M., Ashour, H., Ammar, W.: Automatic scoring of online discussion posts. In: Proceedings of the 2Nd ACM Workshop on Information Credibility on the Web, New York, NY, USA, pp. 19–26 (2008)
24. Feng, D., Shaw, E., Kim, J., Hovy, E.: Learning to detect conversation focus of threaded discussions. In: Proceedings of the Main Conference on Human Language Technology Conference of the North American Chapter of the Association of Computational Linguistics, Stroudsburg, PA, USA, pp. 208–215 (2006)

Informal Mathematical Discourse Parsing with Conditional Random Fields

Raúl Ernesto Gutierrez de Piñerez Reyes[(✉)] and Juan Francisco Díaz-Frías

EISC, Universidad del Valle, Cali, Colombia
{raul.gutierrez,juanfco.diaz}@correounivalle.edu.co
http://eisc.correounivalle.edu.co/

Abstract. Discourse parsing for the Informal Mathematical Discourse (IMD) has been a difficult task because of the lack of data sets, partly because the Natural Language Processing (NLP) techniques must be adapted to informality of IMD. In this paper, we present an end-to-end discourse parser which is a sequential classifier of informal deductive argumentations (IDA) for Spanish. We design a discourse parser using sequence labeling based on CRFs (Conditional Random Fields). We use the CRFs on lexical, syntactic and semantic features extracted from a discursive corpus (MD-TreeBank: Mathematical Discourse TreeBank). In this article, we describe a Penn Discourse TreeBank (PDTB) styled End-to-End discourse parser into the Control Natural Languages (CNLs) context. Discourse parsing is focused from a discourse low level perspective in which we identify the IDA connectives avoiding complex linguistic phenomena. Our discourse parser performs parsing as a connective-level sequence labeling task and classifies several types of informal deductive argumentations into the mathematical proof.

Keywords: Discourse parser · Support Vector Machines · Informal Discourse Mathematical · Controlled Natural Language · Connectives · Arguments · CRFs · Sequence labeling

1 Introduction

Discourse parsing for the Informal Mathematical Discourse (IMD) has been a difficult task because of the lack of data sets, partly because the Natural Language Processing (NLP) techniques must be adapted to informality of IMD. Various complex linguistic phenomena have been treated in relation to discourse parsing deep level in the IMD context. In this sense, we note that many works have been developed such as anaphoric resolution, disambiguation of mathematical structures, narrative structures of mathematical texts processing and a complete mathematical language processing on the syntactical, morphological, semantic and pragmatic level [5,9,17,18]. In this paper, IMD processing is described like a PDTB-styled End-to-End discourse parser into the CNL context in which discourse parsing is performed under a discourse low level perspective. We avoid complex linguistic phenomena such as resolution anaphora and focus in the use

© Springer International Publishing Switzerland 2014
L. Besacier et al. (Eds.): SLSP 2014, LNAI 8791, pp. 259–271, 2014.
DOI: 10.1007/978-3-319-11397-5_20

of rhetorical relations within the informal mathematical discourse. These rhetorical relations are automatically modeled using machine learning for classifying IDA types. We also think that linguistic phenomena in the mathematical proof can be reduced by including CNLs, given that they allow to eliminate ambiguity and reduce the complexity [4]. Specifically, our approach treats discourse parsing of paraphrased mathematical argumentations (IDAs) like two subtasks of classification. First, we focus on explicit connectives and the identification of their arguments (Arg1 and Arg2), and second, we proceed with the classification of the IDA types by using CRFs. We exclusively focus here on the classification of informal deductive argumentations as a sequence labeling problem under the annotation protocol of Penn Discourse TreeBank (PDTB) [9]. In this regard, the discourse parser is designed as a cascade of CRFs trained on different sets of lexical and syntactic features from a Mathematical TreeBank (M-TreeBank) corpus [11]. We also train CRFs on contextual features of connectives from Mathematical Discourse Treebank (MD-TreeBank) corpus following PDTB guidelines [9]. This article is organized as follows: in Sect. 2 we present related work to discourse parsing. In Sect. 3 we present corpus linguistics in the IMD. In Sect. 4, for the considered IDAs, we focus on explicit connectives and the identification of their arguments. In Sect. 5 we detail the classification of informal deductive argumentations. In Sect. 6 we describe the experiments and results; and, finally, we draw our conclusions in Sect. 7.

2   Related Work

In this paper, we address the tasks of automatically extract discourse arguments and classification of mathematical argumentations under a discourse low level perspective. Automatic extraction of discourse arguments is performed for giving IDA by using a explicit discourse connectives. On the one hand, we refer to parsing discourse based on the identification of discourse arguments using the PDTB guidelines. Dines [2] was the first to carry out such an experiment on the PDTB in which were extracted complete arguments with boundaries. Wellner and Pustejovsky [16] extracted discourse arguments using a sophisticated probabilistic model based on argument *heads*. In contrast, Ghosh [3] integrates the argument spans for identifying arguments and defines an end-to-end system for discourse parsing. In addition, Lin [6] defines an end-to-end system which identify arguments using contextual features to implicit and explicit connectives. In this sense, Pitler et al. [8] investigated features ranging from low-level word pairs to high-level linguistic features. In our work, we identify arguments in IDAs using spans and mathematical features under tne PDTB protocol. On the other hand, we indicate the little work related to rhetorical relations in IMD. Zinn [18] was the first introducing anaphoric resolution in mathematical discourse parsing and uses DRT (Discourse Representation Theory) analyses for course-books proofs. Humayoun [4] also uses DRT for mathematical discourse processing. Wolska [17] defines a language for processing more informal dialogues on mathematical proofs including annotation of rhetorical relations. Another important work was defined by Kamareddine [5] in which rhetorical relations are annotated under a logic structure based on graphs. All previous approaches have as their main characteristic the anaphor resolution and also they plan to implement their algorithms informed by a corpus analysis [18]. To date, our approach is the only that has studied rhetorical relations data-driven based under PDTB guidelines. In the broader context, our work is the starting point and a focus for processing of verifying informal proofs (as in [18]).

3   Corpus Linguistics in the IMD Context

3.1   The Mathematical TreeBank

The Mathematical TreeBank (M-TreeBank) is a manually annotated corpus with syntactic structures for supporting study and the analysis of IMD phenomena. The M-TreeBank consists of about 748 sentence/tree pairs in Spanish. The average sentence length is 33 tokens. These sentences are part of the 150 IDAs of our standard corpus. On average, each IDA has 132 tokens and between five or six sentences per IDA. The treebank trees contain information about the constituency and syntactic structure, as well as connective annotations, indefinite and definite descriptions annotation [18] and morphological information. The standard corpus is composed of a standard set of 150 IDAs in which their sentences are well-formed within our CNL. The standard corpus is the transformation of a defined original corpus after a previous experiment in which the students were required to demonstrate two theorems of set theory: (1) if $A \subseteq B$ then $U \subseteq \overline{A} \cup B$; (2) $P(A \cap B) = P(A) \cap P(B)$. The M-TreeBank annotation scheme basically follows the Penn TreeBank II (PTB) scheme [7], human-annotated. Based on PTB, we used the clausal structure, as well as most of the labels. We adapted annotation structure of connectives from UAM Spanish TreeBank [13] as well as the annotation scheme for morphological information. We also adapted the set of morphological features from Freeling 2.2.[1]

3.2   Discursive Annotation of a Standard Corpus

The Mathematical Discourse TreeBank (MD-TreeBank) is a discursive corpus over the standard corpus. This corpus was annotated with 150 argumentations and it was defined in the annotation protocol of Penn Discourse TreeBank (PDTB). As in PDTB, the MD-TreeBank is syntactically supported by the M-TreeBank. In MD-TreeBank, we only follow explicit relations and focus on the types and subtypes of class level (*Temporal, Contingency, Comparison, Expansion*). We define ten semantic classes or senses which specify both functionality and deductive character of connectives, among the classes defined are: *alternative-conjunctive (A-C), alternative-disjunctive (A-D), cause-justification (C-J), cause-result (C-Re), cause-reason (C-R), cause-transitive (C-T), conclusion-result (D-C), conditional-hypothetical (C-H), instantiation (I)*, and

---

[1] http://nlp.lsi.upc.edu/freeling/

restatement-equivalence (R-E). We also focus on attributions of arguments, specifically, the property *type* in which we defined six types of attributions according to CNL. The attribution type labels defined are: AFIRM: Statements, COND: Conditionals, SUP: Assumptions, EXP: Explanations, DED: Deductions, JUS: Justifications. In MD-TreeBank, we annotate discourse arguments (`Arg1` and `Arg2`) like paraphrased logic propositions. These propositions keep a deductive order and share a linear order within mathematical proof. Each IDA is annotated as a sequence of connectives by following the CNL's discourse structure. Note that although this annotation is only linguistic it's not intended for annotating his logic representation. As operating in PDTB, the argument in clause that is syntactically bound to the connective is called `Arg2`; the other one is called `Arg1`. For example, causal relation in (3) belongs to type "Pragmatic Cause" with subtype label "Justification" in which connective <u>porque</u> indicates that `Arg1` is expressing a claim and `Arg2` is providing justification for this claim.

(3) Entonces *el elemento x pertenece al conjunto complemento de A o al conjunto A* [*the element x belows to the complement set A or to set A*] <u>porque</u> [*because*] **la unión de el conjunto complemento de A con el conjunto A es igual al conjunto universal U** [*The union of complement set A with set A is equal to universal set U*].

In MD-TreeBank the statistics of the position of `Arg1` w.r.t. the discourse connective is shown in Table 1. The position of `Arg1` w.r.t. the discourse connective shows that 55 % of explicit relations are same sentence (SS); 45 % are precedent sentence (PS). In PS, the position `Arg1` in previous, adjacent sentence (IPS) is 20 % and previous, non adjacent sentence (NAPS) is 25 %.

Table 1. Statistics of position `Arg1` in MD-TreeBank.

Position	
`Arg1`, in same sentence (SS) as connective	55 %
`Arg1`, in previous, adjacent sentence (IPS)	20 %
`Arg1`, in previous, non-adjacent sentence (NAPS)	25 %

4   Identification of Arguments in an IDA

Before classifying the informal deductive argumentations within mathematical proof, we identified and labeled the connectives and their arguments (propositions). In this section, we present a discourse parser that, given an input argumentation (IDA) automatically extracts discourse arguments (`Arg1` and `Arg2`) linked to each connective of the argumentation. We develop the identification of arguments as a set of steps in which a step output feeds input of the next step based on Support Vector Machines (SVMs) [14]. In this work, SVMs are used for identifying   `Arg1` and `Arg2` given a mathematical argumentation following the annotation protocol of PDTB [9]. As in [3,6], our discourse

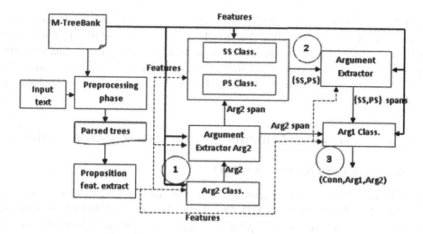

Fig. 1. System pipeline for the arguments classifier

parser labels `Arg1` and `Arg2` for each explicit connective of a mathematical proof. We follow in three steps: (1) identifying the `Arg2`, (2) identifying the locations of `Arg1` as SS or PS, and (3) identifying the `Arg1` and labeling of `Arg1` and `Arg2` spans. Figure 1, shows the pipeline architecture used for such a process. First (1), we identify the `Arg2` and then we extract the `Arg2` span; next (2), we identify the relative position of `Arg1` as same sentence (SS) or precedence sentence (PS). Then that relative position is propagated to an argument extractor and the `Arg1` is correspondingly labeled; finally (3), `Arg1` is identified given (`Arg2,SS,PS`) spans, and arguments (`Arg1` and `Arg2`) are extracted. Before the segmentation and labeling of `Arg1` and `Arg2` arguments of the input text (mathematical argumentation), we must do a preprocessing phase (The left side of Fig. 1). First, we take the input text and segment it into sentences; second, each sentence is tagged; third we use a statistical parser model (Bikel's parser [1]) for obtaining the parsed tree for each sentence. Next, we use a rule based algorithm for building a features vector for each proposition found in each parsed tree. For each proposition corresponds a numerical transformed feature vector which will serve as dataset of SVMs (In Fig. 1, the bolded line represents the test data and the dotted line represents the training data). In this section, we only present the results of `Arg1` classifier in which the `Arg1` and `Arg2` arguments are classified and labeled (For information, see [12]). We trained the `Arg1` classifier under two models; $M1 = CONN + NoS + POS + Arg2\text{-}s + mf$ and $M2 = CONN + NoS + POS + Arg2\text{-}s + mf + SS\text{-}s + PS\text{-}s$. These two models are differentiated because M2 including the corresponding SS and PS spans. We use contextual features to the connective and their corresponding contextual features to the arguments (`Arg1` and `Arg2`). For instance, we define as contextual features the connective string ($CONN$), its syntactic category (POS) and the sentence number (NoS) in which the connective is found. Among contextual features to the arguments we define: the `Arg2` span ($Arg2\text{-}s$), the SS span

(SS-s), the PS span (PS-s) and the mathematical features (mf). We trained $\texttt{Arg1}$ classifier under the two models. For $\texttt{Arg1}$ and $\texttt{Arg2}$ arguments, the M1 model was 97 % F1 score and the M2 model was 98 % F1 score. As it was expected, this small increment is due to adding of SS (SS-s) and PS (PS-s) spans. This result will be used in the next phase for the classification of informal deductive argumentations.

5   Classification of Informal Deductive Argumentations

5.1   Discursive Structure of an IDA

Informal deductive argumentations are mathematical proofs that have been paraphrased and annotated in MD-TreeBank. An informal deductive argumentation type is annotated as a sequence of connectives with their corresponding connective POSs, semantic senses and argument attributions, respectively. For each argumentation type, we define a semantic senses sequence that represents a deductive and linear order according to mathematical argumentations of the standard corpus. This senses sequence represents an argumentation type which is characterized as one of 15 argumentation types annotated in the standard corpus. IDA types were labeled with numbers such as {01,03,04,05,06,07,09, 023,024,025,026,052,115}. In Fig. 2, we show the connectives sequence of IDA type 05 (at the top of figure) with its corresponding semantic senses sequence {C-R, R-E, A-C, C-Re, A-C, R-E, D-C} (at the bottom of figure). The sentences $s1, s2$ and $s3$ have a following linear order $s1 \prec s2 \prec s3$ and therefore their connectives. In Fig. 2, we define the discursive structure for the IDA type 05. This argumentation type consists of three sentences $s1, s2, s3$ in which $s1$ is an assumption with explanation, $s2$ is a statement with explanation and $s3$ is a deduction. Additionally, we define semantic information for each sentence based in semantic sense of each connective (see Sect. 3.2). We define a semantic sense for each explicit connective that depends of relation between $\texttt{Arg1}$ and $\texttt{Arg2}$. For each sentence, we can suppose one or two propositions between each connective. The following excerpt in text $s1$ in the IDA type 005: *Suponga que* **prop1:** *la intersección del conjunto potencia de A y el conjunto pontencia de B contiene al conjunto X*, *es decir*, **prop2:** *el conjunto X pertenece al conjunto A y* **prop3:** *el conjunto X pertenece al conjunto B*. In Fig. 2, the tag *cause-reason ($\widehat{C}$-R)* is used because the connective "Suponga que" indicates that the situation described in **prop1**=$\texttt{Arg2}$ is the cause and the situation described in **prop2+prop3**=$\texttt{Arg1}$ is the effect. The tag *restatement-equivalence (R-E)* is applied because the connective "es decir" indicates that **prop1**=$\texttt{Arg1}$ and **prop2+prop3**=$\texttt{Arg2}$ describe the same situation from different perspectives.

5.2   Classification of IDAs as a Sequence Labeling Problem

In this work, the classification problem of argumentation types is tackled as a sequence labeling task in which are built two sequence classifiers; first, we developed a classifier for automatically annotation of semantic senses which feeds a

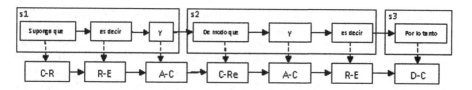

Fig. 2. The IDA type 05 annotated in MD-TreeBank.

second classifier for classification of argumentation types. More formally, the classification task of semantic senses can be treated as a sequence labeling problem. Let $\mathbf{x} = (x_1, x_2, \ldots x_n)$ denote the observations sequence in an argumentation instance, where c_i is a sequence of connectives and x_i is c_i augmented with additional information such as the POS tag (t_i) and the sentence number tag (s_i) of c_i. Each observation x_i is associated with a label $y_i \in \{$ C-R, C-Re, C-J, A-C, A-D, D-C, C-H, I, R-E$\}$ (which are the hidden labels or tags, and these are called hidden states in the general concept) which indicates the semantic sense of c_i. Let $\mathbf{y} = (y_1, y_2, \ldots y_n)$ denote the sequence of labels for $\mathbf{x}$, then for a candidate argumentation instance $\mathbf{x}$, let $\hat{Y}$ denote the valid set of label sequences. Our task consists in finding the best label sequence $\hat{y}$ among all the possible label sequences for $\mathbf{x}$, that means to find an appropriate sequence of tags can maximize the conditional likelihood according to Eq. (1).

$$\hat{y} = \operatorname*{argmax}_{y \in \hat{Y}} p(\mathbf{y}|\mathbf{x}) = \operatorname*{argmax}_{y \in \hat{Y}} \prod_{i=1}^{n} p(y_i | c_{1:n}, t_{1:n}, s_{1:n}, y_{i-1}) \qquad (1)$$

Similarly, we also can formulate the classification of informal deductive argumentations as a sequence labeling problem taking into account the semantic senses annotated in the senses classifier. Let $\mathbf{x^*} = (x_1, x_2, \ldots x_n)$ denote the observations sequence in an argumentation instance, where c_i is a sequence of connectives and x_i is c_i augmented with additional information such as the POS tag (t_i), the sentence number tag (s_i) and semantic sense tag (r_i) of c_i. Each observation x_i is associated with a label $y_i \in \{$01,03,04,05,06,07,09,023,024,025, 026,052,115$\}$ which indicates the argumentation types of c_i. Let $\mathbf{y^*} = (y_1, y_2, \ldots y_n)$ the sequence of labels for $\mathbf{x^*}$, then for a candidate argumentation instance $\mathbf{x^*}$, let $\hat{Z}$ denote the valid set of label sequences. As well as in first classifier the idea consists in finding the best label sequence $\hat{y*}$ among all the possible label sequences for $\mathbf{x^*}$.

$$\hat{y*} = \operatorname*{argmax}_{y^* \in \hat{Z}} p(\mathbf{y^*}|\mathbf{x^*}) = \operatorname*{argmax}_{y \in \hat{Y}} \prod_{i=1}^{n} p(y_i | c_{1:n}, t_{1:n}, s_{1:n}, r_{1:n}, y_{i-1}) \qquad (2)$$

For the models above is very hard to compute both equations as it involves too many parameters. In order to reduce the complexity, we employ two linear-chain CRFs for solving our two sequence labeling problems. In this sense, we use linear-chain CRFs in which all nodes in the graph form a linear chain and each feature involves only two consecutive hidden states [10]. For the first CRF (see

Eq. (1)), we define the general form of a feature function as $f_i(y_{j-1}, y_j, c_{1:n}, t_{1:n}, s_{1:n}, j)$, which maps in a pair of adjacent states y_j, y_{j-1}, the whole input sequence $c_{1:n}$ as well as $t_{1:n}, s_{1:n}$ and the current connective's position. For example, we can define a simple feature function which produces binary values: it is 1 if current connective c_j is "es decir", the corresponding part-of-speech t_j is "CONN-EXP", the sentence number in the IDA s_j is 1 and the current state y_j is the semantic sense "R-E":

$$f_i(y_{j-1}, y_j, c_{1:n}, t_{1:n}, s_{1:n}, j) = \begin{cases} 1, \text{if } c_j = \text{es decir}, \ t_j = \text{CONN-EXP}, \ s_j = 1 \\ \text{and } y_j = R\text{-}E \\ 0, \text{otherwise} \end{cases} \quad (3)$$

Another example of feature function is used for classifying of IDA types:

$$f_i(y_{j-1}, y_j, c_{1:n}, t_{1:n}, s_{1:n}, r_{1:n}, j) = \begin{cases} 1, \text{if } c_j = \text{es decir}, \ t_j = \text{CONN-EXP}, \ s_j = 1, \\ r_j = R\text{-}E \text{ and } y_j = 05 \\ 0, \text{otherwise} \end{cases}$$

$$(4)$$

Where r_j is the semantic sense and the hidden labels y_j, y_{i-1} are IDA types. The feature functions in Eqs. (3) and (4) will be both active with the IDA type 05, if, besides the current connective is "es decir", the POS tag is "CONN-EXP", and it is found in the sentence $S1$ with the semantic sense as R-E. In both classifiers we define a set of training instances $\{\mathbf{x}^{(m)}, \mathbf{y}^{(m)}\}_{m=1}^{M}$ where $\mathbf{y}^{(y)}$ is the correct label sequence for $\mathbf{x}^{(k)}$. For this we need fully labeled data sequences, for instance, for training the semantic senses CRF we define the fully labeled review data as $\{(c^{(1)}, t^{(1)}, s^{(1)}, y^{(1)}), \ldots, (c^{(M)}, t^{(M)}, s^{(M)}, y^{(M)})\}$, where $c^{(i)} = c_{1:n_1}^{(i)}$ is the first connectives sequence and $\mathbf{x}^{(1)} = (c^{(1)}, t^{(1)}, s^{(1)})$ is the first observations sequence.

5.3   Processing of an End-to-End System

Discourse parser is an end-to-end system that has two main components: in Fig. 1, we showed the first component such as the preprocessing phase and the arguments classifier. In Fig. 3, we present the second component which is an architecture of an IDA types classifier. Automatically discursive parsing can range according to application domain and depends on robustness of corpus and extraction methods. However, we applied natural language processing techniques to IMD processing into the low-level discourse framework. In this sense, we develop the three basic phases according to [3,15] such as (1) segmentation of basic units, (2) identification of arguments and relations, and (3) definition of relation types between arguments. In this section, we tackle the third phase as a labeling sequence problem and we develop an end-to-end system based on phases 1 and 2. In this work, we present the pipeline architecture for solving two sequential labeling problems involved in the classification of IDA types. First (1), we identify the semantic sense for each connectives sequence of input text, and next (2), we use this first result for identifying the informal deductive

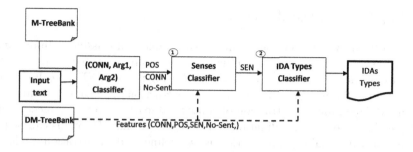

Fig. 3. The informal deductive argumentations classifier.

argumentation types. As shown in Fig. 3, for (1), we extract the POS (POS tag of connectives c_i), CONN (connectives span c_i) and No-Sent (sentence number s_i) from the arguments classifier (see Sect. 4). The senses classifier is trained using the senses features from the MD-TreeBank (the dotted line in figure). Next, for (2) in Fig. 3, we use the same input features of (1) plus of the output sequences from the senses classifier. The IDA classifier is trained using features from the MD-Treebank. Finally, the IDA classifier annotates the argumentation type of an informal deductive argumentation (input text) as output.

Table 2. Results for the senses classifier.

Sense	Prec.	Rec.	F_1	Fre.
A-C	79.31	97.87	87.62	58
A-D	83.33	55.56	66.67	6
C-R	80.85	86.36	83.52	94
C-J	80.00	94.12	86.49	20
C-Re	84.62	75.00	79.52	40
C-T	83.33	83.33	83.33	10
D-C	80.00	94.12	86.49	94
C-H	80.00	80.00	80.00	12
I	90.00	81.82	85.71	12
R-E	85.71	100.00	92.31	42

Table 3. Results for the IDA Types classifier.

Pro. type	Prec.	Rec.	F_1	Fre.
01	42.86	68.85	52.83	294
03	81.82	61.76	70.39	77
04	83.33	19.44	31.53	42
05	78.57	76.24	77.39	98
06	66.67	20.25	31.07	24
07	77.93	82.33	80.07	299
09	84.00	63.16	72.10	100
023	20.27	57.69	30.00	74
024	41.85	86.92	56.50	270
025	90.56	96.45	93.41	180
026	50.59	89.58	64.66	85
052	91.59	95.15	93.33	107
115	73.20	44.09	54.04	153

6   Experiments and Results

In this section, we present the results of two sequential classifiers. Information related to syntactic parser and arguments classifier are presented in [11,12]. Performance of the two classifiers was measured using precision (**P**), recall (**R**) and *F1* measures. We use the MD-TreBank for extraction of semantic features and use the M-TreeBank for extraction of contextual features to the connective. We performed a 10-fold cross-validation, since we ensure that 150 proofs can be partitioned into 10 equal size proofs; that is, we define 15 demonstration models at the 150 proofs. In Table 2, we report performance of the senses sequential classifier given an IDA input. Performance measures (**P** and **R**) were used according to semantic senses where the frequency (Frec.) is the number of times that shows semantic sense in the proposition. A sentence consists of prepositions (`Arg1` and `Arg2`) and an IDA has many sentences. We establish the individual annotation of each tag corresponding to semantic sense for each connective on the input IDA. Semantic sense in IDAs means that a label represents the relational semantic of the connectives according to PDTB. In this regard, it is worth highlighting that results of Table 2 are more focused to statistics of senses per propositions. We use leave-one-out-cross-validation (LOOCV) in which a single sample of test data is tested against the training data partitions. Test data collected were obtained from the performed tests to students (these were in advance instructed over the CNL), 226 IDAs in total, with 1949 propositions of which 1815 were correctly considered, for an overall precision of 93.12 %. In Table 2, we report precision, recall and F1 based on a basic features set (CONN+POS+No-Sent+SEN). As expected, the *R-E (restatement-equivalence)* shows 92.31 % F1-measure because it do not present ambiguity with respect to other senses. The sense *R-E* is the semantic class a single connective "es decir". Others senses such as *I,D-C,A-C* also show a score very similar to *R-E* because these senses have a single connective. Although the sense *D-C* is the semantic class of two connectives. In contrast, the sense *C-Re (cause-reason)* shows the lowest F1 (79.52 %), which means that this sense is very ambiguous being the semantic class of the connectives "*Entonces, por lo tanto, de modo que,* etc.". In this regard, we discard F1 score from A-D *(alternative-disjunctive)* like the lowest because this semantic class has the lowest frequency. We extract all features from MD-TreeBank for training data and we extract all features from the arguments classifier for testing data (see Fig. 3). In Table 3 we show results of IDA types classifier. As with the senses classifier, we collected 256 IDAs as test data, with 2179 propositions of which 1445 were correctly considered, for an overall precision of 63.31 %. We extract all features from MD-TreeBank for training data. For testing data (see Fig. 3), we extract some features such as (POS+CONN+No-Sent) from the arguments classifier. For each IDA type, we manually annotated propositions or arguments (`Arg1` or `Arg2`) with the IDA type. For instance, we annotate ten (10) propositions with the number 01. We did the same with rest of IDA types. Note that each proposition begins by a connective. As seen Table 3, IDA types 01 and 115 show the same F1 score due to these argumentations have a similar discursive structure. Therefore, their connectives sequences also are very similar as well as semantic senses sequence. For instance, we found that many annotated propositions like IDA type 01 were annotated like IDA type 115 and vice versa. It could explain because F1 score of both 01 and 115 have similar F1 scores (52.83 % and 54.04 %, respectively). Precision of IDA type 115 is larger than Precision of IDA type 01 and Recall of IDA type 01 is larger than Recall of IDA type 115. We can infer that the number of bad annotations is equal the number of correct annotations. With respect to lowest F1 scores for types 04 and 023 it is because their frequencies are very low. This can be ratified given that types 04 and 024 have the same discursive structure. In proportion, F1 score for type 04 should be fewer with respect to frequencies number compared with type 024. The highest F1 score was for type 025, in this case, we found connectives sequences such as {*Suponga, es decir, y, De manera que, es decir, y, Por lo tanto*} in which we can note a greater number of unambiguous connectives related to semantic senses, for instance, the connective "es decir" has a single semantic sense. A similar situation applies to the connective "y" which has a single semantic sense.

7   Conclusions

In this work, we presented an algorithm that performs IMD parsing within CNL context in which an end-to-end system is implemented. This system is a discourse parser which was performed in three phases; a first preprocessing phase, a second phase in which we identify connectives and their arguments (**Arg1** and **Arg2**), and a third phase where we classify informal deductive argumentations by using sequential labeling. We define a discursive structure for each IDA type in which both linear and deductive order were defined. This discursive structure was annotated as a connectives sequence in MD-Treebank. For the senses classifier, we trained a CRF classifier with lexical, syntactic and semantic features extracted from MD-TreeBank. For the IDA types classifier, we trained another CRF classifier taking advantage the output labels of senses classifier. We tested a set of IDAs that consists of argumentations performed by students in which these were instructed in advance on the CNL. Finally, we presented results of two classifiers. We think that these results benefited from ambiguity underlying in connectives and senses. We also think that the use of sequential labeling of connectives is a first approach to automation of mathematical proof under a discourse low level perspective. In the light of these results, to intermediate term, the objective will set other CRF structures more contextuals for treatment of ambiguity.

References

1. Bikel, D.: Design of a multilingual, parallel processing statistical parsing engine. In: Proceedings of the 2nd International Conference on Human Language Technology Research HLT'02, pp. 178–182. Morgan Kaufmann Publishers Inc., San Francisco (2002)

2. Dines, N., Lee, A., Miltsakaki, E., Prasad, R., Joshi, A., Webber, B.: Attribution and the (non-)alignment of syntactic and discourse arguments of connectives. In: Proceedings of the Workshop on Frontiers in Corpus Annotations II: Pie in the Sky, CorpusAnno '05, Stroudsburg, PA, USA, pp. 29–36. Association for Computational Linguistics (2005). http://dl.acm.org/citation.cfm?id=1608829.1608834

3. Ghosh, S., Johansson, R., Riccardi, G., Tonelli, S.: Shallow discourse parsing with conditional random fields. In: Proceedings of the 5th International Joint Conference on Natural Language Processing, Chiang Mai, Thailand, pp. 1071–1079 (2011)

4. Humayoun, M., Raffalli, C.: Mathabs: A representational language for mathematics. In: Proceedings of the 8th International Conference on Frontiers of Information Technology, FIT '10, pp. 37:1–37:7. ACM, New York (2010). http://doi.acm.org/10.1145/1943628.1943665

5. Kamareddine, F., Maarek, M., Retel, K., Wells, J.B.: Narrative structure of mathematical texts. In: Kauers, M., Kerber, M., Miner, R., Windsteiger, W. (eds.) MKM/CALCULEMUS 2007. LNCS (LNAI), vol. 4573, pp. 296–312. Springer, Heidelberg (2007). http://dx.doi.org/10.1007/978-3-540-73086-6_24

6. Lin, Z., Ng, H.T., Kan, M.: A PDTB-styled end-to-end discourse parser. Comput. Res. Repository (2011)

7. Marcus, M.P., Marcinkiewicz, M.A., Santorini, B.: Building a large annotated corpus of english: The penn treebank. Comput. Linguist. **19**(2), 313–330 (1993). http://dl.acm.org/citation.cfm?id=972470.972475

8. Pitler, E., Nenkova, A.: Using syntax to disambiguate explicit discourse connectives in text. In: Proceedings of the ACL-IJCNLP 2009 Conference Short Papers (ACLShort 2009), Stroudsburg, PA, USA, pp. 13–16. Association for Computational Linguistics (2009). http://dl.acm.org/citation.cfm?id=1667583.1667589

9. Prasad, R., Dinesh, N., Lee, A., Miltsakaki, E., Robaldo, L., Joshi, A., Webber, B.: The penn discourse treebank 2.0. In: Proceedings of the 6th International Conference on Languages Resources and Evaluations (LREC 2008), Marrakech, Marocco (2008)

10. Qi, L., Chen, L.: A linear-chain CRF-based learning approach for web opinion mining. In: Chen, L., Triantafillou, P., Suel, T. (eds.) WISE 2010. LNCS, vol. 6488, pp. 128–141. Springer, Heidelberg (2010)

11. Gutierrez de Piñerez Reyes, R.E., Díaz Frías, J.F.: Preprocessing of informal mathematical discourse in context of controlled natural language. In: Proceedings of the 21st ACM International Conference on Information and Knowledge Management, CIKM '12, pp. 1632–1636. ACM, New York (2012). http://doi.acm.org/10.1145/2396761.2398487

12. Gutierrez de Piñerez Reyes, R.E., Díaz Frias, J.F.: Building a discourse parser for informal mathematical discourse in the context of a controlled natural language. In: Gelbukh, A. (ed.) CICLing 2013, Part I. LNCS, vol. 7816, pp. 533–544. Springer, Heidelberg (2013). http://dx.doi.org/10.1007/978-3-642-37247-6_43

13. Ruesga, S.L., Sandoval, S.L., León, L.F.: Spanish treebank: specifications version 5. Technical report, Universidad Autónoma de Madrid (1999)

14. Vapnik, V.: The Nature of Statistical Learning Theory. Springer, New York (1995)

15. Wellner, B.: Sequence models and ranking methods for discourse parsing. Ph.D. thesis, Brandeis University (2009)

16. Wellner, B., Pustejovsky, J.: Automatically identifying the arguments of discourse connectives. In: Proceedings of the 2007 Joint Conference on Empirical Methods in Natural Language Processing and Computational Natural Language Learning (EMNLP-CoNLL), Prague, Czech Republic, June 2007, pp. 92–101. Association for Computational Linguistics (2007). http://www.aclweb.org/anthology/D/D07/D07-1010

17. Wolska, M.: A language engineering architecture for processing informal mathematical discourse. In: Towards Digital Mathematics Library, pp. 131–136. Masaryk University (2008)

18. Zinn, C.: Understanding informal mathematical discourse. Ph.D. thesis. Universität Erlangen-Nürnberg Institut für Informatik (2004)

Corpus-Based Information Extraction and Opinion Mining for the Restaurant Recommendation System

Ekaterina Pronoza[✉], Elena Yagunova[✉], and Svetlana Volskaya

Saint-Petersburg State University, 7/9 Universitetskaya Nab.,
Saint-Petersburg, Russia
{katpronoza,iagounova.elena,svetlana.volskaya}
@gmail.com

Abstract. In this paper corpus-based information extraction and opinion mining method is proposed. Our domain is restaurant reviews, and our information extraction and opinion mining module is a part of a Russian knowledge-based recommendation system.

Our method is based on thorough corpus analysis and automatic selection of machine learning models and feature sets. We also pay special attention to the verification of statistical significance.

According to the results of the research, Naive Bayes models perform well at classifying sentiment with respect to a restaurant aspect, while Logistic Regression is good at deciding on the relevance of a user's review.

The approach proposed can be used in similar domains, for example, hotel reviews, with data represented by colloquial non-structured texts (in contrast with the domain of technical products, books, etc.) and for other languages with rich morphology and free word order.

Keywords: Information extraction · Opinion mining · Restaurant recommendation system · Machine learning

1 Introduction

In this paper information extraction (IE) and opinion mining (OM) for the restaurant recommendation system are considered. Our goal is to introduce an effective corpus-based restaurant IE and OM method for Russian using machine learning techniques. We try to combine thorough language analysis, adopted from corpus linguistics, with fully automatic machine learning techniques.

The system is built in several steps. First, we define a set of restaurant aspects; the most frequently occurring ones are to be extracted using machine learning techniques. We conduct corpus analysis and construct dictionaries and sentiment lexicon (these procedures are described in our earlier paper [27]). Second, we compare several classification techniques in combination with various feature sets to determine the best classifier for each of the restaurant aspects defined earlier [27].

There are two types of restaurant aspects extracted from reviews: those which are evidently subjective and depend on the user's tastes (staff amiability, food quality, etc.)

© Springer International Publishing Switzerland 2014
L. Besacier et al. (Eds.): SLSP 2014, LNAI 8791, pp. 272–284, 2014.
DOI: 10.1007/978-3-319-11397-5_21

and those which are more or less objective (crampedness, noise, etc.) – users usually agree on them. Thus, our task is similar to IE or OM depending on a restaurant aspect. In this paper we do not distinguish these two tasks and use the term OM to refer to both of them when conducting our experiments.

2 Related Work

IE and OM system we implement is a part of the restaurant recommendation system. Recommendation systems are usually classified as content-based and collaborative filtering ones [25, 26, 28]. In text-based systems checking items similarity usually demands the application of text mining techniques. However, using linguistic methods or ideas is not common practice in this area: the overview of text mining methods gives an idea of the dominating bag-of-words and key words approaches [11, 16, 21]. Sometimes more advanced techniques are applied, but IE is completely integrated into items ranking algorithm inside the recommendation system [17].

As far as IE and OM are concerned, the corresponding problems are solved using both rule-based and statistics-based methods.

In the early days of IE, hand-crafted rules were used. They were later substituted with automatically extracted ones, and then machine learning-based approach developed [30]. Although traditionally IE is a domain dependent task, machine learning approaches to IE are both domain dependent and domain independent [12, 35]. In OM, rule-based approaches usually include the application of a semantic thesaurus [10], and machine learning ones do not always demand such linguistic resources.

As part of our research, we analyze features and classifiers commonly used in OM (and/or sentiment analysis). They are listed in Tables 1 and 2.

As far as unigrams are concerned, it is shown in [22] that occurrence-based unigram features generally perform better than frequency-based ones. Some researches consider higher order n-grams (non-contiguous ones) not useful [22] while others argue that they improve overall performance for some tasks [7, 36, 37].

Table 1. Features in sentiment analysis

Feature	References
Unigrams	[2, 4, 7, 12, 15, 22, 29, 37]
N-grams (bigrams, trigrams, etc.)	[2, 4, 7, 22, 37]
Unigrams of a certain POS (adjectives, adverbs, etc.)	[2, 7]
N-grams of certain POS	[2, 3]
Token positions	[22]
Emoticons	[8, 19, 31]
Substrings	[7]
Syntactic relations, syntactic n-grams	[19, 23, 33, 34]
Valence shifters	[12, 15]
Semantic classes	[7, 29]
Sentiment lexicon words	[6, 15, 31]

Table 2. Supervised models in sentiment analysis

Model	References
Naive Bayes	[2, 4, 6, 7, 20, 22, 29, 31, 37]
Logistic Regression	[31]
Maximum Entropy	[22, 29]
Support Vector Machines	[2, 4, 7, 12, 15, 22, 23, 31]
Random Forest	[31]
Perceptron	[1, 2]
Neural Networks	[32, 34]

Sometimes position information (i.e., the position of a token in a paragraph) is also included in the feature set [22]. Lower order features, such as substrings, are experimented with in [7].

When dealing with informal language (e.g., tweets or sms), some authors propose taking emoticons into account [8, 19, 31].

As for part-of-speech (POS) information, unigrams of certain POS as well as word combinations of certain POS [2] (e.g., adverbs and adjectives in [3]) are often used.

Deeper linguistic-based features include dependency or constituent-based features [24]. Semantic classes of words are also employed. Thus, words referring to a particular object are replaced with their class labels (and extracted inside n-grams) [7, 29]. In [14] valence shifters (intensifiers, diminishers and negations) are used. As for negation, a common approach to its handling involves attaching "not" to the negated word [2, 6, 7, 19, 20].

As far as classifiers are concerned, it can be seen from Table 2 that Naive Bayes classifier (NB) and Support Vector Machines (SVM) and the most popular ones in sentiment analysis. NB is commonly used as a baseline model, as SVM usually demonstrates better performance. However, it is shown in [4] and [37] that NB can outperform SVM on short-form domain like microblogs.

As stated earlier, our goal is to introduce an effective corpus-based restaurant IE and OM method using machine learning techniques. Since there is lack of linguistic resources for Russian, we heavily rely on corpus analysis and our focus is on the application of machine learning to the problem. Our objectives include the identification of feature sets and models to experiment with, and the automatic selection of the best model and feature set combination. Our tasks are to conduct corpus analysis and construct dictionaries, to define feature sets and models for the experiments, to evaluate the classifiers and to propose the rules for automatic selection of the best classifier.

As far as features are concerned, our choice is dictated by the realities of the Russian language. Since there are no available linguistic sentiment resources for Russian known to us, we employ various lexicons either learnt semi-automatically or constructed manually from the corpus. Thus, we use combinations like "modifier + predicative-attributive word" as one of the features, and this idea is similar to that of "adverb + adjective" pairs and valence shifters. We also experiment with occurrence-based unigrams and contiguous bigrams, emoticons and exclamation marks. As there is lack of parsing tools for Russian, we consider non-contiguous bigrams an alternative to syntactic ones, taking into account free word order and the variety of sentiment expression in Russian. Negation is also to be covered by non-contiguous bigrams.

As for machine learning models, in our research we experiment with NB, Logistic Regression (LogReg), linear SVM and Perceptron [28]. NB appears to be the best one at classifying opinion in our domain for most restaurant aspects. It agrees with the results obtained in earlier papers [4] and [37] for English and with the notion that NB, as a simple generative model, is better at small amount of data.

3 Data

The data consists of 32525 reviews (4.2 millions of words) about restaurants in informal Russian language. The corpus is full of slang, misprints and prolonged vowels (as a means of expressing emotions). The reviews are mostly unstructured and vary from 1 to 96 sentences. A part of the corpus, with 1025 reviews about 206 restaurants from the central part of Saint-Petersburg, is annotated.

We outline a list of restaurant characteristics which are presumably mentioned in users' reviews (see Table 3). Our recommendation system suggests a dialogue with a user based on a predefined list of restaurants aspects, and therefore we do not perform automatic topic clustering described, for example, in [18].

Table 3. Restaurant aspects

Aspect	Value domain	Aspect	Value domain	Aspect	Value domain
Restaurant type	String	Noise level	$\{-2; -1; 0; 1; 2\}$	Dancefloor	{yes; no}
Cuisine type	String(s)[a]	Cosiness	{yes; no}	Bar	{yes; no}
Food quality	$\{-2; -1; 0; 1; 2\}$	Romantic atmosphere	{yes; no}	Parking place	{yes; no}
Company	{large; small}	Crampedness	{yes; no}	VIP room	{yes; no}
Audience	String(s)	Price level	$\{-2; -1; 0; 1; 2\}$	Dancefloor	{yes; no}
Service quality	$\{-2; -1; 0; 1; 2\}$	Average cheque	Integer or Interval	Railway station	{yes; no}
Service speed	$\{-2; -1; 0; 1; 2\}$	Smoking room	{yes; no; area; room}	Hotel	{yes; no}
Staff politeness	$\{-2; -1; 0; 1; 2\}$	Children	{yes; no}	Shopping mall	{yes; no}
Staff amiability	$\{-2; -1; 0; 1; 2\}$	Children's room	{yes; no}		

[a] Multiple valued cuisine type and audience are split into several binary aspects

Table 4. Restaurant aspects distribution in the corpus

Occurrence percentage	List of aspects
[85 %; 100 %]	Food quality (86 %)
[55 %; 85 %)	Service quality (55 %)
[25 %; 55 %)	Staff politeness & amiability, service speed, price level, cosiness
[10 %; 25 %]	Noise level, crampedness, romantic atmosphere, company

Our task can be considered a classification problem. For each aspect the system should either label a review with one of the possible classes or reject it as irrelevant with respect to the given aspect.

The aspects in italics are the most frequent ones in the annotated subcorpus. As most restaurants characteristics are never mentioned in the reviews, we define an empirical threshold frequency value of 10 % and consider aspects mentioned in at least 10 % of reviews frequent. We only train classifiers for the frequent aspects. In Table 4 the 11 selected aspects are divided into groups according to their frequency in the reviews.

4   Methods

The research described in this paper is conducted in several stages, including corpus analysis, features and classifiers identification and the automatic selection of the best classifiers and features.

4.1   Corpus Analysis

The first step is described in [27] in detail. It consists of corpus preprocessing (tokenization, lemmatization, spell checking and splitting into sentences) and dictionaries learning. The dictionaries learnt from the corpus include

- trigger words dictionaries (for service and food frames only),
- predicative-attributive dictionaries (for service and food frames only),
- modifiers dictionary (for aspects taking one of the 5 values from "−2" to "2"),
- key words and phrases dictionary and
- sentiment lexicon (for aspects taking one of the 5 values from "−2" to "2").

Trigger words and predicative-attributive dictionaries are learnt semi-automatically from non-contiguous bigrams (gathered from the corpus) using bootstrapping procedure. Here and further in the paper by trigger words we mean such nominations for service or food that if a trigger word occurs in a review, the review is likely to contain relevant information about the aspect in question. Predicative-attributive words are adjectives and participles which occur in the context of trigger words.

The modifiers are filtered from the adverbs list (collected from the corpus) and key words are written out and annotated manually. Sentiment lexicon consists of adjectives and participles and is also annotated manually.

4.2 Features Identification

We try to incorporate the dictionaries learnt at the corpus analysis stage into our feature sets. There are 11 different feature sets defined (see Table 5).

All the features except for the emoticons and exclamations (taking frequency values) are occurrence-based and binary. Baseline features consist of unigrams and bigrams, and non-contiguous n-gram features are represented by non-contiguous bigrams (with at most two words between the components).

There are predicative-attributive features for each word from the respective dictionaries. They take "1" values when a word from predicative-attributive dictionary occurs within 3 words to the left from any of the trigger words. If a modifier occurs inside such left context, a corresponding feature is taken into account too. Sentiment lexicon features also take "1" values when occurring in a trigger word context. They take form "LEX_label_LEFT" or "LEX_label_RIGHT" depending on their position with respect to the trigger word, and "label" stands for aspects class label e.g., "-2", "-1", etc.).

Table 5. Feature sets

Feature set \feature	N-grams	Non-contiguous N-grams	Emoticons and exclamations	Key words	Predicative-attributive words and modifiers	Sentiment lexicon
Baseline	+					
Extended Distant	+	+				
Extended Distant Emoticons	+	+	+			
Extended Distant Emoticons Lex	+	+	+			+
Extended KWs	+			+		
Extended KWs PredAttr Lex	+			+	+	
Extended PredAttr	+				+	
Extended Lex	+					+
Extended KWs Lex	+			+		+
Extended Emoticons KWs Lex	+		+	+		+
Extended All	+	+	+	+	+	+

Table 6. Best models and feature sets according to Holm-Bonferroni tests series

Aspect	Model: feature set	Priority class	Accuracy, %	Average F1, %
Class selection				
Amiability	MNB:extended_All	3	77,30	76,84
Cosy	MNB:extended_Distant	3	96,00	95,74
Cramped	MNB:baseline	2	87,86	87,52
Level	MNB:baseline	2	65,00	61,82
Noise	MNB:extended_KWs	3	82,67	80,40
Politeness	NB:extended_All	3	79,66	79,20
Service quality	MNB: extended_Distant_Emoticons	3	72,71	71,96
Food quality	MNB:extended_Distant	3	75,05	74,05
Speed	MNB: extended_KWs_PredAttr_Lex	2	69,71	68,72
Relevant vs. irrelevant				
Amiability	NB:baseline	3	82,78	82,76
Company	LogReg:baseline	2	93,24	92,66
Cosy	LogReg:baseline	3	89,91	89,78
Cramped	*LogReg:baseline*	*3*	*92,22*	*91,73*
Level	LogReg:baseline	3	92,96	92,95
Noise	LogReg:baseline	3	93,89	93,68
Politeness	*LogReg:baseline*	*3*	*87,59*	*87,52*
Service quality	*LogReg:baseline*	*3*	*82,87*	*82,79*
Romantic	Prcp:baseline	2	93,98	93,91
Speed	LogReg:baseline	2	88,33	88,30

Since the size of feature space appears to be quite large, we prune irrelevant features using Randomized Logistic Regression implemented in scikit-learn.

4.3   Models

The models we experiment with include NB (Bernoulli, with non-occurring features taken into account, and Multinomial, with non-occurring features ignored), LogReg, linear SVM and Perceptron (with shuffled samples) from scikit-learn[1].

For each of the restaurants aspects there are two classifiers trained: first, to label a review as relevant or irrelevant with respect to the aspect, and then, if relevant, to predict its class. Given the size of our annotated corpus, it should be mentioned that

---

[1] http://scikit-learn.org

while for the relevance/irrelevance task there are only two classes and the whole training data set available, while for further sentiment classification task only relevant reviews are considered. Therefore, when the latter task is concerned, we have a limited amount of data for some of the aspects, especially for the subjective ones.

5   Evaluation: Classifiers Selection

To choose the best combination of model and feature set for each of the restaurant aspects, we conduct a two-step procedure. First, 10-fold cross-validation is held (with models trained and tested on the same random data splits and test size equal to 10 % of the corpus). Then statistical tests are employed to check whether the best combinations are significantly better than the other ones.

As it was mentioned in Sect. 4.3, we choose two classifiers for each of the aspects: the first one is to decide whether a review was relevant or not and the second one – to predict its class label. As far as the former is concerned, it is an intermediate task, and we only try the classifiers on baseline feature set according to the empirically derived conclusion that n-grams-based approach is sufficient to separate relevant reviews from the irrelevant ones. For the latter task we experiment with 11 different feature sets (or 4, for the aspects not belonging to service and food frames). Thus, there are 5 and 55 (or 20) different combinations respectively.

To be able to compare the models, one has to choose some single score. During the cross-validation procedure we calculate average weighted F1 scores for our "classifier + feature set" combinations. These F1 scores are average across all the classes with weights equal to their frequencies in the training data set.

To test whether the best combinations are significantly better than the other ones, we follow the recommendations described in [9]. The tests are conducted in two stages. First, we apply a modified non-parametric Friedman test (proposed by Iman and Davenport in [13]) to see whether there is any significant difference between our models performance scores. Then, in case the difference is significant, we proceed with a series of post hoc Holm-Bonferroni tests (also described in [9]).

Holm-Bonferroni test is quite powerful and can be used for comparing one classifier to the others even for dependent data sets. In our research we not only test whether the best classifier is significantly better than the other ones, but also divide them into groups according to their ranks. The ranks are calculated for each classifier as average ranks for each test set in the cross-validation.

Thus, let us assume there are four types of classifiers: the best one (class 3)[2], those which are significantly worse than the best one (class 0), those which are significantly better than each classifier from class 0, except for the best one (class 2) and all the rest (class 1). As far as the latter (class 1) is concerned, one cannot tell whether there is any

---

[2] It should be noted that the classifier ("model + feature set" combination) with the highest rank does not necessarily demonstrate the highest average weighted F1 score. The classes 0, 1, 2 and 3 assigned to the classifiers in this paper are based on their ranks (according to non-parametric Holm-Bonferroni test) and not F1 scores.

significant difference between their ranks compared to the best rank (class 3) or the worst ranks (class 0).

Having divided our classifiers (i.e., "classifier + feature set" combinations) into the groups as described above (in case there is significant difference according to Friedman test), we choose the classifier for each of the restaurant aspects according to the following rules:

- if there is no statistically significant difference between the classifiers (e.g., null hypothesis is not rejected in Friedman test or all groups belong to class 1), choose the simplest combination[3] among the classifiers with scores within 2 % from the maximal score (in case of ties the classifier with better scores should be chosen);
- if there is statistical difference according to Friedman test, and the groups are class 3 and class 0 only (or class 3 and classes 1 and 0 only), choose the only element of class 3;
- if there is statistical difference according to Friedman test, and all the four groups (0, 1, 2 and 3) take place, choose the simplest combination among those class 3 + class 2 classifiers which are within 1 % from the class 3 F1 score, either higher or lower (if none such classifiers in class 2, choose the only element of class 3; in case of ties choose the classifier with better scores).

Such an approach seems reasonable because it guarantees that significantly worse classifiers (if any) are never chosen and provides a balance between high performance scores and computational effectiveness.

Cross-validation results are shown in Table 6. It contains information only about the aspects for which Friedman test proves significant difference at the 0.05 level between classifiers performance. During the series of Holm-Bonferroni tests we also test the null hypotheses at the 0.05 significance level. For the aspects for which the null hypothesis in Friedman test is not rejected we adopt Multinomial NB classifier and baseline feature set by default.

As far as relevance/irrelevance task is concerned, LogReg appears to be the best classifier. For crampedness, politeness and service quality (in bold) LogReg is significantly better than all the rest classifiers at the 0.05 level. For most of the other aspects it performs better than Bernoulli and Multinomial NB and Perceptron. And indeed, LogReg is known to be better on large training sets, and for the relevance/irrelevance task there is more training data than for the task of classifying relevant reviews.

Thus, we suggest that LogReg could be recommended for the classification of informal unstructured Russian texts into those which contain information or opinion about the specific aspect and those which do not.

---

[3] Baseline features set is considered the simplest one, while Extended_All – the most complex one. MNB and NB models are considered the simplest models, Perceptron – a more complex one, and LogReg and linear SVM – the most complex ones (in fact, they are both similar to Perceptron but their training is more computationally expensive [5]). MNB and NB classifiers are considered similar in the degree of "simplicity" as well as LogReg and linear SVM. A simple model with complex features is considered simpler than a complex model with simple (e.g., baseline) features.

As for deciding on the sentiment or opinion class, NB classifiers are chosen for all the aspects. It can be partly explained by the nature of the classifier itself and the rules which direct our choice. Namely, NB, having high bias, usually behaves better when there is small amount of training data, and, according to the outlined rules, simple classifiers have higher priority. However, with the given rules, for 6 aspects out of 9, NB classifiers, combined with extended feature sets, still have the highest ranks. Therefore it might be suggested that NB is good at classifying sentiment in the informal texts with small training set.

Holm-Bonferroni test series also reveals the following tendency: some of the "model + feature set" combinations are never labeled with class 0 for any of the aspects considered. Such combinations include Naive Bayes (Bernoulli) with the following feature sets: extended_PredAttr, extended_KWs_PredAttr_Lex, extended_Distant, extended_Distant_Emotions and extended_Distant_Emotions_Lex. It suggests that even if we simply pick up one of these combinations for each of the aspects, the obtained scores will not be among the worst ones.

Another observation that can be made is that including emoticons and exclamations into the extended_Distant feature set is not a good idea unless the aspect to be extracted is service quality. For the other restaurant aspects extended_Distant_Emoticons feature set does not improve F1 score or even worsens it.

As for dictionaries, the corresponding features can improve the results for the service frame. However, food quality, one of the most important restaurant characteristics along with service quality, is best extracted using just non-contiguous bigrams which seem to cover a wide variety of the expressions of opinion. Thus, a more elaborate lexicon and dictionaries construction could be one of our future work directions. For example, sentiment lexicon currently includes ambiguous words and thus demands elaborate sense differentiation.

6 Conclusion: Further Work

In this paper we propose a corpus-based method of information extraction and opinion mining for the restaurant recommendation system. It uses machine learning techniques and is based on elaborate corpus analysis and automatic classifier selection.

We have experimented with a number of machine learning models and feature sets with respect to our tasks, and employed statistical tests to select the optimal classifier for each of the restaurant aspects. The features include dictionaries constructed during corpus analysis stage (the latter is described in our earlier paper [27]). Selection procedure is based on a set of rules and classifiers priorities and enables us to choose the most computationally effective combination of a model and a feature set among those which perform best during cross-validation.

As a result of the experiments, Bernoulli and Multinomial Naive Bayes classifiers appear to be the most appropriate ones for the opinion class labeling. For the task of deciding on the relevance of a user's review Logistic Regression, outperforms other models. Thus, these classifiers could be recommended for the tasks similar to those described above where the data is represented by colloquial non-structured texts and its amount is limited.

For service quality frame, the application of dictionaries improves models performance, which confirms the idea of employing preliminary corpus analysis.

Thus, the results of the research verify the effectiveness of corpus-based methods with respect to the problem of information extraction and opinion mining from colloquial non-structured texts (in domains similar to restaurants) inflective languages with rich morphology and relatively free word order (like Russian), especially under resourced ones.

Our further work directions include more sophisticated sentiment and modifiers lexicons construction and annotation with the help of several experts. Since sentiment degree of a word may depend on the restaurant aspect it refers to, the words are to be annotated with respect to every aspect separately; verbs are also to be included in the lexicon. We also plan to extend annotated subcorpus and to conduct another series of experiments according to the method described in the paper to verify that Logistic Regression and SVM which are normally more effective than Naive Bayes on larger data sets will outperform it.

Acknowledgement. The authors acknowledge Saint-Petersburg State University for a research grant 30.38.305.2014.

References

1. Aston, N., Liddle, J., Hu, W.: Twitter sentiment in data streams with perceptron. J. Comput. Commun. **2**, 11–16 (2014)
2. Bakliwal, A., Patil, A., Arora, P., Varma, V.: Towards enhanced opinion classification using NLP techniques. In: Proceedings of the Workshop on Sentiment Analysis where AI meets Psychology (SAAIP), IJCNLP, pp. 101–107 (2011)
3. Benamara, F., Cesarano, C., Picariello, A., Reforgiato, D., Subrahmanian, V.S.: Sentiment analysis: adjectives and adverbs are better than adjectives alone. In: Proceedings of the International Conference on Weblogs and Social Media (ICWSM) (2007)
4. Bermingham, A., Smeaton, A.: Classifying sentiment in microblogs: is brevity an advantage? In: Proceedings of the International Conference on Information and Knowledge Management (CIKM) (2010)
5. Collobert, R., Bengio, S.: Links between Perceptrons, MLPs and SVMs. In: Proceedings of the 21th International Conference on Machine Learning (2004)
6. Das, S.R., Chen, M.Y.: Yahoo! for Amazon: sentiment parsing from small talk on the web. Manage. Sci. **53**(9), 1375–1388 (2007)
7. Dave, K., Lawrence, S., Pennock, D.M.: Mining the peanut gallery: opinion extraction and semantic classification of product reviews. In: Proceedings of the 12th International Conference on World Wide Web, pp. 519–528 (2003)
8. Davidov, D., Tsur, O., Rappoport, A.: Enhanced sentiment learning using twitter hashtags and smileys. In: Proceedings of the 23rd International Conference on Computational Linguistics: Posters, pp. 241–249. Association for Computational Linguistics (2010)
9. Demšar, J.: Statistical comparisons of classifiers over multiple data sets. J. Mach. Learn. Res. **7**, 1–30 (2006)
10. Devitt, A., Ahmad, K.: Is there a language of sentiment? An analysis of lexical resources for sentiment analysis. Lang. Resour. Eval. **47**(2), 475–511 (2013)

11. Emadzadeh, E., Nikfarjam, A., Ghauth, K.I., Why, N.K.: Learning materials recommendation using a hybrid recommender system with automated keyword extraction. World Appl. Sci. J. **9**(11), 1260–1271 (2010)
12. Gatterbauer, W., Bohunsky, P., Herzog, M., Krüpl, B., Pollak, B.: Towards domain-independent information extraction from web tables. In: Proceedings of the 16th International Conference on World Wide Web, pp. 71–80 (2007)
13. Iman, R.L., Davenport, J.M.: Approximations of the critical region of the Friedman statistic. Commun. Stat. **18**, 571–595 (1980)
14. Kennedy, A., Inkpen, D.: Sentiment classification of movie reviews using contextual valence shifters. Comput. Intell. **22**(2), 110–125 (2006)
15. Kotelnikov, M., Klekovkina, M.: The automatic sentiment text classification method based on emotional vocabulary. In: RCDL'2012 (2012)
16. Leksin, V.A., Nikolenko, S.I.: Semi-supervised tag extraction in a web recommender system. In: Brisaboa, N., Pedreira, O., Zezula, P. (eds.) SISAP 2013. LNCS, vol. 8199, pp. 206–212. Springer, Heidelberg (2013)
17. Li, Y., Nie, J., Zhang, Y., Wang, B., Yan, B., Weng, F.: Contextual recommendation based on text mining. In: Proceedings of the 23rd International Conference on Computational Linguistics (Coling 2010): Poster Volume, pp. 692–700 (2010)
18. Liu, J., Seneff, S.: Review sentiment scoring via a parse-and-paraphrase paradigm. In: Proceedings of the 2009 Conference on Empirical Methods in Natural Language Processing, Singapore, pp. 161–169 (2009)
19. Marchand, M., Ginsca, A.L., Besançon, R., Mesnard, O.: [LVIC-LIMSI]: using syntactic features and multi-polarity words for sentiment analysis in twitter. In: Proceedings of the 7th International Workshop on Semantic Evaluation, pp. 418–424 (2013)
20. Narayanan, V., Arora, I., Bhatia, A.: Fast and accurate sentiment classification using an enhanced Naive Bayes model. In: Yin, H., Tang, K., Gao, Y., Klawonn, F., Lee, M., Weise, T., Li, B., Yao, X. (eds.) IDEAL 2013. LNCS, vol. 8206, pp. 194–201. Springer, Heidelberg (2013)
21. Naw, N., Hlaing, E.E.: Relevant words extraction method for recommendation system. Int. J. Emer. Technol. Adv. Eng. **3**(1), 680–685 (2013)
22. Pang, B., Lee, L., Vaithyanathan, S.: Thumbs up? Sentiment classification using machine learning techniques. In: Proceedings of the Conference on Empirical Methods in Natural Language Processing (EMNLP), pp. 79–86 (2002)
23. Pak, A., Paroubek, P.: Language independent approach to sentiment analysis. Komp'uternaya Lingvistika i Intellektualnie Tehnologii: po materialam ezhegodnoy mezhdunarodnoy konferencii "Dialog", vol. 11(18), RGHU, Moscow, pp. 37–50 (2012)
24. Pang, B., Lee, L.: Opinion mining and sentiment analysis. Found. Trends Inf. Retrieval **2** (1–2), 1–135 (2008)
25. Park, D.H., Kim, H.K., Kim, J.K.: A literature review and classification of recommender systems research. Soc. Sci. **5**, 290–294 (2011)
26. Pazzani, M.J., Billsus, D.: Content-based recommendation systems. In: Brusilovsky, P., Kobsa, A., Nejdl, W. (eds.) The Adaptive Web, LNCS, vol. 4321, pp. 325–341. Springer, Heildelberg (2007)
27. Pronoza, E., Yagunova, E., Lyashin, A.: Restaurant information extraction for the recommendation system. In: Proceedings of the 2nd Workshop on Social and Algorithmic Issues in Business Support: "Knowledge Hidden in Text", LTC'2013, (2013)
28. Ricci, F., Rokach, L., Shapira, B., Kantor, P.: Recommender Systems Handbook. Springer, New York (2011)
29. Saif, H.: Sentiment analysis of microblogs. Mining the New World. Technical Report KMI-12-2 (2012)

30. Sarawagi, S.: Information extraction. Found. Trends Databases **1**(3), 261–377 (2007)
31. Shah, K., Munshi, N., Reddy, P.: Sentiment Analysis and Opinion Mining of Microblogs. In: University of Illinois at Chicago, Course CS 583 - Data Mining and Text Mining (2013). http://www.cs.uic.edu/~preddy/dm1.pdf
32. Sharma, A., Dey, S.: An artificial neural network based approach for sentiment analysis of opinionated text. In: Proceedings of the 2012 ACM Research in Applied Computation Symposium, pp. 37–42 (2012)
33. Sidorov, G., Velasquez, F., Stamatatos, E., Gelbukh, A., Chanona-Hernández, L.: Syntactic n-grams as machine learning features for natural language processing. Expert Syst. Appl. **41**(3), 853–860 (2014)
34. Socher, R., Perelygin, A., Wy, J.Y., Chuang, J., Manning, C.D., Ng, A.Y., Potts, C.: Recursive deep models for semantic compositionality over a sentiment treebank. In: Proceedings of the Conference on Empirical Methods in Natural Language Processing (2013)
35. Turmo, J., Ageno, A., Català, N.: Adaptive information extraction. ACM Comput. Surv. **38**(2), 3 (2006)
36. Turney, P.: Thumbs up or thumbs down? Semantic orientation applied to unsupervised classification of reviews. In: Proceedings of the 40th Annual Meeting of the Association for Computational Linguistics (ACL), pp. 417–424 (2002)
37. Wang, S., Manning, C.D.: Baselines and bigrams: simple, good sentiment and topic classification. In: Proceedings of the 50th Annual Meeting of the Association for Computational Linguistics (ACL), vol. 2, pp. 90–94 (2012)

Author Index

Printed in the United States
By Bookmasters